THE GREAT WAR

THIRD VOLUME
THE ORIGINAL GERMAN PLAN AND ITS CULMINATION

BY

GEORGE H. ALLEN, Ph. D., formerly of the University of Pennsylvania, History Department; Fellow in Classical Archæology, American School of Classical Studies, Rome, Etc.,

HENRY C. WHITEHEAD, Captain in the United States Army, Served in Europe, by Official Assignment, for Observation, Etc.,

AND

Admiral F. E. CHADWICK, U. S. N.

PRINTED *AND* PUBLISHED *BY*

GEORGE BARRIE'S SONS

at PHILADELPHIA

CONTENTS

LIST OF ILLUSTRATIONS

FACING PAGE

PREFACE

Odiis etiam prope maioribus certarunt quam viribus. . . . Livy 21, 1.

[*They fought with almost greater animosity than strength.*]

Once in the waning of a summer afternoon I slowly climbed the long acclivity that leads from Spa towards Stavelot, refreshed by the cool shadows of the tall fir-trees and by the balsam-scented atmosphere, and tarried at the hushed, mysterious hour of twilight on a summit near the boundary. The gathering dusk, subduing every contrast, blending all the landscape in evasive, neutral tones, produced a grateful sense of absolute serenity. Behind me stretched the variedly fascinating, rugged tract of eastern Belgium; before, through the filmy mist that half obscured the yellow fields and rolling meadows, the fancy helped the eye to discern the undulating outline of the hills in Germany. The situation lured to meditation.

I seemed to stand upon the common margin of two mighty streams of human culture, my spirit quickened by a sense of intimate communion with the Latin and the Teutonic genius, to be lost in them, and to thrill with consciousness of common heirship in all that the long succession of their yearning, striving, impressionable generations had wrought. In weariness and tribulation the image of that hour still lingers wistfully.

Some of us had yielded to the seductive vision of a world in which the nationalities, each through the medium of its own peculiar usages, without aggression or constraint, should strive in friendly emulation for the benefit of all,—the preservation of the rich diversity of human life without the hard, intolerant spirit that had nourished its vitality.

Our blissful dream was suddenly transformed into a hideous nightmare. Startled, chagrined, outraged in our tenderest emotions, we quivered with impulsive, inconsiderate fury. Our feverish resentment cried for vengeance, sought the author of our anguish, claimed a victim. The most conspicuous factors attracted our inflamed imagination: the anachronism of a military caste intolerant of criticism, arrogantly self-sufficient, disdainfully refractory to popular control, violating our instinctive sense of political propriety. Our impotent rage was deeply stirred by trite, invidious catchwords. We reviled with passionate abuse the Chauvinistic spirit which sows suspicion, emphasizes contrasts, hides its ugly, sordid countenance behind the mask of sacrosanct traditions, the insidious deceit by means of which reactionary parties seduce the multitude from the pursuit of righteous aspirations.

But all our abstract speculations, optimist's reproach and cynic's sneer, are vain beside the one, supremely vital fact, that millions do not hesitate to hazard every good and life itself for these irrational, inveterate conceptions, superbly unconcerned by calculations of utility; that for preposterous motives, men will fight with eager, frenzied zeal, and joyous hatred.

Not for its cruelty, nor for its injuries alone, do we consider war with hate and detestation. But we execrate in it the element that dissipates our fondest dreams and most cherished prejudices. War is the relentless, but transcendent realist, that tries our theories and designs not merely by the test of concrete facts, but by the rigorous standard of consistency with stubborn human nature.

Shattered by the awful cataclysm reason is forced, despoiled of every preconception, to cling with humble confidence to sober observation as its sole support and guide.

Heraclitus taught that strife is the father of all things, and Hegel recognized in it the antidote for torpor and stagnation, the mighty agency for progress. Schopenhauer regarded war as the inevitable collision of the blind, unreasoning forces of volition that compose the universe. Nietzsche found a single means of deliverance from this chaos in the conscious *will to power*. A superior type, a superman, must be developed by the encouragement of the military qualities, courage, boldness, obedience, at the expense of the passive virtues, charity, compassion, humility. In Lasson's theory warlike development is a source of health for every nation, a fountain of youth for decrepit peoples. Von Treitschke declared, "the hope of driving war from the earth is not only senseless, but profoundly immoral; realized, this aspiration would convert the world into a great temple of egotism."

War is the fiery crucible that exterminates the moral dross of indolence and self-indulgence. War is the stern, avenging angel that with flaming sword expels its mournful victims from the languorous paradise of self-complacency, condemning them to cultivate with unrelenting toil the rugged wilderness of disillusionment and rigid self-examination.

In estimating war's stupendous process the sway of sentiment must sternly be repressed. It is a test of fortitude for one who feels inductively the heart-throb of diverse contending nations to await the issue with unperturbed detachment and stoic acquiescence. But despite subjection to the severe supremacy of fact, the emotions recover all their fervid function in the contemplation of an unparalleled tragedy in which whole armies act as single characters, and an organization of unprecedented efficacy, strength, and flexibility,—angel of light or demon as you will,—plays the rôle of hero.

GEORGE H. ALLEN, PH. D.

CHAPTER I

FORMATION OF THE GREAT GERMAN PLAN

Geographical position of Germany in relation to the plan of campaign. Possible advantages of a central position. Napoleon's method. Special factors in the situation of 1914. France to be crushed before Russia is ready. Rapidity necessary in the execution of the German plan. Physical features of the theater of hostilities in the West. Possible routes for a German invasion of France. Reasons for the choice of the route *via* Belgium. The German doctrine of tactics and strategy. Concealment of the real nature of the German plan. The French plan. The German aim is to envelop the opposing army, the French is to pierce it. Successive periods in the development of the plans: (1) August 4-15; (2) August 15-22; (3) August 23-September 5; (4) September 6-10; (5) September 10-23; (6) September 23-October 15; (7) October 16-November 11; (8) November 11-December 31. Subordinate position and character of operations in the East. Minor questions regarding the development of the German plan. The German and French Chiefs of Staff, von Moltke and Joffre.

The Great German Plan of Campaign! For years it had been the object of anxious speculation. Its existence had lurked in the background of the imagination like a vague, but stupendous, apparition. There were indications that beneath the peaceful surface of German society dwelt a latent, unfathomable force which could be invoked to action in accordance with a mysterious design with results that would baffle all human prevision. There was a dim consciousness of the existence of a power which could instantaneously galvanize the whole outward machinery of life with a terrible, frenzied energy. Often at night the great square block of the building of the General Staff with all its windows illuminated stood out in striking contrast to the deep shadows of the quiet Tiergarten, like a great factory working overtime. Whoever is familiar with

German method and thoroughness can conceive how laboriously provisions were perfected in anticipation of every emergency. The general outline may have been some individual's flash of inspiration, or the complex product of the sagacity of a group of eminent authorities. But just as a building is erected gradually brick by brick in accordance with the architect's design, so the general ideas for hypothetical campaigns were diligently worked out and elaborated in every detail by a vast number of individual contributions, ranging in scope from the comprehensive treatises of a General von der Goltz, embodying mature reflections based upon a lifelong experience, down to the myriad of petty dissertations, monographs, and articles written by students and young aspiring officers. Military science had become an obsession in Germany. About seven hundred books were added to the literature on the subject every year.

The most elementary factor in the formation of the German plan has been the system of European alliances.

We are frequently reminded that Germany was shut in between her two most redoubtable probable antagonists; and this situation was a source of continual apprehension to the German military authorities. But a central position is not necessarily an element of weakness in a conflict with foes who possess a collective superiority in numbers. The armies of a central power like Germany, operating on interior lines and with highly developed railway communications across the country, can be employed in very intimate correlation in pursuance of a single, homogeneous plan, whilst the forces of its adversaries are incapable of such combined and united action.

Napoleon's most brilliant successes were his repeated victories gained with a smaller, but undivided, army against opponents whose aggregate strength was very much superior to his own. Mustering all his available forces, Napoleon would fall upon his adversaries individually, crushing the first, leaving a containing force to overawe him, and then leading the bulk of his army to crush the next. Thus by repeated blows with the same weapon he beat down his antagonists one by one. As long as he could keep them asunder their collectively superior numbers did not avail. But his most remarkable maneuvers and illustrious victories were achieved when the isolation of his principal adversaries was far less complete than the separation of France and Russia in the world-war of the present. The supposed disparity in the time required for mobilization in France and Russia offered apparently a very suitable occasion for applying the Napoleonic method in the present situation.

The special factors in 1914 which bore upon the German plan, as expressed by Herr von Jagow before the outbreak of the war and repeated later many hundred times by others, are almost a commonplace. Germany's adversaries would eventually be able to dispose of far greater forces than those of Germany and her allies. But Germany's forces would outnumber those of France alone, and surpass at the beginning of the campaign the available field strength of Russia. It was believed that fully six weeks would intervene before the progress of Russia's concentration of troops on her western frontier would make her a formidable antagonist. In the meantime Austria-Hungary with minor German forces could restrain her while Germany dealt with France. The German General Staff saw their one chance of achieving a speedy, decisive victory in dealing a swift, paralyzing blow at France. France must be crushed before Russia could move; then the greater part of the German forces in the West could be conveyed rapidly to the eastern frontier producing an overwhelming superiority over the Russians. The situation offered an enticing possibility, a chance that might never return. It would be a race of vigor, determination, and endurance against time. The consequences of failure might be appalling, but modern Germany had been schooling herself to "live perilously."

"Germany had the speed and Russia had the numbers," and it seemed indispensable for the security of the former not only to inflict a decisive blow on France in the event of war before the Russian concentration had been completed, but even to initiate hostilities so as to prevent Russia, under cover of alleged pacific intentions, from bringing up great masses of troops from all parts of her vast dominions to neutralize the German advantage of rapidity. Delay would be profitable for Russia, but might prove fatal to Germany.

For Germany rapidity of action seemed to be the essential condition for success. She had to strike down her neighboring enemies in turn by utilizing to a considerable extent identical forces, first on her western, then on her eastern frontier. With this Napoleonic method once adopted, the strategic problem for such a campaign as that of 1914 resolved itself into the two main questions of the order and the method of attack. The fact that France would be ready first made her the suitable recipient for Germany's first and most vehement attentions.

The determination of the general method for the German assault on France was the function of strategy. In consequence of the frequent confusion in the use of the two terms strategy and tactics, a very brief digression may not be out of place at this point for the purpose of defining as clearly as possible the distinction between them—a distinction which, as will presently be shown, the continuity and magnitude of the battles of the first campaign tended to obliterate.

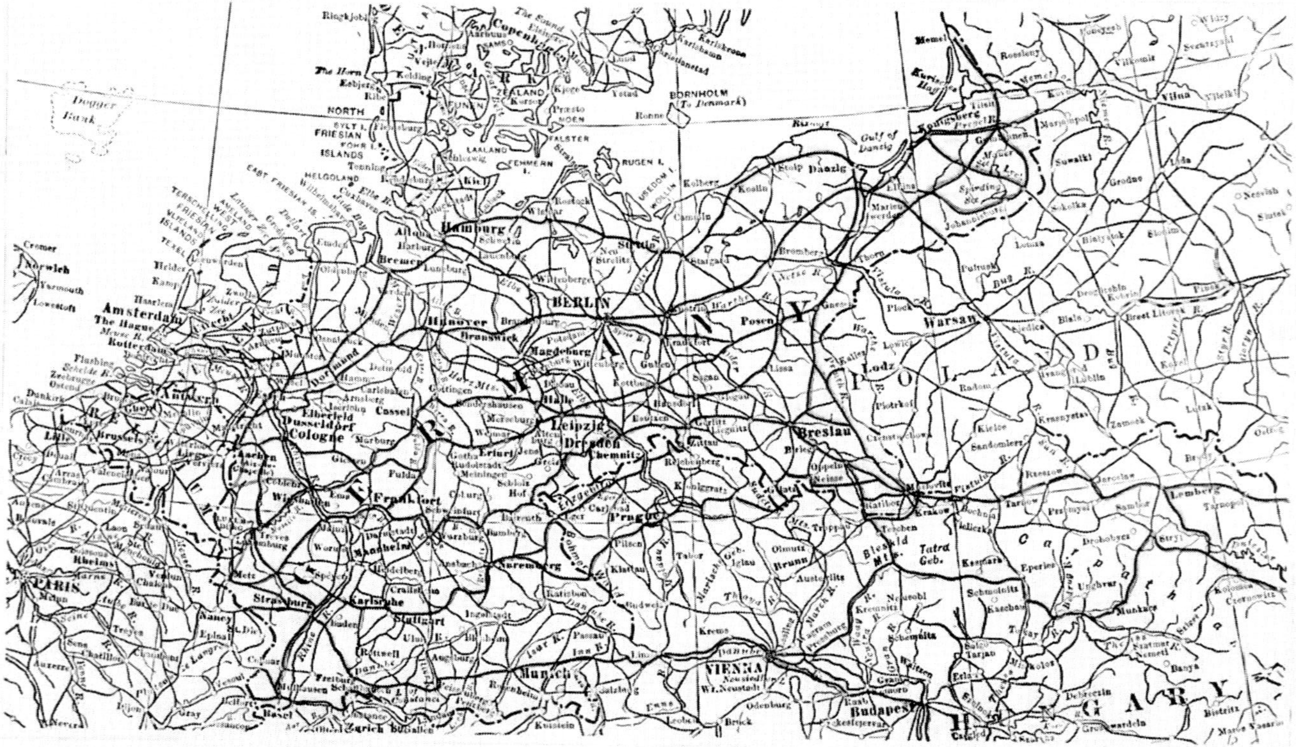

Map showing the highly developed railway syatems across Germany. By means of these trunk lines and their terminating strategic branches, the German General Staff hoped to solve the problem of confronting enemies on both their western and eastern frontiers.

Strategy and tactics together comprise the art and practice of war. The scope of strategy is broader than that of tactics. Strategy is the method of disposing the troops in masses and directing their movements towards the object of the campaign while they are not actually engaged in battle. Tactics is the method of conducting the evolutions of the military forces while engaged in battle. The field of strategy is the whole theater of operations; the field of tactics is confined to the battlefield. Strategy relates to the movement of armies when not in contact with the enemy, tactics to the maneuvers carried on when hostile contact has been established.

A brief survey of the prominent physical features of the territory where a campaign between France and Germany might conceivably develop will illuminate the considerations which influenced the German General Staff in devising their method for delivering the intended staggering blow against France, which determined, in other words, the general lines of their strategy in the West.

Two river systems first claim our attention; the Rhine with its tributaries, the Moselle and the Meuse, which surpass it in immediate strategical importance for this campaign; and the Seine with its tributaries, the Oise (with the Aisne) and the Marne, which likewise surpass in significance for our present purpose the larger stream which receives their waters. Between the Moselle and the Rhine, the Vosges Mountains, rising to an elevation of nearly 5,000 feet, form a natural barrier along about one-half of the common boundary of France and Germany. They are extensively wooded and recall somewhat in appearance the Alleghany Mountains of Pennsylvania.

In its general physiognomy Belgium may be likened to two isosceles triangles of unequal size, the larger one above with its base to the north, the smaller one below with its base against the lower right-hand side of the upper triangle, the line of contact formed by continuous sections of the courses of the Sambre and the Meuse. The generally rugged region of the Ardennes occupies the greater part of the area of the lower triangle. It is relatively arduous for the passage of armies, as compared with the conditions generally in western Europe. But the employment of motor-vehicles for transporting supplies has greatly diminished the difficulties of operations in this region. A glance at the plan of the German strategic railways recently constructed in the region of the Eifel, in Germany, directly eastward of the Ardennes, would seem to indicate that the Germans were not dismayed at the prospective difficulties of provisioning large armies in this region of Belgium. The Meuse runs through a deep ravine, bounded in many parts by rocky, precipitous walls, from Mézières to Namur, and through a somewhat less abrupt depression from Namur to Liége.

The upper triangle is occupied by the Belgian plain, rolling in the east, but gradually becoming absolutely flat towards the west. It is traversed in all parts by the lines of the densest network of railways in Europe. Its lines of communication eastward converge on Liége near the eastern extremity of the triangle. Liége is not only the key to the valley of the Meuse directly above it, but it gathers within itself practically all the railway traffic between Belgium and Germany. It scarcely yields precedence in strategic importance to any of the fortified points which we shall have occasion to consider.

The obvious goal for a German offensive against France was Paris, and three general routes were available to this objective point, as follows:

(1) From Cologne across the Belgian frontier near Aix-la-Chapelle, through Liége, up the valleys of the Meuse

and the Sambre, into France near Maubeuge and thence down the valley of the Oise.

(2) From Coblenz and the valley of the Moselle westward across Luxemburg, into France by Longwy and Stenay, across the Meuse and down the valley of the Aisne.

(3) From Strassburg and Alsace across Lorraine, between the Vosges and Metz, through Nancy, between the great barrier fortresses of Toul and Épinal, and westward down the valley of the Marne.

Of course these indications must not be taken in too narrow or literal a sense. The course of armies advancing in these general directions might be represented on the map more appropriately by broad bands than by lines.

Alleged forays by German cavalry patrols, even before hostilities had actually been declared, near Lunéville and in front of Longwy, might have served as portents of impending invasion by the second and third of these routes.

The invasion of France by the first route involved the violation of the neutrality of Belgium; an advance by the second route, that of Luxemburg.

The German authorities professed to regard it as a matter of life and death to advance into France by the quickest and easiest route so as to strike their overwhelming blow at the earliest possible moment. The strongly fortified eastern border of France was the most palpable argument for the selection of the Belgian route. But the momentous decision of Germany to disregard the neutrality of her friendly but feeble neighbor was based at the same time upon a strategical principle of universal applicability.

In their simplest terms the German doctrines of tactics and strategy were essentially identical. The second was only the expansion of the first to cover the wider field; and the truth is, that, with the vast extent of the battle lines in the present war and the often continuous fighting for long periods, the distinction between the two could not in any case have been considerable. The German doctrine of tactics is based upon a presumption of available numerical superiority. As superior forces naturally advance upon a front more widely extended than the front of their enemy, so as to make their superiority effective, their most natural course of action is to turn their adversary's flank, roll his battle line back upon itself with the almost inevitably ensuing panic and confusion, stifle him in a deadly embrace, and so annihilate him as a combatant. German authorities prescribed the most energetic application of the enveloping maneuver with a formidable pressure at the same time on all parts of the enemy's line to engage and pin down his forces. The enveloping maneuver was the essential feature of official German tactics. The German General Staff employed it in directing the movements of the tremendous forces operating against France with almost the precision of the deployment on a single battlefield. In their unity of design, the movements throughout the greater part of the first campaign in France might almost be regarded as a single battle of enormous dimensions.

Interpreting the spirit of German strategy, we may regard it as the operation of swift, tremendous momentum, of well-considered boldness, of a studied conviction of superiority. German strategists had been attentive students of Napoleon's campaigns, and Napoleon declared that the force of an army, like momentum in mechanics, is the product of the mass multiplied by the velocity. German principles of strategy are based upon the presumption of an immediate, though perhaps only transient, numerical superiority. The natural goal of German strategy is, therefore, to force on decisive action while this superiority is still available at the critical time and place.

The common frontier of Germany and France afforded no room for the great enveloping movement. The French armies completely filled it, supported by their almost impregnable barrier fortresses, their wings covered by neutral Belgium and neutral Switzerland. The characteristic maneuver of German strategical teachings could not be employed without the violation of the neutrality of one of these two states whose territories flanked the French position, and every consideration of expediency demanded that the turning movement should be made across Belgium.

The German General Staff had probably decided as far back as 1904 not to hesitate to take any step which would insure success in the event of war with France and Russia, and particularly to snatch promptly the advantage of traversing Belgium to invade northern France, where the immense German forces could conveniently deploy in the rear of the French armies concentrated along the eastern frontier. This plan had been set forth in a noteworthy memorandum by General von Schlieffen. The irrepressible General von Bernhardi had done his best to advertise it, and many pamphlets and articles had been written in Belgium and France to sound the note of warning against this very project. Yet, strangely enough, the disposition and operations of the French forces, even after the violation of Belgian neutrality had actually taken place, were such as to indicate either an utter failure to suspect that the bulk of the attacking forces would pass through Belgium, or, what is even more incomprehensible, a feeling of confidence that the inadequate garrisons of Liége and Namur and the only half-prepared Belgian field army would be able to stem such a human torrent.

The concealment of the real nature of the great German maneuver in its early stages was a masterpiece of secrecy. While the public in the western capitals was beguiled with the notion that the Germans were hopelessly embarrassed in the execution of their plan, or exulted in tidings which magnified trivial engagements into splendid victories, the human deluge was mounting higher and higher behind the floodgates. Before long the barriers were swept away and a sea of belligerent humanity burst over the Belgian plain. Nobody had fully anticipated the astounding boldness of the German plan, which consisted in throwing a defensive screen of troops along the line of the Vosges, engaging the attention and arousing the apprehension of the French in Lorraine, but gathering the chief strength for the tremendous blow in the north, crossing the Meuse wherever possible between Visé and Dinant, taking Liége and Namur by assault, marching by way of Brussels, and utilizing the various routes from there towards Paris.

A combination of theories and circumstances made the French methods of warfare the opposite of the German.

After the French disasters in 1870-1871 the opinion prevailed in France that the Germans had won by the audacious methods of Napoleon I, and it became a settled conviction that the imitation of his generalship was the key to victory French professors of military science reduced their observations of Napoleon's methods to a practical system. His practice in tactics, according to their teaching, was not based upon a predetermined plan of battle. It consisted in rapping on all parts of the enemy's front to discover the weakest place and, when this vulnerable spot had been found, in directing against it a sudden, shattering blow with all available forces, particularly with strong reserves held in anticipation of this critical movement at some distance in the rear of the battle front. With this theory as its prototype, the offensive is described by the French Field-Service Manual of 1895 as composed of three distinct operations: (1) the preparatory conflict in search of the weak spot; (2) the decisive attack at the feeble point; (3) the entry into action of the general reserve. The purpose of this system of tactics is to perforate the enemy's line and roll the parts thus severed back upon themselves.

With the French, the offensive in tactics and strategy had as the common essential element the piercing of the enemy's front. But in spite of the general precept that the offensive alone will produce decisive results, strategy includes the defensive action also. The situation seemed to demand of the French a more cautious attitude. It made the expediency of the offensive and the occasion for its application debatable points for them. French strategists were confronted with the prospect of a serious numerical inferiority in a war with Germany. The French had constantly before them the problem of compensating for this weakness by the employment of greater flexibility and discernment, by circumspection and economy in the use of their resources in men, and by maintaining a relative degree of compactness in the general disposition of their masses of troops to permit rapid movement to and fro for support.

The conditions seemed to preclude a general French offensive in a war against Germany, except perhaps in very close concurrence with the operations of Russia, and the supposed sluggishness of Russian mobilization and concentration seemed to exclude the hope of an opportunity for such combined action in the early weeks of hostilities.

The normal plan of campaign for the French, as adopted after an initial period of vacillation in 1914, may be described as follows. The forces are grouped in a certain number of separate masses, placed in positions with connection on interior lines, in expectation of the enemy's offensive movement. The mass which first receives the hostile impact retreats fighting, decreasing as far as possible the momentum of the enemy's mass by its resistance while it retires. The compactness of the general disposition of the different masses of the French forces increases in proportion as this fighting mass falls back towards the common center. The capacity of this one mass to withdraw fighting before a vastly superior invading army without losing its organization is crucial for the success of the general scheme, and even for the safety of the entire army. Finally, by a common converging movement of all the masses, the others take their places by the side of the first. Advantage is then taken of the local superiority thus produced to crush and break through the portion of the invading forces directly in contact, and then to overwhelm separately the parts of the enemy's extended line, when, after wheeling about, they approach the critical point of operations, embarrassed and delayed by their change of plan and direction.

A French general plan of strategics was thus opposed to a German one. Each was suited to the circumstances, though not necessarily to the temper of the respective combatants. For the waiting attitude prescribed by the French plan was at variance with the traditional, and generally accepted, theory that the French were temperamentally incapable of fighting at their best on the defensive.

The Germans concentrated their efforts on enveloping their enemy, the French aimed to penetrate the front of their adversary. The Germans wished to push operations to a decisive issue with the utmost possible speed. It was natural that French strategy should aim to delay as long as possible a decisive collision. For time would fight for the side whose antagonist's initial superiority was subject to progressive diminution. The French held large forces in reserve for the decisive moment; the Germans, scorning

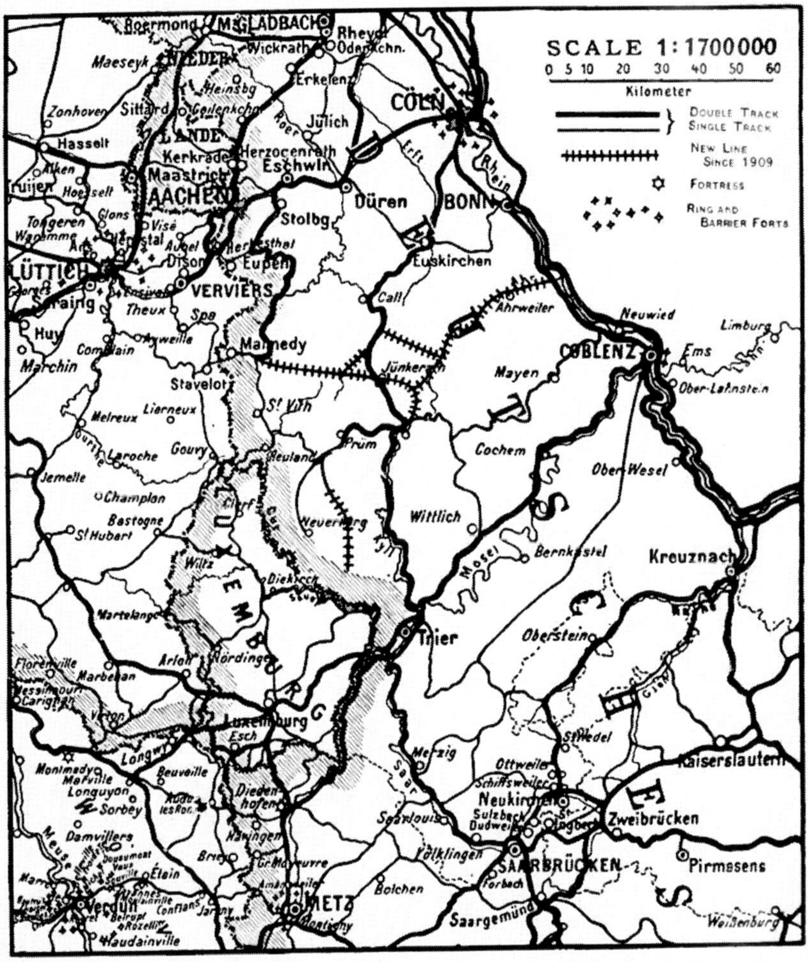

Map showing the German strategic railways leading to the frontiers of Belgium and Luxemburg. Based on a portion of a map published in *Illustrierte Weltkriegschronik.*

such a precaution, brought all their forces into immediate action for the single, tremendous maneuver.

It was naturally to be expected that the ultimate issue of the campaign would depend mainly upon which of the two plans should prove more effective. But curiously enough, before the campaign was many weeks old, the rapid course of events led each antagonist to employ the peculiar maneuver of his opponent's strategy, as will be explained in the appropriate connection.

In the treatment of the period embraced in the present volume it will be convenient to observe a number of subdivisions based upon the successive stages in the development of the principal action, as follows:

(1) August 4-15. The attack on Liége was the natural beginning of the execution of the German plan. Before the swelling tide of invasion destroyed the barrier at Liége, operations in the Belgian plain west and north of the Meuse, growing chiefly out of cavalry reconnaissances, were of slight importance.

(2) August 15-22. The remaining obstruction was swept away on the 15th and the deluge rolled with irresistible momentum across the Belgian plain. In the space of a week the principal masses of the army of invasion advanced without opposition of serious consequence from the Meuse at Liége to the line Namur-Charleroi-Mons, while other bodies of German troops advanced to the upper Meuse.

(3) August 23-September 5. The capture of Namur and the defeat of the Allies on the line of the Sambre and between Charleroi and Mons opened France to invasion. The grayish-green billow rushed forward with awful volume, tearing away every restriction in its course, and broke with angry violence at the very foot of the defenses of Paris and along the banks of the Marne.

(4) September 6-10. A battle of unprecedented magnitude was the turning point in the campaign. The initial momentum of the German invasion was finally checked. A dike was successfully interposed against the flood that was inundating France. The conclusion of the Battle of the Marne and the commencement of the German retreat to the Aisne might seem to be a suitable termination for this present volume. But considerations of proportion in the division of the material as a whole, and the continuity of all the military action in the West in the autumn of 1914 make it appropriate to interpret the operation of the original German plan in a broader sense and include the treatment of events whose dependence on the original impulsion is not quite so immediate.

(5) September 10-23. The receding tide halted after passing the Aisne. Strong positions had been prepared in the hills on the north side of the river, heavy artillery had been mounted, and what the Allies at first regarded as only a rear-guard action to cover the further retreat developed into a general battle, which raged for about two weeks with intermittent periods of exceptional fury, and fixed the limit of the Teutonic inundation for many months.

(6) September 23-October 15. The forces of invasion restrained in front expanded laterally. A formidable concentration of Allied forces at Amiens on September 21st threatening the right flank of the enemy's position and an important movement of German forces to St. Quentin in response mark the beginning of a veritable race in prolonging the two opposing fronts, each party striving to outflank the other, until their progress was halted at the coast of the North Sea. This movement furnished the chief immediate motive for the determined attack upon and capture of Antwerp.

(7) October 16-November 11. The German armies were still impelled by a frenzied resolve to achieve the purpose of the campaign. A tremendous blow was launched at the northwestern extremity of the Allied line. The much-heralded intention of penetrating to Calais was perhaps a somewhat misleading indication of the chief intent of the enterprise, which was probably designed as a renewed attempt to fulfil the original purpose of the campaign by turning the general Allied position. Desperate encounters continuing for about three weeks with terrible bloodshed served only to consolidate the positions of the opposing armies.

(8) November 11-December 31. An equilibrium of forces was thus produced in the West. The fighting became intermittent and finally subsided into comparative calm. The first great plan had reached its culmination without decisive results.

Sound considerations of strategy, as already explained, prescribed for the French army a firm defensive attitude in the early stages of a war with Germany, until a Russian invasion of Germany on a formidable scale had been fairly launched, or until the Germans had been drawn into a position in the West where the French could bring superior strength to bear against them at the strategically critical point. But contrary to this principle by which the operations of the French forces were in general conducted, the French at the very outset yielded to the impulse of invading Alsace and Lorraine, an action which deviated entirely from the regular development of events according to the normal plans of the two adversaries, as outlined above.

In the first period of the campaign French advance guards hastily penetrated both Alsace and Lorraine, but were driven back to the frontier in a few days. A very much stronger French force invading German Lorraine between Metz and the Vosges in the second period was defeated on August 21st and 22d and thrown back into France, suffering considerable loss. A similar invasion of Alsace from the direction of Belfort was likewise fruitless.

From the point of view of the German plan and the operations conducted in the endeavor to accomplish its fulfilment, which we regard as the unifying principle for the treatment of the first campaign, events in the East, in spite of the enormous forces involved in them, are subordinate to those in the western theater. During the most important part of this campaign the Germans possessed the initiative. They were able, in other words, to attack wherever they wished. They chose to direct their supreme effort against France. Russia could evade defeat a long time by reason of her vast size, and therefore the first decisive results had to be sought in France. The destruction of the army of the Allies in the West was doubtless regarded by the German General Staff at the beginning of the war as the necessary antecedent to offensive operations of a decisive character against Russia.

For these reasons it is suitable that the narrative of events in the East should be adjusted to the general scheme of treatment for the entire campaign based upon the continuous course of events in the West. The original plan of the Central Empires provided, doubtless, that Austria-Hungary should dispatch a sufficient army to subdue Serbia and with the rest of her forces invade Russian Poland with the object of embarrassing and breaking up the Russian concentration, which would be in progress chiefly behind the Vistula. This operation would serve to prolong the time available for the great offensive movement in the West. Meanwhile Germany would leave a few army corps on her eastern border as a covering force.

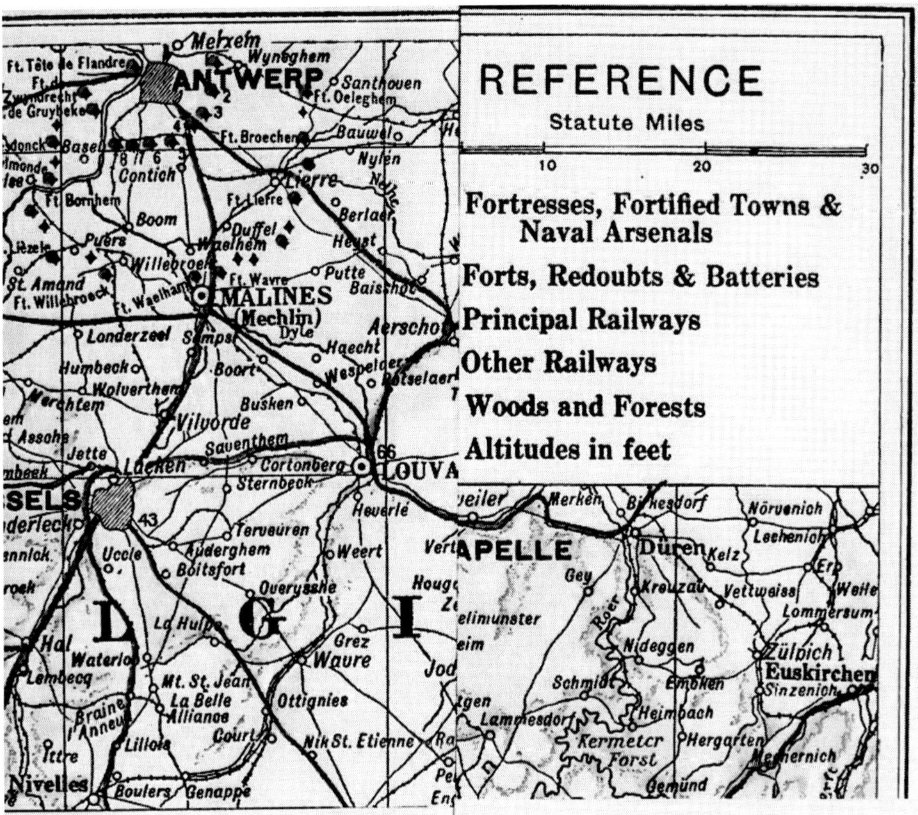
REFERENCE
Statute Miles
10
20
30
Fortresses, Fortified Towns & Naval Arsenals
Forts, Redoubts & Batteries
Principal Railways
Other Railways
Woods and Forests
Altitudes in feet
Merxem
Wyneghem
Ft. Tête de Flandre
ANTWERP
Santhoven
Ft. Oeleghem
Ft. Broechem
Bauwel
Nylen
Contich
Lierre
Ft. Lierre
Berlaer
Ft. Bornhem
Boom
Duffel
Waelhem
Heyst
Puers
Willebroek
Ft. Wavre
Putte
Baisshot
St. Amand
Ft. Willebroeck
Ft. Waelhem
MALINES
(Mechlin)
Dyle
Aerschot
Londerzeel
Haecht
Humbeck
Boort
Wespelaer
Wolverthem
Werchtem
Busken
Vilvorde
Assche
Saventhem
Jette
Laeken
Cortonberg
LOUVA
Sternbeck
SELS
Heverlé
Merken
Düren
Nörvenich
Lechenich
Kelz
Erp
Terveuren
Auderghem
Weert
APELLE
Uccle
Boitsfort
Gey
Kreuzau
Vettweiss
Weile
Overyssche
Roer
Lommersum
L
G
I
La Hulpe
Nideggen
Zülpich
Hal
Waterloo
Grez
Waure
Euskirchen
Lembecq
Mt. St. Jean
La Belle Alliance
Schmidt
Embken
Sinzenich
Ottignies
Heimbach
Braine l'Anneu
Lammesdorf
Lillois
Court
Kermeter Forst
Hergarten
Ittre
Nik St. Etienne
Mechernich
Nivelles
Boulers
Genappe
Gemünd

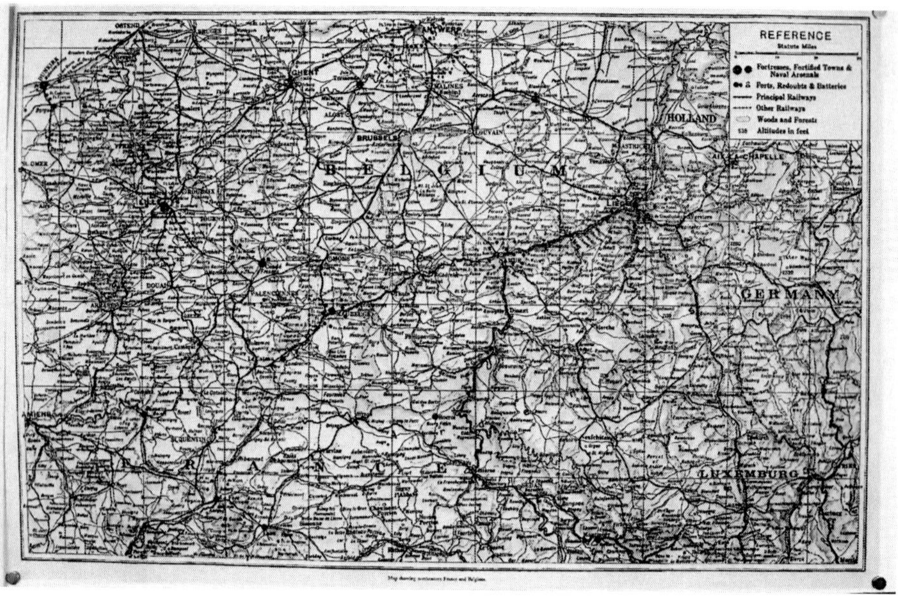

Map showing northeastern France and Belgium.

The prompt concentration of a considerable part of the Russian forces disappointed the expectations of the Teutonic Empires, and contributed to the failure of the Germans to reach the goal of the campaign in the West by compelling them to transfer a number of army corps from France to the eastern theater of war at a critical time.

In the second period, just as the great turning movement in the West had fairly started on its way, about August 16th, the Russians invaded East Prussia and Galicia simultaneously. They overran all the eastern part of Galicia, forced the Austro-Hungarian army in Poland to retreat, and took Lemberg after an eight days' battle on September 3d. But the Russian army invading East Prussia from the south was shattered in the Battle of Tannenberg on August 28th by General von Hindenburg, who routed the other army under General Rennenkampf which was invading the same province from the east on September 10th.

Later, the Germans in the West attained what may be regarded, from the comprehensive point of view of the progress of their general plan, as a partial success in closing their front from Switzerland to the North Sea in such a way as to secure for the time the possession of what they had won and to exclude the danger of a turning movement by the Allies, and so to be able to transfer considerable forces to the eastern front. In this way they were enabled to carry on the campaign in Poland somewhat as originally planned. The striking feature of the remainder of the year in this quarter was the successive desperate attempts of the Teutonic allies to penetrate to Warsaw.

The German peaceful penetration of the Ottoman Empire was described in the first volume, and a forecast was made of its tremendous military, political, and commercial possibilities. Very likely the German General Staff counted from the first on the participation of Turkey in the war, which would be especially serviceable in effecting the isolation of Russia by closing the Dardanelles. But it is very doubtful whether they expected that military operations of major importance in that quarter would be required for their purposes. They were doubtless convinced at the beginning that the war would be decided in Europe. A Turkish offensive against the Suez Canal and Egypt was planned with the concurrence, and probably at the suggestion, of the German officers in Turkey. But the Germans probably regarded this movement, originally, like the insurrections which they were prepared to encourage in the dependencies of their opponents, as a diversion, calculated to distract their adversary's attention from the presumably critical field of operations in France.

Many are the minor questions relating to the formation of the German plan which stimulate our curiosity. But there is space for the treatment of only a few of them here.

A discrepancy as to German intentions is implied in the representations made in London on August 1st and those on August 3d. On the former date Prince Lichnowsky hinted to Sir Edward Grey that Germany might promise to respect Belgian neutrality if Great Britain would engage to remain neutral in the war. On the latter date, to judge by the proposals for British neutrality formulated by Baron Kuhlmann of the German Embassy in London, Germany was prepared only to engage not to make any warlike use of the seacoast of Belgium. We may assume that the final decision of Germany in respect to Belgium would in any case have been made in accordance with the requirements of the General Staff, and it would be very interesting to know whether the attitude of the military chiefs of the German nation underwent any change between the 1st and 3d.

In treating this question we must take several possibilities into consideration. In the first place, Prince Lichnowsky's hint was either made at his own initiative or in consequence of instructions from Berlin. In connection with a matter of such gravity it seems likely that Prince Lichnowsky spoke in accordance with a suggestion from Berlin.

The question next arises whether such a suggestion was made by the German Foreign Office from purely political motives regardless of the plans of the General Staff or with conscious reference to the opinion of the General Staff. In the second case we may assume that it reflects an attitude of indecision on the part of the General Staff as late as August 1st with regard to the Belgian field of operations. The final decision to invade Belgium must then have been made between the conversation on the 1st and the publication of the German proposal to Great Britain on the 3d. Perhaps the decision of Italy to remain neutral influenced the attitude of the General Staff in this matter between these two dates. Italy's decision was undoubtedly a great disappointment to them. But it is much more consistent with all the other evidence of German military plans and intentions to suppose that the design of traversing Belgium had been definitely adopted before August 1st. If this is true Prince Lichnowsky's suggestion does not in any way reflect the military attitude. It might indicate a lingering lack of harmony of view between the German General Staff and the Foreign Office. But it is far more likely that Prince Lichnowsky's inquiry, whether Great Britain would remain neutral if Germany would give an engagement not to violate Belgian neutrality, was merely a *bluff*, a diplomatic maneuver intended to make Great Britain appear as the irreconcilable party.

It is quite natural that the suggestion should have been made that political and economic, in addition to strictly military, considerations were a prominent factor in the deliberations of the German General Staff for determining the direction of the offensive movement in the West; that special circumstances outside the field of strategy invested Belgium with an alluring influence of great potency for the leaders of German military policy.

With the vivid impression in one's mind of Germany's discontent with her inadequate colonial possessions, it is difficult to escape the suggestion that the invasion of Belgium in 1914 was related to the desire to attain control of the great Belgian dependency in the Congo basin.

When the annexation of the Boer republics by Great Britain blighted German hopes of a future Teutonic South Africa, Germany turned her attention to other suitable opportunities for colonial expansion in the Dark Continent. She apparently conceived the idea of a consolidation of territory in Central Africa with the Congo Free State as the core. This project might naturally lead to the extension of German rule across the continent from the Atlantic to the Indian Ocean. The existence of such an aspiration was revealed by the terms for the conclusion of the Agadir incident in 1911, when Germany received as compensation for the strengthening of French influence in Morocco an extension of territory in the French Congo which brought the German Kamerun practically into contact with the coveted Belgian Congo. Railways were constructed both in Kamerun and German East Africa towards the western and eastern boundaries of the Belgian Congo respectively, as if the eventual connection of these two lines by a German line across the present Belgian territory were a part of manifest destiny. Was it not conceivable that the situation created by the passage of German troops across Belgium and the destruction of Belgian neutrality, if judiciously exploited, would give Germany control of Belgium's colonial possessions, and is it not likely that the German General Staff was moved partly by this prospect in planning the campaign of 1914?

We turn our attention for a moment to the conjecture that industrial and commercial considerations were among the factors which concurred in influencing the judgment of the leaders of Germany in their choice of a plan of operations. An imposing array of evidence can be marshalled in support of this opinion. More than ever before, iron and coal are fundamental elements of strength in war and peace alike, while the metallurgical industries are now recognized to be the corner-stone of military as well as manufacturing supremacy.

In the utilization of her coal supplies for the production of mechanical energy and in the manufacturing of iron and steel, Germany had been advancing with giant strides. Her coal production had nearly doubled since 1900, and her production of iron had grown from about 9,000,000 tons to 19,000,000 tons by 1913. Coal is the modern talisman by which treasures are amassed in the localities which possess it. Westphalia contained the leading coal-producing area in Germany and the chief centers of the metallurgical industries. It is the heart of manufacturing Germany. A group of flourishing cities near the junction of the Ruhr with the Rhine, occupying an area not too large with modern conditions to constitute a single urban community, only require to be incorporated to take their place in the census lists as one of the great world-cities, the prospective rival of Berlin and Paris. Their fuel supplies, sufficient for centuries, lie beneath and around them. The same carboniferous basin extends under Belgium into northern France, determining the location of the leading industrial centers in those countries also.

In a war that was to be waged as much by industry as by arms, Germany crippled her opponents from the start by occupying precisely these regions of Belgium and France.

Again, the great industrial center of Westphalia does not enjoy convenient access to the sea in national territory. Rotterdam and Antwerp have grown and prospered amazingly, chiefly because they are its natural seaports.

There are said to have been no fewer than 20,000 German residents in Antwerp before the war, and there were important German colonies in Brussels and other leading cities of Belgium. Ostend and Blankenberghe were becoming German watering-places. Antwerp was regarded as an outpost of Westphalia. The shipping interests of Antwerp were largely in German hands. A large part of the tonnage of imports and exports consisted of wares which were destined for Germany, or originated there, and traversed Belgium in transit. Just as in Italy and elsewhere, Germany was establishing an economic supremacy in Belgium. It is not surprising, therefore, that the belief has been expressed that Germany already intended that this peaceful penetration should some day ripen into an opportunity for annexation and that she regarded the crisis of 1914 as the culmination of this process.

Yet in spite of all this striking evidence it seems probable that the minds of the German General Staff were dominated by the purely military considerations. They could scarcely have been blind to the great possibilities which have been enumerated. They may have regarded them as incidental advantages which a happy chance had placed in their way, a sort of strategical by-product.

The evidence for the theory that the military policy of the German General Staff was determined by the above-mentioned extraneous motives is, after all, entirely circumstantial, while purely military considerations growing out of the problem of confronting enemies on both the western and eastern fronts might suffice to explain the action of Germany in every respect. The evident supposition of the German government that Belgium would offer no real resistance, and therefore no pretext for political subjection, and the obvious desire to restore friendly relations, even after the capture of Liége, contradict the opinion that the German plan of campaign was deliberately manipulated to subserve the ambitions for aggrandizement which have just been described.

It seems likely that the original intention of the German General Staff, exactly as implied in the text of the demands sent to Brussels, was to occupy only enough of Belgium to secure lines of march into France and lines of communication for their armies after their arrival in France. In other words, the expressed intention of respecting the *integrity* of Belgium was probably sincere. If this conjecture is true, it might appear, at least to the Germans, that the Belgians by their obstinate resistance merely played into the hands of the extreme expansionist party in Germany, by forcing the German government to do what in its moderation it would not otherwise have done, namely, to occupy practically the entire country with a plausible excuse for retaining it.

Field-Marshal Count von Moltke, the illustrious Prussian Chief of Staff in the three wars which led to the establishment of the present German Empire, declared in his essays on war:

"In the assembling and placing of the different armies at the beginning of the war, all the various and many-sided political, geographical, and strategical considerations must be taken into account. It is hardly possible during the whole course of the campaign to correct a mistake committed in the initial disposition of the armies. But there is an abundance of time to weigh these dispositions carefully, and if this is done, they must without fail bring about the result desired, provided, of course, the army is ready and the system of transportation perfectly organized."

This utterance conveys a comprehensive impression of the essential function of the General Staff, the brain of the German army. This organ was stamped with its peculiar character by the genius of the great von Moltke, who may be regarded in a certain sense as the originator of the office of Chief of the General Staff, since he brought about the separation of this position from the post of Prussian Minister of War in 1857. To-day the German General Staff is generally regarded as the most thorough body of its kind in the world. It has served largely as the model or prototype for the corresponding organizations of other countries.

General Helmuth Johannes Ludwig von Moltke was Chief of the General Staff at the commencement of the Great War. His appointment to this supremely responsible position in 1906 was severely criticized. It was thought to be a consequence of the fact that he was nephew of the illustrious strategist of the preceding generation, and was regarded by many as a foolish act of deference to an illustrious name. This distinguished relationship has been at once an advantage and a misfortune for the younger soldier. It furnished him an inspiration, an ideal, and an example; but it suggested an ignominious distinction between von Moltke the Great and von Moltke the Less.

The almost superhuman demands upon the Chief of Staff require a man of almost unattainable endurance, amplitude of intelligence, and determination. A certain similarity to the uncle in outward appearance and character suggested that the nephew was not destitute of these essential qualities. He possessed the same sharply-cut features, though his countenance and figure were cast in a

GENERAL JOSEPH JACQUES CÉSAIRE JOFFRE

Commander-in-chief of the French army.

more generous mold. He had the same quiet, unassuming manner, plainness of speech, and aversion for popular display. Reserve and unobtrusiveness had become a principle of conduct at the General Staff.

The younger von Moltke was born in Mecklenburg-Schwerin on May 23, 1848. Determined to devote his life to the career of an officer he joined the 86th Regiment of Fusileers as ensign and received his commission as second lieutenant in time to serve with distinction in the Franco-German War, where he won the Iron Cross. He spent the three years, 1876-1879, at the *Kriegsakademie* and was ordered to duty with the General Staff at the termination of his studies there. He was appointed captain in the General Staff in 1881, and the next year became adjutant of his uncle, with whom he was henceforth in intimately personal relationship until the field-marshal's retirement from the position of Chief of Staff.

As major the younger von Moltke became personal aid on active duty to the Kaiser in 1891, which was the beginning of a long period of personal contact with the sovereign that was of fundamental significance for the remaining course of his career. He was raised to the rank of colonel in 1895 and commanded the 1st Grenadier Regiment of the Guards. As major-general in 1899 he was assigned to the command of the 1st Infantry Brigade of the Guards, and as lieutenant-general in 1902 to that of the First Division of the Guards. In the meantime he had became adjutant-general to the Kaiser.

Count von Schlieffen, who enjoyed a distinguished reputation as an authority on the art of war, then Chief of the General Staff, discerning the ability of General von Moltke, selected him as his first assistant in the post of quartermaster-general, February 18, 1904. From this time his ultimate advancement to the post of greatest importance as the culmination of his career could be regarded as predetermined. He became Chief of Staff on the retirement of Count von Schlieffen, sixteen years after his uncle's death, in 1906, being elevated at the same time to the rank of general of infantry.

The supreme command of the land and naval forces of France is nominally vested in the President of the Republic, and he is not excluded, constitutionally, from the exercise of this function. But in normal circumstances his command is exercised through the minister of war, to whom the Chief of the General Staff is responsible. The latter is automatically designated, as generalissimo, in case of a serious conflict, to wield the full military authority of the president in commanding the armies in the field.

For nearly a century the rule has existed in France that all generals shall retire at sixty-five,—a senseless rule that would have deprived Germany of the services of Count von Moltke before the triumphs which made his name illustrious, if it had been applied in Prussia. As the French Chief of Staff is not a commanding officer, his selection is not subject to seniority of rank, and it was possible to pass over generals who were nearing the age of retirement in favor of one who had time before him for the development of a systematic, progressive policy. In these circumstances General Joseph Joffre was summoned to the post of Chief of the General Staff in 1911 as the choice of the Caillaux ministry then in power. The appointment was received by the press with considerable adverse criticism, like the selection of General von Moltke to the corresponding position in Germany five years before. The fact is, that the qualities which commend an officer for appointment to this position of paramount importance are not those most likely to win notoriety, especially in time of peace. General Joffre was almost as unknown in France in 1911 as he was abroad before the outbreak of the war in 1914.

His conception of the duties of his position and of the function of the General Staff is clearly expressed in the following quotation from an address delivered at a reunion of former students of the École Polytechnique:

"To be prepared in our days has a meaning which those who prepared for and fought the wars of other days would have great difficulty in understanding. It would be a sad mistake to depend upon a sudden burst of popular enthusiasm, even though it should surpass in intensity that of the volunteers of the Revolution, if we do not fortify it by a complete preparation. To be prepared to-day we must assemble all the resources of the country, all the intelligence of her children, all their moral energy, and direct them towards a single goal: Victory!

"We must have organized everything, foreseen everything. Once hostilities have begun, no improvisation will be worth while. Whatever is lacking then will be lacking for good and all; and the slightest lack of preparation may involve disaster."

General Joffre brought to his position a spirit of method, thoroughness, and efficiency which is not unlike the prevailing atmosphere in the General Staff in Berlin. His appointment has been regarded as an expression of the new feeling of patriotism which was beginning to take hold of the French people in 1911. His ardent republicanism doubtless recommended him to the Caillaux government. The fact that he was a Protestant enabled him to suppress the disruptive anti-clerical agitation in the army without exciting suspicion or diminishing his prestige. He combines indomitable energy with thorough technical information and a comprehensive practical experience, the

CHAPTER II

The Die is Cast

The German occupation of Luxemburg. Casting the die; the crossing of the Belgian frontier on August 4th. Seizure and destruction of Visé. Desperate resistance of the Belgians. Participation of civilians in the fighting and its consequences. Liége and its defenses. Assault and capture of Liége. Bombardment of the forts. Consequences of the resistance of Liége on the subsequent course of the campaign.

The beginnings of great events are bathed in an atmosphere of solemnity when viewed in historical perspective, even though they may be comparatively insignificant in themselves. The incidents even which attended the initial movements of the Great War in the diminutive Grand-duchy of Luxemburg assume this illusory appearance. As the clouds grew heavier in the murky, midsummer days of 1914, the leading individuals of the Grand-duchy of Luxemburg were chiefly concerned about the continuation of their fuel supply. A war between the Great Powers might very likely cause an interruption in the importation of coal, upon which their principal industry depended. The closing of the smelting works in consequence of lack of fuel would throw a large part of the laboring class out of work and fill the grand-duchy with want and misery. Their supply of coal was derived almost exclusively from Germany. Apparently they felt even less apprehension than the Belgians about a possible violation of their neutrality.

Just as in Belgium, the shock of consternation came in the night. Between one and two o'clock in the morning of August 2d the first body of German troops crossed the Moselle at Wasserbillig and set foot on the soil of the grand-duchy. An automobile containing German officers appeared in one of the suburbs of Luxemburg about five the same morning, but withdrew at sight of Major van Dyck, commander of the forces of the grand-duchy, the gendarmerie of 155 men, who had taken his stand on the famous viaduct by which the road from Trier enters the city.

The Belgian minister learned of the violation of the territory of the grand-duchy about six and immediately telegraphed the ominous news to Brussels.

A German armored-train of nine box-cars, and a flat-car loaded with rails, drew up in the station of the capital about nine in the morning. The captain of the company of engineers who arrived in this train announced that he had orders to occupy the station and railways. Shortly afterwards the viaducts leading into the city were occupied and German sentinels were posted before the post office and other public buildings. Other trains loaded with German troops arrived in great numbers during the day from all points.

On the 4th M. Mollard, French Minister at Luxemburg, was dismissed at the dictation of Herr von Buch, the German Minister, and the latter proposed that French interests in Luxemburg should be entrusted to the care of the Belgian minister for the duration of the war. This rather obscure circumstance testifies to the belief of the German government that Belgium would comply with their demands, or would not, in any case, carry her resistance so far that a complete rupture in diplomatic relations between the two countries would be necessitated. For of course Germany would be no more disposed to tolerate in Luxemburg a diplomatic representative of a hostile Belgium than of a hostile France.

On the 8th the German military authorities in Luxemburg demanded that Count de Jehay, the Belgian Minister, should leave the grand-duchy, and in compliance with this command, as reluctantly transmitted by the Minister of State, President of the Government, M. Eyschen, he departed the next day by way of Coblenz, Cologne, and Holland.

Germany endeavored to justify her action in Luxemburg by the specious argument that the state of war made it necessary to provide for the safety of the railways in the grand-duchy, which were operated by the German government as a part of the imperial railway system of Alsace-Lorraine. In formulating the terms of the lease by which the grand-ducal government turned over its railways to the German imperial management every precaution had been taken to prevent the use of the lines for any military purpose incompatible with the neutrality of the country. The section of this agreement, which was dated November 11, 1902, bearing upon this point, article 2. is drawn up with unusual explicitness, as follows:

"The Imperial Government binds itself never to employ the railways of Luxemburg, as managed by the German Imperial Directorate of the Railways of Alsace-Lorraine, for the transportation of troops, arms, material of war, or munitions, and not to use them during a war in which Germany shall be involved for provisioning the troops in a manner incompatible with the neutrality of the Grand-duchy, and in general not to institute or tolerate in the management of these lines any act which should not be in strict accord with the duties incumbent upon the Grand-duchy as a neutral state."

It is scarcely necessary to remark that Germany violated the spirit as well as letter of this compact in every single respect.

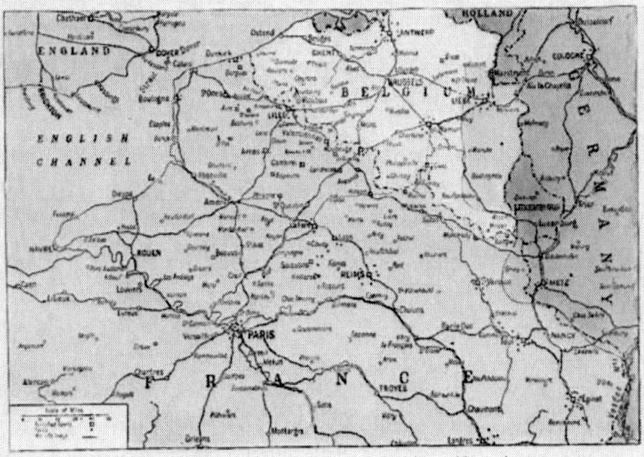

Map showing by the shaded portion the extent of the German advance into Belgium and France, August 10, 1914

The whole subsequent course of the campaign of 1914 is evidence,—if further evidence is needed,—of the hollowness of German professions regarding Luxemburg and Belgium. The German government represented the occupation of these territories as an absolutely indispensable and, at the same time, very distasteful measure of self-protection, and then at once made Luxemburg as well as Belgium the base from which to launch the most stupendous aggressive movement recorded in history. Can anyone frankly believe that the direction of all this mighty force was merely secondary, incidental to Germany's concern for the safety of her flank? Can anyone contemplate the human avalanche which swept from the Meuse to the Marne, premeditated and elaborately prepared in all its parts, the culmination of years of indefatigable toil and forethought, the supreme effort to which all the varied resources of Germany contributed, and then admit for a moment that this tremendous phenomenon, proceeding as it did, was circumstantial in the sense that any degree of innocence or discretion, any attitude of candor or conciliation on the part of these two neutral states, any precaution short of the creation of a military establishment which would have made Germany's enterprise clearly unprofitable, could have shielded Luxemburg and Belgium from the violation of their territories?

The German General Staff drew up their plan of campaign, no doubt, with supreme indifference for all but strategical considerations, leaving to the civilian chiefs of the state the thankless task of palliating the violence and injustice incidental to its execution, as best they could, in the eyes of the world. We may safely assume that the military leaders dictated the traverse of Belgium as the indispensable condition, without which they refused all responsibility for the safety of Germany.

The events in Luxemburg which have been narrated were only the prelude to the drama whose real action began on August 4th.

We considered in Volume II the circumstances of Germany's brief hesitation before venturing into a game of fortune in which the stakes were believed to be world-power or downfall, a gamble with destruction. The fatal die was cast on August 4th and the German boundary near Aix-la-Chapelle (Aachen) became a second Rubicon in world-history.

In several instances there is a striking coïncidence in date between important events in the first campaign of the present conflict and those of the Franco-German War. Thus the German forces crossed the Belgian frontier on precisely the same date and at almost the same hour that they commenced the invasion of France forty-four years before. One is tempted to believe that such concurrences were not entirely accidental. There is little doubt, for instance, that the Germans put forth every effort to strike a blow on September 1st or 2d which should repeat the decisive victory of Sedan. We cannot suspect that the General Staff, or even the Kaiser, was actuated by superstition or a subjective sentimental impulse in arranging the program of the great offensive drive. But in their zeal for arraying every favorable condition on their side, the military authorities probably recognized, and tried to profit by, the common human weakness for the association of dates.

The first requisite in the accomplishment of the great German venture was the capture of Liége, the key to the valley of the Meuse, by surprise, if possible, before the Belgian army was concentrated or the Allies had penetrated the nature of the German design or had had time to come up. General von Emmich, Commander of the Ninth Army Corps, was doubtless a very fortunate selection as leader of this first operation in the West, a stroke upon which so much depended.

The appearance of General von Emmich in the early hours of August 4th among the German troops which had been hastily concentrated on the Belgian frontier was greeted with the most animated expressions of enthusiasm. The hour of destiny had sounded. The crossing of the boundary at Gemmenich in the dim gray light that precedes the dawn must have thrilled with the suggestion of mysterious possibilities the imagination of the youthful soldiery, convinced as they were of speedy victory, but probably somewhat confused as to the immediate purpose of their march. The hour in which this expedition was set in motion had not yet taken on the character of awful gravity with which the gradual revelation of its significance has subsequently invested it. We now know that the immense travail with which the ever prolific civilization of ancient Europe was to bring forth a new era in the life of man commenced at that fateful moment. Who could realize at that time that the mere crossing of an invisible boundary, while humanity still slumbered, would transform the whole aspect of life? The old appearance of Europe was rolled together as a scroll, and men awoke on the morning of the 4th to changed impressions and a different outlook.

The territory directly eastward of Liége is hilly, broken, and wooded in parts. It may be regarded as the northern extremity of the region of the Ardennes. The principal railway line connecting Liége with Germany ascends the valley of the Vesdre, passes through Verviers, the center of the woolen industry of Belgium, traverses the more rugged parts of the section by means of tunnels, and reaches the German boundary at Herbesthal, proceeding thence to Aix-la-Chapelle and Cologne. Northward the valley of the Meuse expands and the elevations towards the German frontier are less abrupt.

The German columns advanced towards Liége by several different routes. A flying column was dispatched by the easier northerly route, from Gemmenich directly westward, to seize with the least possible delay a bridge across the Meuse at Visé ten miles below Liége. From there forces could proceed along the left bank of the river to coöperate in the attack on Liége, and a cavalry screen could be extended westwards to intercept reinforcements for the embattled city, spread panic far and wide across the Belgian plain, and prepare the way for the further march of the main armies. An advance-guard of 1500 men hurried forward to Visé in 150 motor-cars, followed by the marching column.

The Germans encountered resistance from the first in their advance on Liége. In many places the Belgians cut down the trees which stood in rows along the highways, causing them to fall across the roadway, and appropriated whatever other material came to hand for the erection of barricades. Attempts were made to blow up the railway tunnels, and when one such effort failed near Verviers, the Belgians started seventeen locomotives into it at full speed from opposite directions hoping that the ensuing collisions would fill the opening with an inextricable mass of twisted wreckage. But the German engineers cleared out the passage in a single night, finding several of the locomotives practically unharmed.

The inhabitants of some of the Belgian towns and villages appear to have taken part in resisting the invaders. Who can be surprised? Later, notices were posted up by the Belgian authorities warning the people not to take any part in hostilities. But the Germans had invaded Belgium in the night of August 3-4 without any declaration of war. The civilian population in the villages first traversed by the

General von Falkenhayn, German Minister of War.

King Albert and General von Emmich at Belgian army manœuvres before the war.

Germans had not been warned of the serious legal aspect and grave consequences of their participation in acts of hostility. Germany's final notification of her intentions was not presented in Brussels until about 6 o'clock on the morning of the 4th. The people, who were surprised and startled by this sudden irruption, regarded themselves as victims of a treacherous outrage. They believed that they acted in righteous defense of their homes, but the Germans retaliated upon them with the severity which was soon to become proverbial.

In the midst of their intricate and generally accurate calculations, the German strategists had failed to anticipate the heroic resistance of the Belgian nation, which occasioned a very annoying and perplexing delay in the execution of their great plan. The Germans had industriously collected a vast store of knowledge about the material resources and political conditions of their possible antagonists. But very likely those whose duty it was to assimilate and utilize as a guide to policy this mass of intelligence were embarrassed by a plethora of information. The German government displayed an incapacity to appreciate the more subtle, emotional forces which actuate the conduct of nations in grave emergencies, the revulsion of national feeling, and the outburst of patriotism. They failed to attribute to their adversaries the same spiritual qualities upon the existence of which in their own nation they so largely depended for the success of their bold design. Afflicted with the characteristic myopia of national egotism they could only perceive the lofty impulses of their own people. They assumed that the German nation would alone be impelled by an irresistible outburst of enthusiasm; while the spirit of its enemies would remain comparatively cold.

In the moment of elation, in eager expectation of a rapid, brilliant consummation of their immediate enterprise against France, the German leaders encountered a bitter, irritating, determined resistance where it had not been expected. The civilian population, roused to a fury of resentment, unfortunately took part in some instances in this resistance. The invaders were exasperated, inflamed with a spirit of unreasoning indignation. The strictly limited time at their disposal for the rush to Paris, the exacting marching schedule, and consequently frantic concern for promptness of movement intensified their irritation at any unforeseen cause for delay. It is not surprising, however outrageous it may seem in the tranquil remoteness of abstract reflection, that they retaliated most harshly upon the civilian population, making the innocent suffer with the guilty, interpreting the conventional rules of war in the most relentless fashion.

A German official communiqué characterizes the operations leading to the capture of Liége as follows:

"Our difficulties lay in the very unfavorable character of the hilly, wooded country, and in the treacherous participation of the whole population, even the women, in the conflict. They fired from ambush, from the cover of villages and forests upon our troops, upon the surgeons who tended the wounded, and even upon the wounded themselves. There was severe, desperate fighting; whole villages had to be destroyed to crush the resistance, until finally our gallant troops forced their way through the girdle of forts and took the city." Making reasonable amplification to compensate for the probable reserve of such an official description, the imagination unfolds a lurid tragedy of hatred, rage, and fury, and depicts the melancholy spectacle of a devastated, ruined countryside.

A discussion of the methods employed for repressing the supposed intractability of the civilian population in Belgium must be postponed to a subsequent chapter, where the general subject of the alleged atrocities will be treated. One observation may be made here with regard to the activity and punishment of *franc-tireurs*, as the irregular combatants are commonly called. To whatever conclusion the sifting of the evidence will lead, the majority of the American people will, for the present, ascribe to the discussion in relation to Belgium a merely speculative significance. For, until the Germans can prove their right to be in Belgium at all, until they can divest their forcible presence of the appearance of outlawry, unbiased opinion will not admit that the unqualified guilt of the Belgian *franc-tireurs* can be established on the basis of the charges preferred against them.

It is always difficult to secure the concurrence of neutral opinion in the condemnation of civilians for participation in acts of warfare, because, by the very nature of the situations from which such accusations originate, they are always made by invaders, oppressors, or alien rulers, so that those who are thus incriminated enjoy the advantage of an assumption in their favor of justifiable motives, as defenders of their own homes. Humanity is inclined to condone their guilt even when it is sufficiently authenticated. It may be remarked that the same German press which denounced the resistance of Belgian civilians in a tone of righteous indignation in 1914 greeted similar action by the civilian population of the Boer republics as deeds of gallant heroism at the time of the South African War in 1899-1902.

Visé was a tranquil little town of about 4,000 inhabitants, two miles south of the Dutch boundary, lying pleasantly along the eastern bank of the Meuse. The Belgian engineers from Liége blew up the iron bridge at Visé before the Germans could seize it. For some time the Belgian artillery on the left bank of the Meuse, and notably the guns from Pontisse, the nearest of the forts of Liége, which was about five miles away, prevented the construction of pontoon bridges at this point, but finally the Germans effected a crossing. Later, some of the inhabitants of Visé were charged with firing on the Germans, and as retribution the town was pillaged and burned, earning the melancholy distinction, heralded throughout the world, of being the first town to be destroyed in the track of the invaders.

Liége, the "Birmingham of Belgium," with its population of about 185,000 souls, is the center of the metallurgical industries of the country. Its fuel supply has been accumulated by nature beneath and around it. The ugly, cone-shaped mounds of refuse in every direction reveal the location of the pit-mouths of the collieries. The city is stretched out in the winding valley of the Meuse at the point where it receives the waters of its tributaries, the Ourthe and the Vesdre, from the right. The heart of Liége lies on the left bank of the Meuse, where the level space by the river is quite narrow, and the city seizes every opportunity to climb the irregular, confining hillsides. There is an extensive tract of level land between the rivers where the buildings were located for the international exposition of 1905, held in commemoration of the seventy-fifth anniversary of the establishment of Belgium as an independent state. At Seraing, above the city on the Meuse, are the famous John Cockerill Works, founded by an Englishman in 1817, famous steel and iron manufacturers, well-known makers of firearms.

We are too apt to think of Liége as a single fortress, with close communication between all its parts, and consequently to feel astonishment at the apparent confusion of statement as to the date of its capture. The truth is that Liége as a stronghold was a combination of entirely separate defensive units. The city itself possessed no immediate

TO THE BELGIAN PEOPLE!

It is with my greatest regret that the German troops find themselves compelled to cross the frontier of Belgium. They act under the compulsion of an inevitable necessity, the violation of Belgium's neutrality having already been made by French officers, who, in disguise, have traversed Belgian territory in an automobile, in order to penetrate Germany.

Belgians! it is my greatest desire that means may yet be found to avoid a combat between two peoples who have been friends up to this time, formerly even allies. Recall the glorious day of Waterloo, when German arms contributed to found and establish the independence and prosperity of your fatherland.

But we must have a free way. The destruction of bridges, tunnels, and railways will be considered hostile acts. Belgians, the choice is yours.

I hope that the German Army of the Meuse will not be required to fight you. A free path to attack him who would attack us, is all we ask.

I give **formal guarantee** to the Belgian people that they shall suffer none of the horrors of the war; that we will **pay in gold money** for the provisions that it will be necessary to take from the country; that our soldiers will show themselves the best friends of a people for whom we feel the highest esteem, the profoundest sympathy.

Upon your wisdom and a well understood patriotism depends the escape of your country from the horrors of the war.

The Commandant-General of the Army of the Meuse,

Von Emmich.

PROVINCIAL GOVERNMENT OF NAMUR

VERY IMPORTANT WARNING

The Civil Governor directs the most serious attention of the people of the province to the very grave danger that may result to civilians through employing arms against the enemy.

It is their duty, and moreover it is proper, that in this respect they observe the most complete abstention.

The defense of the territory belongs solely to the public force.

Non-compliance with this warning would serve to provoke reprisals, conflagrations, etc.

Namur, August 7, 1914.

BARON DE MONTPELLIER.

Examined and approved:

Namur, August 7, 1914.

The Military Governor,

MICHEL.

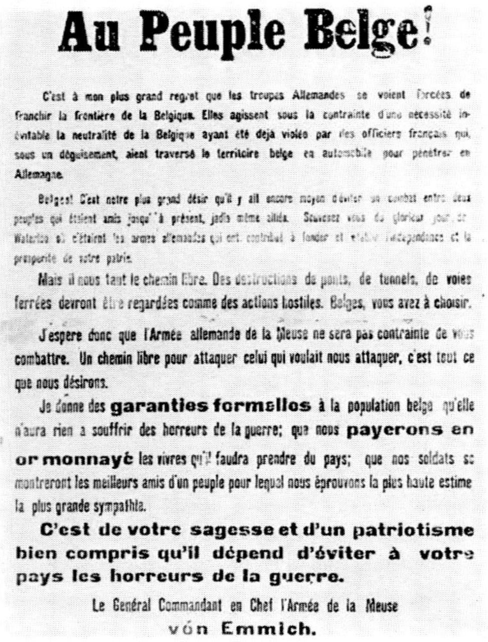

Au Peuple Belge!

C'est à mon plus grand regret que les troupes Allemandes se voient forcées de franchir la frontière de la Belgique. Elles agissent sous la contrainte d'une nécessité inévitable la neutralité de la Belgique ayant été déjà violée par des officiers français qui, sous un déguisement, aient traversé le territoire belge en automobile pour pénétrer en Allemagne.

Belges! C'est notre plus grand désir qu'il y ait encore moyen d'éviter un combat entre deux peuples qui étaient amis jusqu'à présent, jadis même alliés. Souvenez vous du glorieux jour de Waterloo où c'étaient les armes allemandes qui ont contribué à fonder et établir l'indépendance et la prosperité de votre patrie.

Mais il nous faut le chemin libre. Des destructions de ponts, de tunnels, de voies ferrées devront être regardées comme des actions hostiles. Belges, vous avez à choisir.

J'espère donc que l'Armée allemande de la Meuse ne sera pas contrainte de vous combattre. Un chemin libre pour attaquer celui qui voulait nous attaquer, c'est tout ce que nous désirons.

Je donne des **garanties formelles** à la population belge qu'elle n'aura rien à souffrir des horreurs de la guerre; que nous **payerons en or monnayé** les vivres qu'il faudra prendre du pays; que nos soldats se montreront les meilleurs amis d'un peuple pour lequel nous éprouvons la plus haute estime la plus grande sympathie.

C'est de votre sagesse et d'un patriotisme bien compris qu'il dépend d'éviter à votre pays les horreurs de la guerre.

Le Général Commandant en Chef l'Armée de la Meuse

vón Emmich.

Proclamation prepared in advance and distributed by the first of the German cavalry which invaded Belgium.

Gouvernement Provincial de Namur

Avis très important

Le Gouverneur civil attire la très sérieuse attention des habitants de la province sur le très grave danger qui pourrait résulter pour les civils de se servir d'armes contre l'ennemi.

Ils doivent, à cet égard, observer, comme il convient du reste, l'abstention la plus complète.

C'est à la force publique seule qu'il appartient de défendre le territoire.

Toute inobservation de cette recommandation serait de nature à provoquer, le cas échéant, des représailles, des incendies, etc.

Namur, le 7 août 1914

Baron de Montpellier.

Vu et approuvé :

Namur, le 7 août 1914

Le Gouverneur militaire,

MICHEL.

Notice posted by the Belgian authorities warning the people not to take part in the hostilities.

fortifications. But a series of twelve isolated forts, situated at distances varying from 6,500 to 10,000 yards from the center of the city, formed a girdle around Liége. The perimeter thus outlined was about thirty-three miles in circuit. The fall of the city did not necessarily imply the fall of the forts, nor did the loss of any individual fort inevitably involve the loss of its neighbors.

The forts were either triangular or trapezoidal in design. They were constructed chiefly in concrete on the cupola system, steel domes protecting the guns and gunners. The forts were divided into two classes according to their size, the larger ones alone being capable of very effective resistance. Of these, Forts Barchon, Fléron, and Boncelles were situated on the left side of the Meuse, and Forts Pontisse, Loncin, and Flémalle on the right. The larger forts except Flémalle were roughly triangular in plan with the base towards the city and the apex in the supposed direction of attack. Each consisted of a huge concrete mass and was surrounded by a moat. According to the normal scheme, there was a steel turret in the center with two 15-centimeter guns, and there were four other turrets, placed in such a way as to form the corners of a quadrilateral enclosing the first, which mounted 11-centimeter guns. There were also disappearing turrets at the corners of the forts with smaller quick-firing guns. As we shall presently observe, the guns in the forts were in the end hopelessly outclassed by the siege-artillery brought up by the Germans. In the subterranean chambers, formed in concrete, the ammunition was stored and the garrison could take refuge. There were also outer defenses whence infantry with rifles and machine-guns could repel the attacks of storming parties.

Most of the forts were invisible from the interior of the city. In fact a stranger would scarcely have suspected their existence. The intervals between successive forts in the girdle varied from 2,800 to 7,000 yards. These intervening spaces had never been fortified in any way. It is reported that more than 50,000 men were hastily set to work, shortly before the arrival of the Germans, preparing earthworks to facilitate the defense of these open spaces; but the work was far from complete and of little value when the attack began.

The garrison of Liége and the forts was altogether inadequate. The total Belgian forces in the intrenched camp of Liége did not exceed 30,000 at any time. They were manifestly too weak to defend the broad intervals between the forts. They should have been at least twice as numerous. Most of them withdrew after the Germans occupied the city. The garrisons left in the forts were small, and many of the gunners had received no adequate training.

The storming of the city was a distinct performance in the series of operations carried on by the Germans in the vicinity of Liége. A few of the forts were taken at the same time. The Seventh, Ninth, and Tenth German Army Corps had been hastily concentrated on the Belgian frontier without waiting to be mobilized. Light columns were thrown forward without siege-artillery or extensive supplies. They expected to take the city by surprise, where abundant stores of provisions would presumably be found. The German soldiers who first arrived before Liége doubtless suffered considerable hardship from lack of food in consequence of the prolongation of the resistance of the city beyond the prevision of their commanders.

At most about 40,000 Germans were engaged in the operations which culminated in the capture of Liége itself. General von Emmich afterwards declared that his forces did not exceed this number, a statement which is not inconsistent with the official announcement that Liége was taken by six brigades on a peace footing, with cavalry and artillery.

On the evening of the 4th troops of the Seventh Corps, who had advanced by the direct route from Aix-la-Chapelle *via* Herve, attacked Forts Fléron and Evegnée. By August 5th German columns were advancing into Belgium wherever the routes were practicable. On the 5th and 6th the German infantry made repeated attempts to capture some of the forts of Liége by rushing forward in dense masses to reach the supposed zone of safety inside the range and below the trajectory of the heavy guns. But there they were received by the defenders of the outworks with volleys from rifles and machine-guns and repulsed with serious losses. On the 5th Prince Frederick Charles, grandson of the victor of Orléans and Le Mans, won his first laurels by penetrating momentarily into the city with a small detachment of cavalry. An airship, Zeppelin VI, coming from Cologne flew over Liége on the night of the 5th. It hurled a bomb from an elevation of about six hundred yards, and then, descending to a height of three hundred yards, it dropped twelve others in succession, setting fire to the city in several places and causing temporary consternation.

Portions of the Tenth Army Corps, approaching by way of Spa, probably from Malmedy, arrived before Liége on the morning of the 6th, and attacked Chaudfontaine and Boncelles to the southeast and south of the city. The Ninth Army Corps received the task of attacking the forts to the north, probably on both sides of the Meuse. As on subsequent occasions, the Germans exhibited in their operations before Liége a predilection for night attacks. Some of their fiercest assaults on the forts were delivered in the darkness, when the fitful gleam of searchlights, the glare of bursting shells, and the reflection of conflagrations invested an actual Inferno with the added terror of a weird, fantastic setting.

The severest fighting took place on the 6th. The German attacks were repelled in several quarters, particularly on the left bank of the Meuse, with heavy loss. But at nightfall General von Emmich forced his way at the head of the 14th brigade into the obsolete fort and barracks of La Chartreuse, situated on a commanding elevation directly east of the city. General von Emmich won enormous popularity with his soldiers by his personal bravery, dash, and vim, and by his active participation in the fighting before Liége where he was frequently exposed to fire. During the night of the 6th the Germans advancing from the east took possession of three of the bridges spanning the river within Liége, and in the morning the occupation of the city was accomplished. Several of the forts were stormed or silenced on the same day.

The capture of Liége was officially announced on the 7th by one of the Kaiser's aides-de-camp in the Lustgarten opposite the palace in Berlin. This apparently signal achievement on the sixth day of mobilization was the cause of great rejoicing. But the populace undoubtedly made no distinction between the forts which were still holding out and the city itself, and believed that every obstacle to the further progress of German arms at this point had been overcome. The Order *pour le mérite* was conferred by the Kaiser upon General von Emmich in recognition of his service at Liége.

After the occupation of Liége there was a lull in the operations until the heavy siege-artillery was brought up. It was reported that the Kaiser himself gave orders that no more German lives should be sacrificed in storming attacks on the remaining forts. The Germans had apparently reaped what advantage they could gain by rapidity of action. The forts still commanded the railway communications westward. But they could be silenced one by one

Bridge at Liége destroyed by the retreating Belgians and pontoon bridge constructed by the Germans.

Dinant: The citadel on the heights, with the city close to the River Meuse. All the houses, with the Church of Notre-Dame and the bridge on the right, were destroyed by the German shell-fire and incendiarism.

as soon as the heavy artillery arrived. The Germans would then have the advantage of bombarding the forts from the reverse, the inner side, whence attack had not been anticipated in their design. In the meantime, the German concentration for the main advance across Belgium was not complete.

The forts on the right bank of the Meuse were the first to be taken. With the German siege-artillery which was brought up some of the new 42-centimeter (16.8-inch) Krupp mortars made their first appearance. Vague rumors of such mammoth engines of destruction were already afloat. A temporary track was laid to a public square in the heart of Liége by which some of these pieces were brought up and set in position. They deliver projectiles weighing about a ton and a quarter. The violence of the discharge broke all the windows in the vicinity, caused the buildings to quiver and vibrate, and made ceilings collapse. Five projectiles launched from these fearful monsters are said to have silenced two of the girdle forts.

When compelled to retire from the city, General Léman, the heroic commander of Liége, withdrew to Fort Loncin, which dominated the railway westward to Louvain and Brussels. The bombardment of Loncin by 28-centimeter howitzers was begun on August 13th and continued about twenty-six hours at the rate of six shells a minute. When the fort was finally taken General Léman was found lying in one of the passages below the surface, unconscious from the effect of the poisonous gases diffused by the explosion of the German shells. The steel turrets of the forts were smashed and the solid concrete was fairly pulverized by the masses of metal discharged by the German siege-artillery.

The Germans took about 4,000 prisoners in the operations around Liége. Otherwise the losses of the Belgians were probably not very heavy.

The fact that the capture of Liége inaugurated the greatest war in history and the most dramatic movement in that war; the sensation created by the appearance of the new German siege artillery, the greatest surprise of the war; and the intimate association in the minds of so many people in neutral countries with the feeling of indignation aroused by a course of action which outraged their most elementary conceptions of right and justice, emphasize, perhaps unduly, the impression produced by the events described in the present chapter. The Germans proclaimed the capture of Liége as a unique event in military annals, which is probably saying too much. Admirers of Belgium, deeply moved by the self-immolation of the Belgian people on the altar of national honor, and eager to acclaim the benefits of this sacrifice, have declared that the embarrassment in the execution of the German plan occasioned by the heroic resistance at Liége, by giving time for the concentration of the armies of the Allies on the northern border of France, saved the French and British from an irreparable calamity. This statement is clearly incapable of positive proof.

It is erroneous to measure the extent of the delay in the execution of the German plan occasioned by the Belgian resistance as equivalent to the number of days consumed by the Germans in capturing Liége and its forts, the period from August 4th until August 17th. For the Germans would in no circumstances have been ready to advance in force as early as the 4th and the principal masses of their field army did pass Liége before the 17th in spite of the continued resistance of some of the forts. If we assume, as a conservative basis, that the concentration of the German forces would not have proceeded far enough before the 11th, the tenth day of mobilization, for the commencement of the general offensive movement westward, the extent of the delay in the execution of the German plan of invasion caused by the resistance at Liége would be represented by the interval of four days between the 11th and the 15th, when the masses of the German armies were actually set in motion, after the capture of Loncin, which commanded the railway to Brussels. To appreciate fully the very great importance of a four days' respite at this very critical period for the cause of the Allies, we need only recollect that almost all the available forces of France had been concentrated along the eastern frontier with practically no provision to meet the situation created by the German movement in overwhelming strength across Belgium. It is said that one hundred and fifty-nine trains were immediately set in motion to effect the modification in front required by Germany's strategy. But every day's time was precious beyond all computation.

With the vast and complicated organization of service required by modern warfare the advantages of the offensive are even greater than formerly. The party who assumes the offensive translates into action his own premeditated plan, perfected in months or years of prevision and careful preparation. And unless his adversary correctly divines the direction and nature of the coming blow he confronts it in a state of frenzied agitation with arrangements hastily adjusted and in imminent peril of falling into hopeless confusion. The feverish energy of a few days in the face of a mortal crisis can never compensate for a long period of forethought and anticipation. In the actual situation in August, 1914, there was the special danger that flying squadrons of the German Crown Prince's army, which had invaded France through the Gap of Tiercelet about August 10th, would intercept the more direct railway lines between the eastern and northern frontiers of the country and thus greatly impair the French communications.

The world awaited breathlessly, and in part impatiently, the arrival of French and British forces on Belgian soil. We now know how hopeless were the predictions at that time current in England of an impending decisive battle somewhere near the field of Waterloo. But assuming that the respite due to Liége was of four days' duration, without this impediment in their course the German army would have reached the line of the Sambre on the 18th instead of the 22d, when the concentration of the French forces in that quarter was still far from complete and the British Expeditionary Force had not made its appearance. The entire value of the service of the British force in the memorable first days of the retreat from Mons was involved in the defense of Liége, and this was at least considerable.

CHAPTER III

The Deluge Released
(*August 15-22, 1914*)

Renewed proposals by Germany rejected by Belgium. Disposition of the German forces in the West about August 10th. The situation of the Germans at Liége. German cavalry foreshadowing the main advance of the German army through Belgium. Dramatic contrast in the feelings of assurance and consternation in the Belgian towns; Tirlemont, Aerschot, Louvain. The Belgian capital transferred to Antwerp. The German march through Brussels. Reasons for the occupation of Brussels. The German front swings to the left pivoting on Namur.

There is every reason to suppose that the German government thought it probable that Belgium would yield under protest to the requirements presented in the ultimatum of August 2d, perhaps after offering a semblance of resistance. The German authorities still hoped that Belgium, after vindicating her honor by the determined defense of Liége, would be disposed to comply with their demands on the original terms or similar stipulations. After the occupation of Liége, Germany repeated her offer of friendship and the guarantee of the eventual independence and integrity of Belgium on the condition of the free passage of the German armies through the country.

At the request of the Dutch Minister of Foreign Affairs, the Belgian Minister at The Hague, Baron Fallon, forwarded to the Belgian Minister for Foreign Affairs on August 9th the renewed proposals of the German government for a peaceful settlement. Germany had first solicited the mediation of the American minister at Brussels, apparently with the concurrence of Mr. Gerard, the American Ambassador at Berlin. After Mr. Whitlock, who undertook the protection of German subjects in Belgium, declined this special mission, pleading lack of instructions from Washington, the German government through its representative at The Hague begged the Foreign Minister, Jonkheer Loudon, to act as its intermediary in the negotiation.

The text of the German document forwarded by Baron Fallon to M. Davignon was as follows:

"The fortress of Liége has been taken by assault after a brave defense. The German Government most deeply regrets that bloody encounters should have resulted from the Belgian Government's attitude towards Germany. Germany is not coming as an enemy into Belgium. It is only through the force of circumstances that she has had, owing to the military measures of France, to take the grave decision of entering Belgium and occupying Liége as a base for her further military operations. Now that the Belgian army has upheld the honor of its arms in the most brilliant manner by its heroic resistance to a very superior force, the German Government begs the King of the Belgians and the Belgian Government to spare Belgium the horrors of war. The German Government is ready to make any compact with Belgium which can in any way be reconciled with its contest with France. Germany gives once more her solemn assurance that she has not been actuated by the intention of appropriating Belgian territory for herself, and that such an intention is far from her thoughts. Germany is still ready to evacuate Belgium as soon as the state of war will allow her to do so.

"The United States ambassador here concurs in this attempt at mediation by his colleague in Brussels."

The final sentence remained in the original document which was evidently forwarded to The Hague by the American Legation in Brussels.

On August 13th M. Davignon sent the following reply to the Belgian minister at The Hague for transmission to the German government through the medium of the Foreign Minister of the Netherlands:

"The proposal made to us by the German Government repeats the proposal which was formulated in the ultimatum of August 2d. Faithful to her international obligations, Belgium can only reiterate her reply to that ultimatum, the more so as since August 3d her neutrality has been violated, a distressing war has been waged on her territory, and the guarantors of her neutrality have responded loyally and without delay to her appeal."

The general disposition of the German forces in the West was first discernible about August 10th. There were at the time apparently nineteen German army corps, and one Austrian corps which had come to reinforce them. The latter was the Fourteenth Corps which shared with the Fifteenth German Corps the defense of Alsace. The Twenty-first Corps was posted in Lorraine on the line between Strassburg and Metz. Almost the entire extent of the German frontier on the side of France was seemingly guarded only by these three army corps. Notwithstanding the elaborate preparations for the rapid concentration of troops throughout the Reichsland, particularly the strategic railways terminating near the French border and extensive detraining facilities, the bulk of the German forces lay northward of Metz. The available evidence indicated that the forces between Metz and the Swiss frontier were scarcely sufficient for defensive purposes.

On the contrary, as might be expected from the discussion of the German plan, prodigious masses of troops were being concentrated along the western front from Diedenhofen (Thionville) northward. The Thirteenth Corps and First Bavarian Corps were at, or near, Saarbrücken. These were afterwards pushed forward to the frontier of German Lorraine opposite Nancy. The Sixteenth Corps and Second Bavarian Corps were at Diedenhofen and Metz. The Eighth and Twelfth Corps were in Luxemburg and the Third Bavarian Corps either in Luxemburg or at Diedenhofen. The Fifth and Nineteenth Corps were in the Ardennes, the former probably at Rochefort, the latter at Bastogne. The Third and Eleventh Corps had crossed the frontier further north, and occupied positions south of the Meuse, in touch with the Seventh, Ninth, and Tenth Corps which still lay about Liége. The Fourteenth and Eighteenth Corps and the Guard had apparently not reached their final destination at the front. The Fourteenth Corps eventually made its appearance in Lower Alsace on the left wing of the German army operating in Lorraine. The Eighteenth Corps was probably sent to Diedenhofen. The Guard took part in the operations in the Ardennes.

Thus the Allies were confronted with twenty army corps, and probably eight divisions of cavalry in the West. If we reckon a single reserve division for every corps of the first line, a minimum assumption, we shall have an aggregate strength of about 1,200,000 for Germany's mobile forces in the western theater at the close of the period of mobilization or a little later. The fighting strength would consist of about 850,000 rifles, 65,000 sabers, 4,400 guns and field howitzers, and 1,500 machine-guns.

The position of the German forces at Liége remained somewhat precarious even after the city had fallen into their hands. There were three army corps, about 120,000 men, in and around Liége by the 7th or 8th. Thousands

German engineers mining a bridge.

Germans digging out a tunnel which had been blown up by the retreating Belgians.

were daily pouring into this part of Belgium. The reserve formations followed quickly upon the heels of the army corps of the active army. Most of the detached forts were still holding out and their guns commanded the principal lines of communication leading from Germany to Liége, and from Liége westward. The Germans who first arrived before Liége had brought with them only very scanty supplies of food. The resistance of the forts embarrassed the conveyance of supplies and hampered the progress of the Germans westward where their forces could expand and gain elbow-room. There was a threatened congestion in the narrow valley of the Meuse. The Germans would soon have become embarrassed by their own numerical strength. The situation was like forcing a stream of water at high pressure through a narrow spout partly blocked by a sieve.

The resistance at Liége did not entirely prevent an advance westward. General von Emmich remained only a day and a half in Liége and then pushed on up the Meuse in the direction of Namur with the bulk of his corps, not waiting to witness the action of the heavy siege guns upon the recalcitrant forts. The first German cavalry patrols appeared before Namur on the 14th.

The Germans threw out a screen of cavalry, infantry, and light artillery across the Meuse to the westward, to cover the siege operations at Liége, interrupt communication with the forts, reconnoiter the Belgian positions, and, in general, prepare the way for the main advance of the army by spreading consternation as far as possible.

Two divisions of cavalry, the First and the Fourth, crossed the Meuse on the 8th or 9th, chiefly at Lixhe, close to the boundary of Holland, as far as possible from the guns of the northern forts of Liége. They proceeded westward through Tongres and St. Trond, in the direction of Hesbaye, scouring the country and engaging in numerous skirmishes. They were followed, and at times supported by detachments of infantry and field-artillery. The appearance of the 35th Uhlans caused a panic in Tongres Sunday morning, the 9th. On the 11th the Germans seized the station at Landen, twenty-four miles west of Liége on the line to Louvain and Brussels, thirty-eight miles east of the capital. On the 12th there was skirmishing near Tirlemont in which the Germans were supported by their field-pieces.

The novel use of cavalry by the Germans in the present war was no less a surprise, although its effects were somewhat less startling, than the heavy siege-artillery. In their maneuvers the largest organization had been the cavalry brigade of 1,600 sabers. But in the campaign of 1914 each brigade was frequently accompanied by a battalion of infantry with field and machine-guns mounted in armored automobiles, so that it could move independently with far greater assurance.

In the meantime several German army corps were advancing westward through the Ardennes. As the reader has probably observed, the Meuse offers a convenient and natural line of defense for French and Belgian forces repelling a common attack from the eastward. But the Meuse by its course between Verdun and Liége makes a deep depression in this defensive front. It gave the Germans an opportunity of occupying at once a very useful salient. The strategic advantages of the Ardennes for an enemy taking the offensive against the Franco-Belgian front on the Meuse made ample compensation, in the special circumstances, for the relative difficulties of transport in this region. The contour of the Ardennes is, in general, favorable to the movement of an army from east to west. The streams descending to the Meuse flow generally westward or northwestward, so that the valleys extend in directions which correspond with the German lines of advance, and lie athwart the direct line from Verdun to Liége. The Germans encountered practically no resistance in their advance through the Ardennes until the French counter-offensive was launched about August 20th. They pushed westward as far as the line formed by the Meuse directly above Namur as readily as they reached the Meuse below the forts of Liége. The advance-guard of the German army in the Ardénnes came in contact with the French at Dinant as early as August 15th, before the forts at Liége had all been taken. The description of this engagement may be postponed until a convenient point is reached for the consideration of the really significant operations in the region of the Meuse which took place late in August. We may observe, however, in anticipation, that the occupation by the Germans of the salient formed by the Ardennes was not only one of the factors which hindered the timely coöperation of the French forces in Belgium, but it afforded an opportunity of attacking the flank of the Allied armies in southern Belgium and so turning their entire position and endangering their retreat. The capture of the fortress of Huy, midway between Liége and Namur on the Meuse, on August 12th, gave the Germans undisputed mastery of a crossing-place on the river.

The operations of the German cavalry westward of Liége foreshadowed the march of the main invading army.

Either the intelligence department of the Belgian army was entirely inadequate or the Belgian authorities consciously allowed the people to be deluded by untrustworthy information regarding the nature and prospects of the military operations in progress between Louvain and Liége during the ten days following the entry of the

German forces into the latter city. For there was nothing in the accounts published in Belgium to indicate that a disaster impended, or that the progress of operations was other than satisfactory. Stranger still, the belief seemed to be quite universal that the encounters, which were being daily recorded, were really significant battles in which considerable forces of the enemy were taking part.

It is incredible that the progressive concentration of a million men in eastern Belgium could have remained concealed to airmen of ordinary skill or to a military intelligence department of moderate capacity for discernment. Perhaps the Belgian authorities were misled by the hope that the French and British would come to assist them in time to offset the tremendous disparity in numbers; that the French would extend their own front on the left, linking it with the Belgian right, so as to present a continuous front to the invaders and hold them to a contest on parallel lines. This expectation, if really ever cherished by the Belgian Staff, was vain. The hope that the French could arrive in sufficient strength to offset the superiority of the Germans could scarcely have been justified excepting on the assumption that the invasion in the north was only a secondary feature of the German plan of campaign. As soon as it became evident that the Germans were throwing the bulk of their prodigious forces into Belgium, the notion that the French could make the necessary fundamental change in their entire plan of campaign and in the disposition of their forces in time to meet the Germans in Belgium on an approximately equal footing became absurd. It is not to the purpose to compare the distances from the French and the German frontiers respectively to the battle-lines near Louvain. The problem requires us to compare the distance from the French eastern front with that from the Belgian frontier of Germany to the prospective

Imperial
German Legation
in Belgium

Brussels, August 2, 1914.

Very confidential.

"Reliable information has been received by the German government to the effect that French forces intend to march on the line of the Meuse by Givet and Namur. This information leaves no doubt as to the intention of France to march through Belgian territory against Germany.

"The German government cannot but fear that Belgium, in spite of the utmost good-will, will be unable, without assistance, to repel so considerable a French invasion with sufficient prospect of success to afford an adequate guarantee against danger to Germany. It is essential for the self-defense of Germany that she should anticipate any such hostile attack. The German government would, however, feel the deepest regret if Belgium regarded as an act of hostility against herself the fact that the measures of Germany's opponents force Germany for her own protection to enter Belgian territory.

"In order to exclude any possibility of misunderstanding, the German government makes the following declaration:

"1. Germany has in view no act of hostility against Belgium. In the event of Belgium being prepared in the coming war to maintain an attitude of friendly neutrality towards Germany, the German government binds itself, at the conclusion of peace, to guarantee the possessions and independence of the Belgian kingdom in full.

"2. Germany undertakes, under the above-mentioned condition, to evacuate Belgian territory on the conclusion of peace.

"3. If Belgium adopts a friendly attitude, Germany is prepared, in coöperation with the Belgian authorities, to purchase all necessaries for her troops against a cash payment, and to pay an indemnity for any damage that may have been caused by German troops.

"4. Should Belgium oppose the German troops, and in particular should she throw difficulties in the way of their march by resistance of the fortresses on the Meuse, or by destroying railways, roads, tunnels, or other similar works, Germany will, to her regret, be compelled to consider Belgium as an enemy.

"In this event, Germany can undertake no obligations towards Belgium, but the eventual adjustment of the relations between the two states must be left to the decision of arms.

"The German government, however, entertains the distinct hope that this eventuality will not occur, and that the Belgian government will know how to take the necessary measures to prevent the occurrence of incidents such as those mentioned. In this case the friendly ties which bind the two neighboring states will grow stronger and more enduring."

Kaiserlich
Deutsche Gesandtschaft
in Belgien

Brüssel, den 2 August 1914

Très confidentiel.

Der Kaiserlichen Regierung liegen zuverlässige Nachrichten vor ueber den beabsichtigten Aufmarsch französischer Streitkräfte an der Maas-Strecke Givet-Namur. Sie lassen keinen Zweifel ueber die Absicht Frankreichs, durch belgisches Gebiet gegen Deutschland vorzugehen.

Die Kaiserliche Regierung kann sich der Besorgniss nicht erwehren, daß Belgien, trotz besten Willens, nicht im Stande sein wird, ohne Hülfe einen französischen Vormarsch mit so großer Aussicht auf Erfolg abzuwehren, daß darin eine ausreichende Sicherheit gegen die Bedrohung Deutschlands gefunden werden kann. Es ist ein Gebot der Selbsterhaltung für Deutschland, dem feindlichen Angriff zuvorzukommen. Mit dem größten Bedauern würde es daher die deutsche Regierung erfüllen, wenn Belgien einen Akt der

Feindseligkeit

First page of the ultimatum delivered by the German Minister to the Belgian Minister of Foreign Affairs on August 2, 1914.

battlefield in Belgium, and to bear in mind not only that the French, in shifting their forces to the Belgian theater of operations, had to avoid the Ardennes, but that even the railway following the valley of the Meuse down to Namur was menaced or even interrupted as early as August 15th. It was necessary, therefore, that the French should make an extensive detour westward.

It is probable that the Belgian Staff resigned itself as early as the second week of hostilities to the humbler task of merely delaying as long as possible the irresistible progress of the enemy without any expectation of shielding their capital or the greater part of their country from speedy submersion. They could only hope in this way to furnish their allies a few valuable days' respite for consolidating their position, and to make their contribution in this way to the final victory which should eventually redeem their own country from subjection. Considerations of expediency might naturally recommend that the prospect for the immediate future should be concealed. The announcement a week before that Brussels would probably be abandoned to the Germans might have occasioned great consternation and panic without any corresponding advantage for the national cause. The charge brought by the Germans against the leading journals of Brussels of having *deliberately* beguiled the Belgian people with their glowing accounts of fictitious victories is probably ungrounded. Numerous examples may be cited to prove the amenability of the press, even with the most faultless intentions, to the deceptive influence of enthusiasm at the commencement of all great wars. It may be recalled that war was an absolutely unique experience to the Belgians of the present day. With the exception of the insignificant visitation in 1830-1831, the war-god had spared for ninety-nine years the soil of Belgium, once the cockpit of

Europe. The Belgians only felt at second hand the disillusionment of the French in 1870 as a sobering experience.

There were reports that bodies of French troops were present with the Belgian army before the latter's withdrawal from before Brussels in the direction of Antwerp. Such coöperation was probably limited to the French cavalry which entered Belgium as early as the 14th. But the Fifth French Cavalry Division was repulsed by German cavalry in an encounter at Perwez, north of Namur, on August 19th.

The Belgian forces were concentrated mainly on the line of the Dyle with Louvain as headquarters. The rapid development of the campaign, after the principal mass of the invading forces started westward from Liége, reveals the trifling character of the engagements in the Belgian plain which had taken place before that time. It is not likely that the resistance of the Belgians after Liége occasioned any perceptible delay in the advance of the principal masses of German troops. There was some sharp fighting before the Germans entered Brussels. But only the advance-guard of the German armies was engaged in it. The main body of their forces had no occasion to deploy for action. The rate of progress of the bulk of the forces seems to have been determined solely by the marching capacity of the soldiers. Those who arrived in the vicinity of Brussels on the evening of the 19th probably left Liége on the morning of the 16th. They probably reached positions opposite the Charleroi-Mons line of the Allies by August 22d.

A remarkable feeling of buoyancy and assurance prevailed in the towns on the line of the approaching invasion until the enemy was at their very gates, when this impression of confidence was suddenly transformed into a frenzy

Map showing by the shaded portion the extent of the German advance into Belgium and France and the French advance into Alsace and Lorraine, August 20, 1914.

of dismay by the roar of hostile artillery. The contrast was impressively dramatic. The spectacle of the diminutive Belgian army persisting confidently in its hopeless struggle with an adversary whose might was as invincible as destiny, and exhibiting noteworthy examples of individual heroism, stirs the deepest sense of pathos.

On the morning of the 17th the distant rumble of cannon was first audible in Brussels. On the same day the seat of government was transferred to Antwerp, a large part of the archives having been transported thither in motor-vehicles during the night.

Soon after noon on the 18th the Germans began shelling Tirlemont, and immediately the roads leading westward were thronged with thousands of the panic-stricken inhabitants who were leaving all their possessions behind them. Towards evening the Germans, chiefly cavalry and artillery, occupied the town.

The Germans, who had been repulsed before Aerschot on the 18th, resumed the attack early on the 19th. There was a spirited contest for about two hours, when the German infantry, supported by machine-guns, assailed the defenders on the right flank and forced them to withdraw in the direction of Louvain.

Fugitives from Tirlemont and Aerschot, and from many villages in the vicinity of these places, brought terrifying news to Louvain, and the Germans followed closely upon the grim tidings of their approach. There was great excitement in Louvain and many of the inhabitants swelled the streams of fugitives which were filling the roads that led westward. This movement was constantly augmented by the populations of numerous villages which lay in the track of the relentless invaders and had either been burned or were threatened with destruction. Each group of country-folk contributed its tale of brutalities which were exchanged and repeated with inevitable exaggeration and even distortion.

The Belgian General Staff left Louvain for Antwerp on the 19th, the Second and Third Divisions covering the withdrawal of the headquarters and the bulk of the army. There was some sharp fighting near Louvain with the troops of the German vanguard.

Brussels was unusually exuberant during the early days of the war. It retained its cheerfulness and equanimity until the 18th. Every afternoon and evening the principal boulevards were thronged, and a great multitude stood in the Place Rogier, in front of the Northern Railway Station (Gare du Nord), in eager expectation of news. Crowds sat before the brilliantly lighted cafés until late at night discussing the situation and devouring the latest bulletins, or seizing hopeful bits of information from persons who strayed in from the front.

Life for a few days was invested with a peculiar zest, as though grim Fate moved by a fitful impulse of compassion was treating with exceptional indulgence those whom she had condemned to so many months of gloom.

A change in the temper of the capital is said to have become apparent about three in the afternoon of the 18th. The truth about the situation was being revealed by startling evidence, the train-loads of hopeless refugees pouring into the city, the piteous troops of fugitives, some of them bringing a tithe of their possessions in carts and wheelbarrows, others plodding along with a few belongings in sacks thrown over their shoulders, others who had abandoned everything in the feverish haste of their flight. This picture of speechless despair contrasted strangely with the idle gaiety of the city.

The last train left Brussels for Ostend on the night of the 18th and people fought for places in it. Then the

Adolfe Max, Mayor of Brussels.

The white flag used by the municipal authorities of Brussels preliminary to the meeting between Adolfe Max and General von Arnim, at which was arranged the capitulation of the city.

General Sixtus von Arnim, commander of the German forces which entered Brussels.

locomotives and rolling-stock were withdrawn as far as possible from the lines between Brussels and Ghent. By the 19th communications were severed on all sides. Somber crowds stood in the principal squares, their feelings benumbed, deadened, by the crushing weight of anxious foreboding.

During the greater part of the 19th almost uninterrupted processions of automobiles traversed Brussels fleeing westward before the storm of war, while the movement of humbler refugees presented its never-ending spectacle of misery and despair. Some barricades were erected and there was talk of defending Brussels to the last. The Civic Guard, said to have numbered about 20,000, were posted at the principal approaches to the city. The American and Spanish ministers, who remained in Brussels, urged upon the municipal authorities the futility of resistance, which would only result in useless destruction of lives and property, and endanger, particularly, the city's priceless treasures of art. During the following night the order came from the king's headquarters that Brussels should be surrendered without resistance.

About ten on the morning of the 20th shops and houses in Brussels were closed and shutters were closely drawn as the news spread throughout the city that the foe was at the gate. An interview took place just outside the city between the Burgomaster Adolfe Max and General Sixtus von Arnim, commander of the German forces, in which the capitulation of Brussels was arranged on terms which may be briefly stated as follows:

1. The free passage of the German troops through Brussels.

2. The quartering of a garrison of 3,000 men in the barracks of Lailly and Etterbeek.

3. Military requisitions to be paid in cash.

4. Respect for the inhabitants and for public and private property.

5. The management, free from German control, of the public affairs by the municipal administration.

The booming of cannon and the sound of military music conveyed the information to the people of Brussels that the triumphal progress of the German forces through their city had commenced. The Germans made their entry at about two o'clock in the afternoon, after the Civic Guard had been disarmed and disbanded. Although the German march through Brussels was undoubtedly determined by strategic considerations, those who managed its details evidently proposed to secure all the incidental advantages or satisfaction from an impressive display of the varied might and efficiency of the German military machine. All branches of the service and all departments of military activity were represented or suggested while the German host defiled through the streets of Brussels.

Squadrons of Uhlans headed the march followed by one hundred motor-cars mounting quick-firing guns. Every regiment was preceded by its own band. The soldiers sang their national anthems, *Die Wacht am Rhein* and *Deutschland, Deutschland über alles*, to the sledge-hammer time-beat of their heavy boots on the stone pavements. There were cavalry, infantry, and artillery, rapid-fire guns, howitzers, and even heavy siege-guns. The columns kept on one side of the roadway with unfailing precision, leaving the other side unimpeded for the passage of automobiles, the service of communication between different parts of the line by mounted orderlies, motor-cyclists, etc. Portable kitchens accompanied the columns, and food and refreshments were served to the soldiers on the march. Even letters and postcards were collected and distributed by those appointed for this service while the column was in motion.

The Germans entered Brussels by the Louvain Road, descended past the Botanical Garden to Place Rogier in front of the Northern Railway Station, and then passed through the principal boulevards to the heights near Koeckelberg, where many of them encamped. All wore the new German field-uniform of a greenish-gray color, a most neutral, evasive tone, which makes all objects blend completely with their natural environment at a very short distance. The helmets of the soldiers were covered in this same color. The entire material equipment of the army,—gun-carriages and caissons, wagons, motor-vehicles, and the like,—was painted to match. It was an example illustrating the comprehensive scope of German efficiency and prevision. An eye-witness compares this color to the "gray of the hour just before daybreak, the gray of unpolished steel, of the mist among the trees."

Artillery was posted at important positions to command the principal streets and squares, before the ministries and railway stations. The Germans suspected that fully 10,000 Belgian soldiers were concealed in the city in civilian costume. But the people of Brussels maintained a quiet, dignified bearing. Crowds watched the German host as it defiled through the streets with silent, unbending amazement. But Brussels gave no cause for reprisals.

Hour after hour, day after day, the greenish-gray column wound like a serpent through the city, until its duration seemed unearthly, uncanny. Probably about 300,000 Germans passed through Brussels and its vicinity within three days.

The Germans had hoped to overpower the forts of Liége by their unexpected, impetuous assault and straightway open the road into France by way of the valleys of the Meuse and the Sambre without occupying any more Belgian territory than was necessary to afford a passageway across and a covering for their lines of communication. The occupation of Brussels was regarded at the time by the press of the Allies as a spectacular *coup* intended to overawe the Belgians, impress the world at large, and foreshadow the triumphal entry into Paris. It has also been said that the resistance of the Belgians forced the Germans to take a somewhat longer route leading through Brussels. Namur barred the direct route from Liége to Paris. The fact that the Germans appeared on the line of the Sambre above Namur as early as the 21st, while many of the forts of Namur held out until the 25th, shows that they had to pass northward of Namur.

But the occupation of Brussels by the Germans was probably not due solely to the opposition of the Belgians, even the resistance of Namur, although this in itself necessitated a detour northward. It is absolutely essential for the provisioning and orderly progress of large armies that they should advance by many parallel roads. The tremendous forces which traversed Belgium from the vicinity of Liége westward must in any case have utilized all the main highways between the line of the Meuse and the Sambre and a line extending through Brussels. The actual effect of their march was like drawing a great comb across the country, with the teeth moving in the principal lines of communication. The Germans advanced from east to west in parallel columns with a common front extending north and south. When the southern extremity of this front reached Namur about the 18th it was delayed until the 25th or 26th by the resistance of the forts. Meanwhile the front of the advancing columns from Namur northward wheeled to the left, pivoting on its left extremity which remained stationary near Namur, until it faced almost to the south and confronted the Allies on the line of the Sambre and of Charleroi and Mons, in a position extending at an angle

The last telegram received from Brussels. On the morning of August 20, 1914, the Amsterdam telegraph office received the above message which states, "The Germans have arrived, we are retiring, good-bye."

The German forces marching through Brussels, August 20, 1914.

of ninety degrees with its original one. The forces on the German right, at the flying extremity of the wheeling masses, those for instance which passed through Brussels, had to cover a comparatively great distance while those at the pivot were merely marking time. Troops arriving in the vicinity of Mons on the 22d, after passing through Brussels, must have marched a hundred miles from Liége in about seven days. In fact the extent of continuous marching performed by some of them must have been considerably greater; for it is reported that many who passed through Brussels on the 20th had been marching continuously all the way from Aachen (Aix-la-Chapelle). The divisions nearer the axial point at Namur reached their positions on the new front sooner than those on the flying wing, as is shown by the fact that the French on the line of the Sambre were attacked at least a day before the British near Mons.

The possession of Namur, while eventually almost indispensable, was not immediately necessary for the progress of the German plan of campaign. The excellent highways and railway system of Belgium enabled the Germans to disregard for the time the obstruction at Namur.

CHAPTER IV

THE FRENCH COUNTER-OFFENSIVE

Criticism of the French counter-offensive. The French incursion into Alsace; entry into Mülhausen, August 8, 1914, French defeated west of the city two days later. The clearing of the Vosges by the French. Mülhausen retaken by the French on August 20th. The French offensive in Lorraine, beginning August 12th, French advance to Saarburg. Turning of the tide, August 20th. Battle in Lorraine, August 20-24. General collapse of the French offensive. German invasion of France and occupation of Lunéville. Evacuation of Mülhausen by the French. Invasion of France from Luxemburg by the army of the German Crown Prince. Defense of Longwy. French defeated at Neufchâteau. The battle at Dinant, August 15th.

As was explained in the first chapter, a general doctrine for the offensive had obtained the official stamp of approval in France. It did not follow from this that the French should in all circumstances take the offensive from the first. In fact, considerations of prudence, based upon relative conditions of strength, seemed to recommend a defensive attitude for the French in the early stages of the Great War. As the event showed, the Allies would have done better to receive the shock of the German offensive in prepared positions in France, since they were too late to intervene decisively in Belgium, and not to pass to the attack until the initial force of the enemy had been considerably diminished by attrition and the Russian invasion of Germany had been fairly started. But it was difficult to believe that any Fabian policy would commend itself to the French with their impetuous temperament.

While in the main the first campaign was actually conducted by the French in accordance with this more prudent policy, there was at the commencement a period of uncertainty and confusion, perhaps of divided counsels, as to the suitable moment for applying the offensive. In the early weeks of the war the French troops advanced with impetuous eagerness beyond the borders of their own country throughout almost the entire extent of the contested frontier, even before their mobilization had been completed. The French invaded Alsace by the Gap of Belfort, dislodged the Germans from their positions on the Vosges Mountains, penetrated into Lorraine as far as Saarburg, made their way into the Ardennes to Neufchâteau, and took up positions along the line of the Sambre in southern Flanders.

These movements are not all to be judged by the same criterion, although they may all have been regarded by the French General Staff as closely related parts of the same general offensive action.

The incursions into Upper Alsace were hastily conceived and rather carelessly executed with insufficient forces. They were dictated by political rather than strategical motives. The government was naturally eager to inflame national enthusiasm at once by a brilliant stroke appealing to popular sentiment, by taking definite steps towards the redemption of Alsace, and the realization of the hope so passionately cherished for forty-four years. The expeditions here referred to left no permanent results of any consequence. They involved many persons in Mülhausen in the charge of treason, some of whom were brought before a court martial and executed by the Germans after their return. They dissipated uselessly the energy of the French.

The occupation of the summits and passes of the Vosges, while an almost indispensable preliminary step for the conquest of Alsace, might also be justified as a defensive measure.

The principal invasion of German Lorraine from the direction of Nancy was apparently planned with the object of cutting the railway line between Strassburg and Metz, of isolating Metz and Diedenhofen (Thionville) and threatening the communications of the army invading France from Luxemburg, perhaps also of menacing Strassburg and Lower Alsace. It is only fair to judge this project according to the intention and outlook of those who devised and directed it. Strange as it may seem at present, at the time when this maneuver was commenced, the French General Staff probably believed that the Germans would deliver their principal blow through Luxemburg and that the invasion by way of Liége was only secondary. On this assumption there were very good grounds for a vigorous offensive in Lorraine even as a defensive measure; for there was a fair chance that it would paralyze the German effort through southern Luxemburg. On the other hand, the French may possibly have intended this movement as the execution of the characteristic paramount act in their offensive, the penetration of the vulnerable point in the enemy's front. In the first weeks of the war the frontier of Lorraine had probably seemed weakly guarded. The concentration was proceeding at some distance to the rear. The failure of the French invasion of Lorraine was largely due to inadequate intelligence and a miscalculation of the available German defensive forces in this section. The convex contour of Lorraine on the side of France gave the Germans the advantage of maneuvering on inner lines. It was comparatively easy, for instance, for the Germans to shift their forces to and fro between Diedenhofen and Saarburg, the extreme points of military importance in the province. The Germans, moreover, were

Frederick William, Crown Prince of the German Empire and Prussia, with two officers of his personal staff.

well intrenched against the French counter-offensive in Lorraine as well as in Alsace.

Even though the main effort of the Germans was to be made far from Lorraine, on the extreme right wing of their line of concentration, it was scarcely conceivable that they would have denuded this portion of their exposed territory of sufficient forces for effective resistance. The remarkable character of the railway system of the Reichsland reflects, no doubt, in part an earlier German project of penetrating their opponent's lines by way of Nancy, between Toul and Épinal. Several railway lines cross from Alsace-Lorraine into France, notwithstanding that the range of the Vosges bars half the common frontier, while eight or nine other lines end abruptly at the French border. These lines were provided with very extensive sidings and there were eighty unloading platforms in Lorraine alone, each more than 1,500 feet in length, capable of accommodating the longest troop-trains. It was estimated that Germany could detrain between 150,000 and 200,000 men daily between Metz and Strassburg.

The same criticism may be made of the French advance into the Ardennes as of the movement into Belgium generally. It was too late to have any fundamental effect. It drew the French further from their base, away from their prepared positions. It deprived them of the inherent advantages of the defensive without giving them the special advantages of the offensive.

With reference more especially to the French advance to the positions on the Sambre, it may be said that the movement into Belgium was probably made in response to a popular and parliamentary demand that the French should coöperate directly with the Belgian forces in the defense of Belgium. The French troops were first reported on Belgian soil August 14th, the day before the first encounter at Dinant. The advance into Belgium would have been a judicious measure if the French could have arrived early enough in sufficient strength to reinforce and prolong the Belgian front at Louvain and impose upon the Germans a parallel battle. But for this, preliminary arrangements elaborated before the war would probably have been necessary. Either the Belgian and French General Staffs had not formulated any such arrangements for common action in case of the invasion of Belgium, or, if they had, the French authorities, acting in accordance with the letter of Belgium's communication declining their proffered assistance on August 3d, directed the concentration of their forces to follow its normal course, as if the main attack were expected on their own German frontier.

Commendable as it seemed in itself, the French movement northward into Belgium, in the actual circumstances, undertaken some time after the outbreak of hostilities as an afterthought, did not accord with the general plan as already adopted for the French armies. The stupendous movement of the Germans through Belgium inevitably necessitated a fundamental readjustment of the disposition of the French forces at great expense of time and energy. The possibility of this unfortunate necessity was inherent in the general situation and in the relative strength and efficiency of the opposing forces in the West. Unless the French foresaw the direction of the chief blow of their opponents and concentrated their forces against it from the first, renouncing their favorite maneuver of piercing the enemy's weakest point, they ran the risk of being compelled to shift a large part of their forces at a critical period when every hour was immeasurably valuable. The advance into Belgium, at the time when it was made, merely served to increase this inevitable dissipation of energy. The position on the Sambre, a compromise after the failure to coöperate with the Belgian field army, was seemingly fortunate in view of the support of Namur and the supposed capacity of this fortress for an effective resistance.

But the defeat of the French army on the line of the Sambre, even before the forts of Namur had fallen, as well as at Neufchâteau, shows that the French forces were entirely inadequate for successful offensive operations in this direction. It is probable that General Joffre sanctioned the advance into Belgium against his better judgment. Very likely he would have preferred to await the German attack on the Lille-Valenciennes-Hirson line. While this line did not possess the unparalleled strength of the barrier in eastern France, it nevertheless afforded several quite appreciable advantages for a defensive action, such as the fortifications of Lille and Maubeuge, several detached forts between these two cities, and the Scheldt running like a moat along part of the front.

A few days after the collapse of the French counter-offensive there were rumors of a misunderstanding between the French government and General Joffre and of the latter's resignation. A divergence of view regarding the principal aims of French strategy may very likely have contributed to this state of friction, if it really existed.

A confusion in the counsels of the French at the beginning of the war had been anticipated, at least by their adversaries. It is a condition which has been repeated in several crises when the destiny of the nation has not been guided by the strong arm of a military dictator. In a country where parliamentary control is so complete, while the balance of power among the various parliamentary factions is so unstable, it is very remarkable that the situation in this particular instance was not more serious. One of the surprising phenomena of the war has been the comparatively prompt establishment in France of a steadfast leadership with a clear, consistent policy.

General Joffre issued the following proclamation to the people of Alsace on the occasion of the French invasion of the province:

"Children of Alsace!

"After forty-four years of sorrowful waiting French soldiers once more tread the soil of your noble country. They are the pioneers in the great work of revenge. For them what emotion and what pride!

"To achieve this work they have made the sacrifice of their lives. The French nation unanimously urges them on, and on the folds of their flags are inscribed the magic words, 'Right and Liberty.'

"Long live Alsace! Long live France!"

The first incursion into Upper Alsace from the direction of Belfort was merely a reconnaissance in force, as spectacular as it was unimportant. It began on August 7th. The French soldiers who crossed the frontier in the radiant sunshine of the summer morning were thrilled with the conviction that the hour for the fulfilment of the national yearning was at hand and elated by the thought of their own participation in this glorious achievement.

The expedition, consisting of about a division of the covering troops, was led by General d'Amade, the conqueror of Fez. Towards evening the advance-guard, one brigade, appeared before Altkirch and drove the Germans from their field-works at the point of the bayonet, suffering very slight losses. They received an enthusiastic welcome from the people of the town.

On the next day, about five in the afternoon, the French entered Mülhausen, the Germans retiring towards Neu-Breisach, a fortified town about eighteen miles away.

Waldweise
Metrich
Königsmachern
Diedenhofen
1188
Metzerwiese
Dalstein
Bous Battery
Freisdorf
Ay
Bettlainville
Vigy
Antilly
Ft. St. Barbe
Les Étangs
St. Eloy
Ft. Manteuffel
METZ
Ft. Zastrof
Colombey
Fouligny
Ft. Goeben
Peltre
Pange
Courcelles
Elwingen
Ft. de Sorbey
Ft. d'Orny
Forest of Remilly
Ft. Wagner
Verny
Remilly
Luppy
Solgne
Baudrecourt
St. Jure
Lucy
Clemery
Delme
Aulnois
Jallaucourt

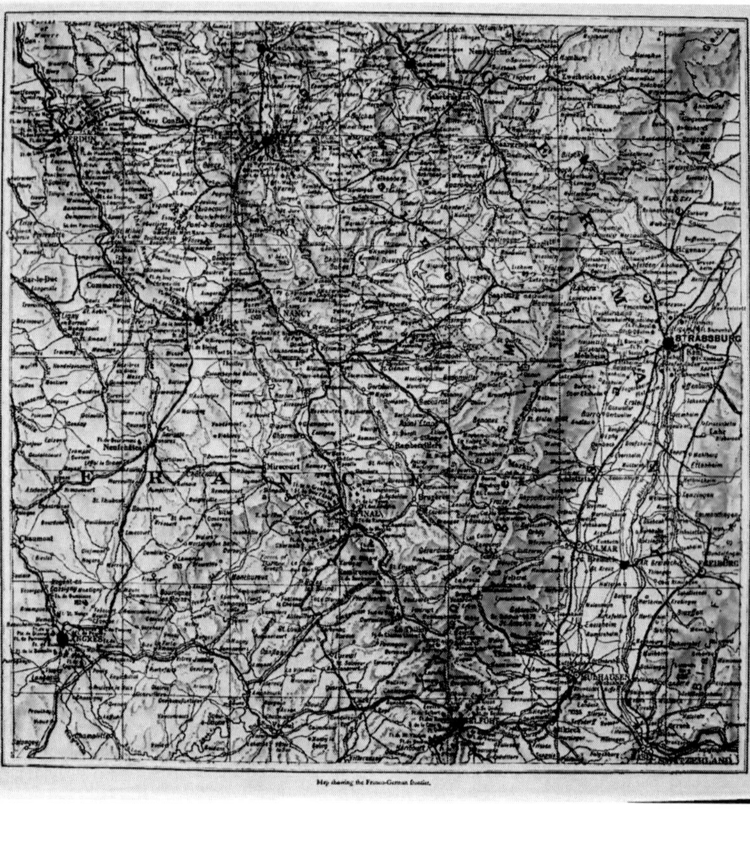

Map showing the Franco-German frontier.

Mülhausen, situated on the Ill, fifty-eight miles south-southwest of Strassburg, a great center for the cotton and woolen industries, with its 100,000 inhabitants, rich and prosperous, set in the midst of a smiling, fertile plain, was one of the jewels of the Reichsland. The upper classes favored the French, while among the laboring class it is said that a sentiment of loyalty to Germany more generally prevailed.

Jubilant crowds filled the streets of Mülhausen, eager to behold again the tricolor and the once familiar red pantaloons of the French soldiers in their midst. They regaled their transitory liberators with wine, beer, and food, and the women pelted them with flowers.

It was a triumphant revelry of short duration. On the next day, Sunday, the 9th, General von Deimling, commander of the Fifteenth Army Corps at Strassburg brought up superior forces to an intrenched position north of Mülhausen and threatened the French outposts. A battle ensued the following day, just outside the city to the west, and the French, who were only a single brigade, were overpowered by the superior effectiveness of the German artillery and forced to retire. It was futile to endeavor to defend Mülhausen, which is an open town. Nevertheless the battle was renewed in the streets during the night; and after the French had been driven away, a number of citizens occupying houses from which shots had been fired upon the German troops were sentenced by court martial and promptly led before a firing squad.

The Germans took 523 prisoners, including ten officers in the fighting at Mülhausen. By the 11th the French were back again on their own frontier. The sudden collapse of the expedition was apparently due to the inadequacy of the French forces, the superiority of the German artillery, and the deficiency in the French intelligence service.

A similarly impulsive incursion into Lorraine was inaugurated by a mixed brigade of the Fifteenth French Army Corps. They encountered the German covering troops near Lagarde on the 11th and were defeated and driven back across the frontier with the loss of about 1,000 prisoners, about one-sixth of their entire number, and found refuge in the Forest of Parroy, northeast of Lunéville.

By the 13th the German authorities reported that Alsace-Lorraine had been entirely cleared of the French.

The French claim that the observance by themselves of a neutral zone along the border within their own territory just before the outbreak of the war was utilized by the Germans as an opportunity for occupying important positions on the crest of the Vosges Mountains. One of the first undertakings of the French after war had been declared was to gain control of these eminences. In this endeavor they were favored by the fact that at least in the south the ascent from the French side of the mountains is comparatively easy. As the forces involved were relatively small in consequence of the rugged character of the ground, these operations have not received as much attention as they merited. Generally not more than a battalion or a regiment, at most a brigade, were engaged. But the difficulties in some places were considerable and much skill was exhibited in surmounting them.

Starting in the south the French captured successively the Ballon d'Alsace, Col de Bussang, Honeck, and Schlucht. Operations against the Col du Bonhomme and Col de Sainte Marie-aux-Mines were much more arduous. In this quarter the slope on the French side is steep, the mountains are densely wooded, and the ridges are mostly so narrow that it is difficult to plant artillery on their summits. The Germans had mounted their guns in well-chosen positions which had already been artificially strengthened. The French Alpine troops and mountain batteries here rendered valuable service. There were serious encounters before these points were taken, but finally the French secured positions on the flank and in the rear of the Germans, who were forced to retire.

By the 17th the passes and summits of the Vosges and the slopes on the side of Alsace were generally in the hands of the French. At this time the general forward movement was inaugurated throughout the entire front of the French armies, in Belgium and on the eastern frontier.

The French renewed their invasion of Upper Alsace. The two regiments which lost Mülhausen on August 10th retook it on the 20th, capturing the German batteries in the outskirts at the point of the bayonet. By the 23d Colmar also fell into the hands of the French.

The Great War has taught the sobering lesson that the day for spontaneous popular uprisings has past, that the passionate enthusiasm of the untrained multitude is of no avail against regular armies, and that a discontented province is bound hand and foot by the mobilization and removal of its military youth. Thus the supposedly seditious peoples of the Austro-Hungarian Monarchy made not the slightest move when the crisis came, which was supposed to be their opportunity.

The French invasion of the lost provinces was actuated in part, no doubt, by the expectation that the smoldering fires of discontent in Alsace-Lorraine would flame up in open rebellion, and that the population would eagerly join forces with their professed deliverers. Several sensational occurrences just before the war had tended to provoke the resentment of the nationalists in Alsace. The initial act in the most recent of these irritating episodes took place in Colmar. Johann Jacob Waltz, a German by birth and citizenship, but a French Alsatian in sentiment, known by the pseudonym "Hansi" through his books and sketches, was brought before a tribunal in Colmar and bound over for trial before the Supreme Court at Leipzig on the charge of treason. His offense consisted in the publication of a book entitled *My Village* (*Mon Village*), with drawings lampooning the Germans in Alsace, caricatures depicting the supercilious attitude of officers, the petty arrogance of the police, the officious pedantry of the schoolmasters. In Leipzig "Hansi" was condemned to a year's imprisonment on the lesser charge of inciting class hatred and libelling public officials. He was granted three days in which to take farewell of his relatives, but he promptly took farewell of Germany. The report of his arrival in France came almost on the eve of the international crisis. Without his prosecution the fame of the Alsatian "Hansi" would scarcely have transcended the limits of his own province. Governments are often stupidly insensible to the fact that persecution only serves to sharpen the edge of ridicule.

General de Castelnau, who commanded the imposing concentration of French forces on the rivers Moselle and Meurthe with Nancy as center, started a general advance along the entire front from the Moselle to the slopes of the Vosges on August 12th.

The French forces along this extensive front may have amounted to eight army corps or their numerical equivalent. The Germans were driven back by the French left near Pont-à-Mousson and Pagny. The French right attacked one of the Bavarian corps intrenched in front of Blamont on the evening of the 14th. The battle was resumed on the next day. Blamont and Cirey were taken by the French at dawn, and the Germans were outflanked and driven from the heights to the northward.

The children and schoolmaster gathered to welcome the stork whose coming heralded the arrival of spring. The German policeman is seen goose-stepping in the background.

The Fête of Messti cannot officially commence until the German policeman has inspected the booths to see that there is nothing displayed which carries the French colors. "For," says Hansi, "it appears that the German Empire, with her thousands of soldiers, fortresses, Krupp cannons, cuirassiers, and Zeppelins, runs an immense danger if at the fête of my village one small boy blows a tri-colored whistle."

Illustrations from *Mon Village*, "For the little children of France," by Hansi, for which he was tried and sentenced by the German authorities to one year in prison. *Published by H. Floury, Paris.*

On the 14th the French occupied Mt. Donon near the northern extremity of the Vosges, a commanding position at the angle between the boundary of Alsace and that of Lorraine on the side of France, at the extremity of a salient in the outline of the French territory. In consequence of their temerity two battalions of German fortress troops from Strassburg exposed themselves to the fire of the French artillery on Mt. Donon the same day in a pass near Schirmeck. They were put to rout, abandoning their guns and evacuating Schirmeck.

In the center of their battle-front the French poured into Lorraine from the valley of the River Seille and the cavalry occupied Château Salins on the 17th. On the right the French advanced towards Zabern, occupying Saarburg on the 18th, and severing railway communication by the direct line between Metz and Strassburg. The left had advanced as far as Mörchingen, nineteen miles southeast of Metz.

The French were pressing upon the entire front of Alsace-Lorraine and the Germans were apparently everywhere retreating, and already a considerable zone of territory on the German side of the frontier was in French hands. But the rapidity of their progress, which was more apparent than real, may have served to put the French off their guard. Apparently they were not fully aware of the formidable concentration of German forces in progress at some distance from the frontier on a line roughly indicated by Mörchingen, Finstingen, and Pfalzburg.

The futility of the French counter-offensive is fully realized if we reflect that the ephemeral achievements of the French invasion were all made before the completion of the German mobilization, while the movement collapsed at once as soon as the concentration of the German field-armies was complete.

The army of Crown Prince Rupprecht of Bavaria which encountered the French offensive in Lorraine consisted probably of five of the regular army corps, and contained in addition a large number of troops in reserve formations. In fact there was probably no great numerical disparity in the French and German forces which confronted each other in this field of operations.

The course of events suggests that the systematic retirement of the Germans from the 13th to the 19th of August may have been in complete accord with a definite intention of luring the French into a position where both their flanks would be open to sudden attack while their front would be exposed to the fearful pounding of the heavier German artillery in concealed positions. The turning point was about eleven o'clock on the morning of the 20th, when the First Bavarian Corps took the offensive near Saarburg.

Apparently the French were surpassed in generalship and in the efficiency of their intelligence service. They had invaded a region with which all the German officers had become thoroughly familiar in the course of recent maneuvers. A division of the French Fifteenth Army Corps was surprised by the German shrapnels as it was preparing its noon meal near Saarburg on the 20th. It was subsequently stated that the defeat of the French in the general conflict which followed was due to the confusion and flight of this division. But it is doubtful whether the failure of a single division to maintain its position was the sole decisive factor in the issue of a battle which raged almost continuously for several days along a very extensive front.

The possession of Saarburg and Dieuze was very hotly contested. In spite of the occupation of the summits of the Vosges, and especially of Mt. Donon, by the French, which would naturally render their position on the battle line in Lorraine more secure, the Germans by crossing the Rothe Saar executed a turning movement against their right flank. Partly in consequence of this maneuver the French were forced back to the frontier on the 21st.

On the 23d the French army received reinforcements from Nancy, but their renewed attack on the German right wing failed, and they retired at evening to the Lunéville-Blamont-Cirey line within the French border.

On the 22d, 23d, and 24th there was desperate fighting along the front of all the armies, on the Sambre and the Meuse, in the Ardennes and in Alsace-Lorraine. Perhaps we may regard these terrific encounters as a continuous battle raging along a line 350 miles in extent, from Flanders to the Jura. The concentration of the first-line troops was now complete and both sides looked forward eagerly to significant action within a very few days. The natural converging point for the hopes and speculations of the whole past generation seemed to have been reached. The world awaited decisive results with tremulous expectancy. Later, after many monotonous months of almost stationary trench fighting in the West, the palpable effect of this first tremendous collision of the opposing forces seemed astounding by comparison. But the results were nowhere decisive.

On the 20th the French press was announcing the sensational progress of the invasion of Germany. A few days later it took comfort in the reflection that the defenses of the country were still intact.

The Allies were everywhere overpowered and beaten back, but without being broken or crushed. The German Twenty-first Corps entered Lunéville on the 24th. The defeat of the French in Lorraine necessitated the withdrawal of their forces from the position in the northern Vosges, particularly from Mt. Donon and the Col de Saales.

The French abandoned their offensive in Upper Alsace on August 25th. In consequence of the failure of their offensive movement in the north and northeast, and the invasion of France, it became necessary to contract their lines and to send all available forces to the threatened regions. Mülhausen was again evacuated. The French retained possession of only a small section of the province adjacent to the French border opposite Belfort as a sort of protruding bastion, from which future sorties might conveniently be made.

Skirmishing was reported in the corner of France that borders on Luxemburg as early as August 3d. The German army advancing through Luxemburg under command of the German Crown Prince began the actual invasion of France about August 10th. It encountered at once the resistance of Longwy on the railway line penetrating France from the direction of Luxemburg and Trier. This resistance, conducted with unexpected stubbornness, was prolonged beyond all prevision. It must have been a veritable thorn in the flesh for the Germans, embarrassing their communications and the movement of supplies for the large army of invasion in this quarter. The defense of Longwy, although comparatively insignificant in respect to the strength of the garrison and the means with which it was carried on, not being one of the major events in the evolution of the larger strategic plans of the war, is one of the most noteworthy episodes of the conflict. The Germans doubtless tried to mask it and continue on their route, but its presence in their rear must have been a continual cause of annoyance.

On the 22d the army of the German Crown Prince repulsed strong French forces which had been pushed forward from Verdun to the vicinity of Longwy, in accordance no doubt with the general offensive movement along the

General de Curières de Castelnau, commander of the French forces which invaded Lorraine.

General Foch, commander of the Ninth French Army.

entire French front at just this time. The old fortress, Longwy, designed by Vauban, once regarded as the gateway of France, had been captured by the Germans on three previous occasions, on August 23, 1792; September 11, 1815; and January 25, 1871. Garrisoned in 1914 by only a single brigade, it held out until August 26th although subjected to continual bombardment for at least five days.

The prolonged resistance of Longwy was only possible, of course, because the German siege-artillery was employed elsewhere. Nevertheless, the defenders gave ample proof of resolution and courage by holding out so stubbornly in an antiquated fortress against the fire of the heaviest artillery which accompanied the German armies in the field. The garrison capitulated only when nearly the whole town had been destroyed, all but one of their guns had been rendered useless, and they themselves were threatened with immediate annihilation. The governor, Lieutenant-colonel d'Arche, was made Officer of the Legion of Honor as a reward for his valor.

This northeastern border of France, vulnerable at any time because the barrier forts do not extend northward of Verdun, was an especially sensitive point during the shifting of the French forces from the east to the north.

Partly, no doubt, for the purpose of threatening the flank of the Crown Prince's army the French launched an offensive movement from the Meuse below Verdun into the Ardennes. The French army crossed the Semoy, encountered the army of Duke Albert of Württemberg near Neufchâteau, and suffered a defeat. The victorious Germans advanced and crossed the Semoy and the Meuse, penetrating into France.

The vanguard of the French army which had been concentrated for operations in the north probably entered Belgium on August 14th, a part of it advancing along the left bank of the Meuse. At the same time the advanced forces of the Germans were pushing westward through the Ardennes. The first collision took place at Dinant, situated eighteen miles south of Namur, at the foot of almost perpendicular cliffs on the left bank of the Meuse.

Early in the morning of the 15th a cavalry division of the Prussian Guard, the Fifth German Cavalry Division, and several battalions of infantry with machine-gun companies occupied Dinant and drove the French from the citadel. Later, in crossing the Meuse, they were attacked by two French infantry regiments, six batteries, and a regiment of cavalry. The French artillery was brought into action on the heights commanding the river on the west and checked the advance of the Germans. The French infantry in a violent onslaught forced the Germans back over the bridge, driving some into the river at the point of the bayonet and dispersing the others. They quickly cleared the town and recaptured the citadel.

The subsequent progress of events growing out of the movement of the French forces into Belgium was merged with the tremendous course of the German rush towards Paris, from which its treatment cannot be separated.

CHAPTER V

THE FLOOD-TIDE OF GERMAN INVASION
(*August 23-September 5, 1914*)

The Kaiser goes to the front. Disposition of the forces in the West, the seven German and five French armies and the British contingent. Concentration of powerful German forces against the Meuse-Sambre salient, far outnumbering the Allies. Defenses of Namur. Operations against Namur: first attack, August 20th; occupation of the city, August 23d; bombardment of the forts, August 22-25. Effect of the sudden fall of Namur. The Germans drive the French from the Sambre. Arrival, significance, and composition of the British Expeditionary Force. The Germans attack the British, August 23d. The "retreat from Mons." The night encounter at Landrecies. The critical situation on August 26th. The real objective of the German operations. Situation on August 29th. The moral crisis in France and Joffre's decision. Continuation of the German advance. Frantic exodus of civilians from Paris. Transference of the French government to Bordeaux. Alarming progress of the Russian invasion of Galicia.

The Kaiser took leave of his capital on August 16th to join the headquarters at the front. The concentration of the German forces had reached a point where great events were expected almost immediately. Before leaving Berlin the Kaiser issued the following farewell decree:

"The course of the military operations compels me to transfer my headquarters from Berlin. My heart requires that I should address to the citizens of Berlin a farewell and my deepest thanks for all the demonstrations and proofs of affection which have been so abundantly given to me in these great days fraught with destiny. I trust firmly in God's help, in the bravery of the army and navy, and the unshakable unanimity of the German people in the hours of danger. Victory will not fail our righteous cause."

The general forward movement followed the completion of German concentration with the automatic precision and promptness of the thunderclap that succeeds the flash of lightning. Tremendous blows were delivered all along the front with the bewildering regularity of action of intricate machinery. It requires an unusual effort of the imagination to correlate the movements of so many parts and to perceive in all the varied operations of these most strenuous days the unity as of a single battlefield expanded to the geographical dimensions of a whole campaign.

We have already taken note of the intense activity beginning with August 20th that manifested itself in different parts of the German front, the hurling back of the invading forces in Lorraine, the defeat of the French near Longwy and at Neufchâteau, and the general advance of the German forces towards the Meuse below Verdun. There remains to be described the most spectacular phase, as it is the most important feature, of this, the first of the gigantic combats of the world-war, namely, the German movement on the extreme right, which had been gathering momentum for a week, as we have already seen.

The great design of the General Staff was growing out of the confusion of preliminary vanguard skirmishes as the walls of a great edifice rise and are gradually distinguished from the welter of surrounding scaffolding. Their purpose was revealed in the general disposition of their forces in the West.

The German field forces in the West were grouped in seven different armies numbered consecutively from right to left along the German front, that is, from its northwestern to its southeastern extremity. The First Army under Colonel-general von Kluck was advancing against the British Expeditionary Force on the extreme left wing of the

DUKE ALBERT

Heir presumptive to the throne of Württemberg.
In field uniform.

GENERAL ALEXANDER von KLUCK

Commander of the first German army in the invasion of France.

French front. The Second Army, commanded by Colonel-general von Bülow, and the Third, by Colonel-general von Hausen, were directed against the salient formed by the Sambre and the Meuse. The Fourth Army under the Duke of Württemberg, after driving the French out of the southern Ardennes in Belgium, threatened the front on the Meuse between Mézières and Montmédy. The Fifth Army, led by the German Crown Prince, was advancing into France from the Grand-duchy of Luxemburg on a line of operations passing between Montmédy and the fortress of Verdun. The Sixth Army under the Crown Prince of Bavaria, after repulsing the invasion of German Lorraine confronted the French on the line Nancy-Lunéville. The Seventh Army, commanded by Colonel-general von Heeringen, after the evacuation of Southern Alsace by the French, was engaged chiefly in watching the northern passes over the Vosges.

Opposed to these seven there were at first only five French armies and the British Expeditionary Force. They were numbered from right to left along the French front, in other words, from southeast to northwest. It follows that the First French Army under General Dubail, in the neighborhood of St. Die, confronted the Seventh German Army in the region of the Vosges. The Second Army under General de Castelnau occupied the position from Lunéville to Nancy. The Third and Fourth Armies, commanded by Generals Ruffey (afterwards succeeded by Sarrail) and Langle de Cary, guarded the line of the Meuse from Verdun to Mézières, or somewhat below, facing the German Crown Prince and the Duke of Württemberg. The Fifth Army under General Lanzerac (subsequently succeeded by Franchet d'Espérey) held the angle between the Meuse and the Sambre, particularly the crossing places on the rivers. Its front along the Sambre was prolonged on the left, where the river no longer formed a natural obstacle, by the British contingent.

Throughout the greater part of the western front single German armies were matched against single French ones. But the two armies which the Germans now possessed in excess of the French were added to the right wing, the northwestern extremity. This was not all. For, while the average effective strength of the individual armies on both sides was probably between 200,000 and 250,000 men, the German armies varied greatly in size, the First and Second being especially powerful.

About 500,000 Germans traversed the plain of Belgium in the great movement westward, while from 150,000 to 200,000 crossed the Ardennes in the direction of Namur and the section of the Meuse directly above Namur. All these threatened the salient formed by the line of the Meuse and the line of the Sambre as extended westward by the front of the British Expeditionary Force.

The Germans possessing the initiative distributed the weight of their attack as they chose, baffling their adversaries by the mysterious concealment of their intentions. It presently became evident that the most strenuous effort would be made to turn the Anglo-French left flank, and that therefore the British force occupied the position of gravest danger, since it was posted at the extreme left of the line of the Allies. Von Kluck was leading the First Army composed of five army corps, probably the Second, Third, Fourth, and Ninth of the active army, and the Fourth of the reserve, against the British, who were numerically scarcely the equivalent of two normal army corps. To the Second German Army Corps from Stettin was allotted the all-important task of turning the British left from the direction of Tournai. This corps, which made its first appearance at this time, must be added to those already enumerated, as the twentieth German, or twenty-first Teutonic, active army corps in the western theater. General von Bülow was bringing up the Second German Army composed of four corps, probably the Seventh, Tenth, and the Guard, and the Tenth of the reserve, against the Fifth French Army under General Lanzerac, which consisted of only three corps. Moreover, the Third Army, commanded by General von Hausen, probably four army corps, was directed mainly against the Fifth French Army, although a portion of the three army corps of the Fourth French Army may have received a part of its pressure. We should probably not be very far wrong in assuming that in this section of the front the Germans outnumbered the Allies, including the Belgian division at Namur, in the proportion of nearly two to one.

The Allies grossly underestimated the numerical superiority of their adversaries. They doubtless calculated that the investment of the forts at Namur would absorb an appreciable part of the enemy's forces. The Meuse, moreover, was a considerable obstacle protecting their right flank against attack from the east.

The defenses of Namur, like those of Liége, with which their construction was contemporaneous, consisted of a ring of detached forts, in this case nine in number, built of concrete with armor-plated turrets. These forts were likewise armed with 15 and 11-centimeter guns and 21-centimeter mortars. General Michel, who commanded the Belgian Fourth Division forming the garrison of Namur, about 22,500 men, endeavored to close the intervals between the forts by means of trenches with barbed-wire covering.

German patrols made their first appearance before Namur on August 14th. The main bodies of the Germans began to arrive in the vicinity about the 18th. They did not repeat the infantry assaults in dense formation which had characterized the early operations at Liége. This method of attacking the forts, which had cost them so many lives, was not required at Namur, partly because this stronghold did not constitute so serious an obstacle as Liége, and partly because the heavy siege-artillery was available soon after the arrival of the main bodies of troops.

On the evening of Thursday, August 20th, the Germans first opened fire against a section of trenches two and one-half miles in length between the Forts Cognelée and Marchovelette, northeast of the city. Without exposing themselves the Germans swept the improvised Belgian field-works with a tremendous fire of artillery, against which the feeble armament of the Belgians was useless. The Belgians held out for about ten hours, crouching in their trenches, and then withdrew with difficulty to escape annihilation, permitting the German infantry to penetrate the space within the girdle of the forts on the morning of the 21st. The Germans did not force their way into the city until the 23d, when six of the forts were still holding out.

The bombardment of the forts was mainly carried on by the 15-centimeter heavy field-howitzers and by the 21-centimeter (8.4 inch) and 28-centimeter (11.2 inch) siege-artillery, discharging shells weighing 250 and 760 pounds respectively. These pieces can be transported by automobile tractors and fired from the wheeled-carriages upon which they are conveyed. One of the 42-centimeter (16.8 inch) mortars is said to have been employed also in the bombardment. The 42-centimeter pieces cannot be fired from their travelling carriage. They are usually transported by rail, although it is reported that they were sometimes conveyed by highway in four separate parts, each part being hauled by three broad-wheeled traction-engines

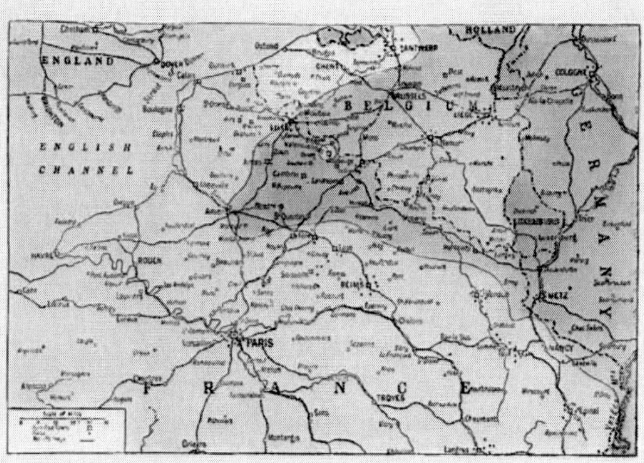

Map showing by the shaded portion the extent of the German advance into Belgium and France, August 30, 1914.

of steam-roller type, an extra engine preceding the convoy to test the road. As has aleady been mentioned, their enormous projectiles weigh 2,500 pounds.

The German artillery took up positions beyond the effective range of the 6-inch guns of the forts. The intensity of the bombardment may be appreciated by the record of shots fired into Fort Suarlee, 600 on the 23d, 1,300 on the 24th, and 1,400 on the 25th. The turrets of this and the other forts were wrecked and the concrete structure shattered. Fort Suarlee surrendered at five o'clock on the afternoon of the 25th, and all the forts were in the hands of the Germans by the next day.

The Germans overpowered Namur suddenly and unexpectedly. The reason, as we have seen, was simple. Their siege-artillery, which greatly surpassed in range the guns in the forts, was placed in positions where it subjected the forts to the full effect of its fire without receiving any damage in return. The facility with which the capture of the fortress was accomplished was an alarming fact for the Allies. The possession of Namur was of almost vital importance for the safety of the French forces in the angle formed by the Meuse and the Sambre, and even for the British force which continued the line of their front westward. The fact that the Allies expected to maintain their position in this salient can only be explained on the assumption that they were confident that Namur would make a much more prolonged resistance. Vastly superior forces were advancing against them with furious momentum, urged on by commanders of reckless determination. Namur at the point of the angle was like the prow of a ship, which cleaves the billows, but receives the full fury of the blast.

If the forts of Namur had held out, the position of the Allies between the two rivers might have been an excellent one, since they could shift their forces to and fro on interior lines and strike in either direction. The position of the German forces attacking on the two fronts would have been an awkward one, their communication being hampered by the control of the adjacent river-crossings by the forts at Namur.

But the failure of the Belgians to support the pressure at Namur precipitated the whole course of events that led down to the Battle of the Marne. It inaugurated the most critical period of the whole campaign in the West for the cause of the Allies. The last barrier to the invasion of France went down with the submersion of Namur. The human tidal-wave rushed forward with eager certainty of engulfing its prey in its tremendous sweep. The invasion did not wait for the reduction of the forts. It went around and percolated between them. The Germans had gathered their strongest force to shatter the Allies on this part of the line at the first blow. With the resistance crushed at Namur the salient in the Allied front was immediately flattened out beneath the terrific impact.

A shell crashing through the roof of the railway station in Charleroi at 7.20 on the morning of the 21st was the prelude to the bombardment of the town by the Germans. On the 22d the German infantry forced its way into the city and at the same time seized the crossing-places on the Sambre above and below. There was desperate fighting in the streets of the city. The Germans gradually advanced to the railway station where the French made their last stand. The canal passes in front of the station and the Germans only gained possession of the canal-bridge after an encounter in which the French resisted desperately for two hours. After retiring from the lower town of Charleroi the French bombarded it in their turn. Meanwhile, the panic-stricken inhabitants took refuge largely in cellars.

On the next day, Sunday, the 23d, the French after being reinforced returned and attacked the Germans in the streets of the town and drove them back over the Sambre. A striking feature of the engagement was a terrific hand-to-hand encounter between the Turcos and a part of the Prussian Guard. The French colonial troops inflicted heavy losses on the Germans, but were finally repulsed by rapid-fire artillery which had been held in concealment. The French were finally driven from the city.

The crossing points on the Sambre were now in the undisputed possession of the Germans. At once the importance for the Germans of their occupation of the salient formed by the Meuse in its deviation westward became manifest. Just at the time when the Belgians were losing hold at Namur, and the Germans were pouring across the Sambre at every available point in front, the French army was threatened on its flank and even in its rear by the German forces which had advanced through the Ardennes to the right bank of the Meuse.

A hasty retirement of the army of General Lanzerac between the Meuse and the Sambre had become imperative by the evening of the 22d. But the departure of this army exposed the right flank of the British Expeditionary Force, compelling it to retire the next day. It endangered, moreover, the position of the French along the entire left bank of the Meuse as far as Verdun.

The British Expeditionary Force originally dispatched to the continent attracted from the first a degree of attention out of all proportion to its comparatively diminutive size. There were several reasons for this. The presence of a friendly British army on French soil was an almost unique phenomenon in history, and it was greeted in Western Europe as a hopeful augury for permanent concord and the progress of democratic ideals. The reports of the operation of the British force, submitted from time to time by General French, its commander, were unusually clear and comprehensive. The intense feeling of bitterness against Great Britain prevailing in Germany magnified the significance of the presence of the British Expeditionary Force in the German imagination, so that victories over it completely cast into the shade successes achieved in action against the French where three or four times as many men were involved.

The first encounter between the most highly developed professional army and the most efficient national army was everywhere awaited with intense, impatient curiosity.

The London *Times* published on October 1, 1914, the text of an alleged order of the day, said to have been issued by the Kaiser to the officers of the German First Army advancing against the British, on August 19th, as follows:

"It is my Royal and Imperial command that you concentrate your energies, for the immediate present, upon one single purpose, and that is that you address all your skill and all the valor of my soldiers to exterminate first the treacherous English and walk over General French's contemptible little army."

The authenticity of this document was afterwards denied by the German government; but whether it is true or false, it may serve as a fair expression of the common sentiment in Germany.

The British Expeditionary Force was commanded by Field-marshal Sir John Denton Pinkstone French, the most distinguished soldier in Great Britain with the exception of Earl Kitchener. Born of Irish extraction at Ripple Vale in Kent, September 28, 1852, he was nearly sixty-two years old at the outbreak of the war. Though destined by his parents for the church, he chose the navy at the age of

Lieutenant-general Sir Douglas Haig, commander of the First Corps of the British Expeditionary Force.

Major-general Edmund H. H. Allenby, commander of a division of cavalry, British Expeditionary Force.

General Sir Horace L. Smith-Dorrien, commander of the Second Corps of the British Expeditionary Force.

fourteen and served as cadet and midshipman, but passed over to the army in 1874. He performed active service with the 19th Hussars during the Soudan campaign of 1884-1885 and became their commander in 1889. After serving on the staff and at army headquarters from 1893 until 1897, he became a brigadier, commanding the 2d cavalry brigade.

The South African War was the great opportunity for the establishment of General French's reputation. For he was the one British general who may be regarded as having been uniformly successful throughout the struggle. He commanded a cavalry division in Natal in the autumn of 1899 with the rank of major-general. Later, his conspicuous merit, as displayed in a position of very great responsibility in charge of the operations around Colesberg, November 10, 1899, to January 31, 1900, in his brilliant dash to the relief of Kimberley, and in his leadership of the cavalry division in Lord Roberts's campaign which ended in the capture of Bloemfontein and Pretoria, was recognized by promotion to the rank of lieutenant-general in 1900 and to command of the First Army Corps at Aldershot in 1901. He served as Chief of the Imperial General Staff from 1911 until 1914 and received the rank of field-marshal in 1913. Military action being threatened owing to the critical conditions arising out of the Home Rule for Ireland Bill, Sir John French and other military officers resigned their commissions in March, 1914. On August 4th the British government having decided to dispatch the Expeditionary Force to France, Sir John was appointed to its command. Short and stocky, though wiry, in build, his appearance is enlivened by his clear, penetrating eyes and his alert, sensitive countenance. His coolness and sound judgment are associated with initiative and well-calculated intrepidity. Of modest, unobtrusive personality, he is ever ready to praise the merit of his subordinates.

Field-marshal Sir John French arrived on the continent and visited the French headquarters on the 14th, and went to Paris to pay his respects to the president the next day. The first British troops reached the position at Mons assigned to them in the French general plan of campaign on August 19th. The concentration of the forces originally dispatched to the theater of operation was completed by the evening of August 21st. They were composed of two corps; the First Corps, commanded by Lieutenant-general Sir Douglas Haig, consisting of the First and Second Divisions, and the Second Corps, under General Sir Horace L. Smith-Dorrien, made up at this time of the Third and Fifth Divisions. General Smith-Dorrien had just succeeded General Grierson, who died of heart failure on the 17th. A cavalry regiment was attached to each of the divisions, and there were also five cavalry brigades, four of which formed a division under Major-general Edmund H. H. Allenby.

The center of the British position was at Mons. The First Corps occupied a front extending eastward as far as Binche near the western extremity of the French Fifth Army. The Second Corps was assigned to a position westward along the canal from Mons to Condé.

British aëroplanes, and cavalry scouts reconnoitering as far as Soignies, reported no overwhelming force of the enemy near on the 22d. But about three on the afternoon of the 23d the situation changed. The Germans were attacking some portions of the British front with increasing violence. The severity of the fighting during the afternoon of the 23d is illustrated by the fact that one company of the Coldstream Guards lost twelve killed and seventy-two wounded out of a total of 120 effectives. At five in the afternoon General French received General Joffre's alarming communication that at least three German corps, a reserve corps, the Fourth Corps, and the Ninth Corps, were moving against the British position in front; that the Second Corps was engaged in a turning movement from the direction of Tournai threatening the British left flank; and that the retreat of the Fifth French Army on the British right had commenced because the Germans had gained possession of the passages of the Sambre between Namur and Charleroi the day before.

After testing by means of his aëroplane scouts the accuracy of this information, General French, who had already withdrawn his forces a little way from Mons and the line of the canal, decided to retire to the line Maubeuge-Bavai, where positions had already been reconnoitered. This movement, the first stage in the famous "retreat from Mons," was commenced at dawn on the 24th, and was successfully accomplished by ten at night, although the retiring British were almost constantly assailed by two corps on their receding front and one on their left flank.

The terms right and left as applied to bodies of troops will always be used in their significance with reference to the position of a member of the respective body in confronting the enemy. Thus we shall do every army engaged in a retrograde movement the formal honor of assuming that it retreated with its face to the foe.

The Second British Corps bore the brunt of the fighting on the 24th, threatened as it was with envelopment. On the night of the 24th the British occupied the position from Maubeuge westward. General French very wisely resisted the temptation of seeking cover behind the defenses of the fortress of Maubeuge.

The continued retirement of the French army necessitated the further retrograde movement of the British;

otherwise they would have been outflanked at both extremities and surrounded. Therefore, the British retreated all day on the 25th to the line Landrecies-Le Cateau-Cambrai, where positions had been in part prepared. Landrecies is situated on the Sambre and Cambrai on the Scheldt. The Fourth Division of the Expeditionary Force which had detrained Sunday, the 23d, at Le Cateau, advanced to coöperate with the retiring forces on the 25th and was incorporated with the Second Army Corps, relieving the pressure on its flank. The Second Corps arrived at its appointed destination about six in the afternoon. Meanwhile the First Corps following parallel routes skirted the eastern border of the great forest of Mormal, which contains 22,000 acres, pursued all day by the Ninth German Corps which was advancing by a route leading through the forest, endeavoring apparently, to intercept communication between the two parts of the British army.

The First Corps completed its appointed march between nine and ten in the evening, after sixteen hours on the road, and was immediately the object of a furious attack by the Ninth German Corps. The Germans debouching from the woods north of Landrecies assailed the 4th Guards brigade in the town. Crowding into the narrow streets they were met by the deadly fire of the British machine-guns and repulsed with the loss of nearly 1,000 in a very short time. Meanwhile the First Division was heavily engaged south and east of Maroilles. The assistance of two French reserve divisions and the skilful generalship of Sir Douglas Haig extricated the troops from a very difficult situation, so that the corps was able to resume its march in the morning.

General French decided that it was necessary for the British forces to continue the retreat until they could find a temporary opportunity for repose and reorganization behind the Oise or some other important barrier. The 26th was the most critical day in the whole retreat. The First Corps was directed to retire towards Guise, the Second on St. Quentin. The First Corps set out at daybreak, but the Second Corps was attacked in such overwhelming force by almost the entire army of von Kluck that it was incapable of disengaging itself. The artillery of four army corps was concentrated upon it. The Germans were making a supreme effort to accomplish the design of crushing their adversary's left wing. The most far-reaching consequences hung upon the issue. General Sordêt with a French cavalry corps of three divisions retiring by routes lying eastward of the British was unable on the 26th to lend any assistance. After maintaining the unequal contest with remarkable stubbornness and pluck until 3.30 in the afternoon, the British commenced their retreat, which was covered by the cavalry and artillery with the utmost courage. Nearly a whole battalion of the Gordons was cut off in the evening. But the preservation of the Second Corps, even with serious losses, in the face of such unparalleled dangers, an achievement little short of marvellous, would have been impossible without a commander of exceptional coolness and determination.

The First Corps experienced less difficulty on the 26th. But on the morning of the 27th the Munster Fusileers on its extreme right were surprised and surrounded by the Germans as they were breaking camp and were forced to surrender.

Henceforward the pressure on the British was very much relieved by the coöperation of General Sordêt's cavalry and of the Sixty-first and Sixty-second French Reserve Divisions under General d'Amade, which withdrew from the neighborhood of Arras. The retreat was continued through the 27th and 28th, bringing the British to a position on the line Noyon-Chauny-La Fère in alignment with the general front of the French armies.

From the 23d to the 27th the British Expeditionary Force had retreated about sixty-four miles in four days, an average distance of sixteen miles a day, fighting every step of the way with characteristic stubbornness in rear-guard actions, counter-attacks, and even desperate pitched battles, against forces that outnumbered it more than two to one. It had suffered serious losses; but had probably inflicted heavier losses on its pursuers. Its organization and spirit were still unbroken. The incidents in the retreat thus far, one of the most brilliant episodes in the annals of the British army, form an indivisible feature of the narrative of the first campaign of the Great War.

In the meantime, the main part of the French Fifth Army was retiring in a direction roughly parallel with that of the British contingent,—and in accordance with the general retrograde movement of all the French forces eastward as far as Verdun,—by way of Philippeville and Chimay to Hirson and later to the Aisne. The retreat of the French Fifth Army commenced at least a day before that of the British Expeditionary Force, as we have seen, but after a day or two the British recovered contact with the French on their right. The Fifth Army was engaged with superior forces led by von Bülow near Avesnes and Chimay at the same time that the British were fighting on the line Landrecies-Le Cateau-Cambrai.

On the evening of the 28th, when the British were in the position La Fère-Noyon along the Oise, the Anglo-French front was practically continuous. The French Fifth Army extended from La Fère to Guise along the Oise, and thence eastward. The French armies had everywhere retired before superior forces, their own inferiority being due to their somewhat slower concentration and their failure to foresee the direction of their adversary's principal attack and to dispose their forces accordingly.

The retirement of the British exposed the heart of industrial France to the German invasion. As soon as the retreat of the Allies left it isolated, the Germans invested Maubeuge, the most important fortress in the North, which commands the principal railway line from Belgium to Paris.

Thousands of the people in the thickly-populated northern departments were driven from their homes by fear of the Uhlans. Hunted from place to place they suffered great privation. The bewildering rapidity of the German advance deprived them of permanent safety for many days. The railways were overflowing with the streams of refugees. The French evacuated Lille, and later, Amiens. The advance of German detachments in the direction of the coast cut off the communications of the British army with the channel ports nearest England. Consequently Field-marshal General French changed his base to St. Nazaire with an advance base at Le Mans.

Von Hausen with the Third German Army inaugurated his movement southward by driving the French from Dinant on the 22d and afterwards destroying the town. After this encounter the French forces in this part of the valley of the Meuse retreated southward. There was almost continuous fighting on both sides of the river, but the main body of the French retired along the left bank, while the Germans advanced as rapidly as possible on the right bank, eager to gain control of a serviceable crossing place. The French in their retreat destroyed no fewer than thirty-three bridges between Dinant and Charleville.

In one of the delaying actions of the French rear-guard, a body of 5,000 French at Marville is said to have held in check four times its number for twelve hours. The immediate objective of the German advance in this quarter was the important center Mézières-Charleville, the most convenient crossing-point on the Meuse. The two towns are separated only by the river, which was here spanned by three bridges. Charleville, situated on the left bank, is by far the more important place.

The deep rumbling of cannon, becoming gradually more distinct, and even more unmistakably the inevitable crowds of fugitives from points in the valley further north, brought to the people of these two towns the ominous warning of the approaching storm of conflict. Public criers announced in the streets at eight o'clock on the evening of the 24th that Charleville had to be evacuated by its inhabitants at once, since it was going to be bombarded in two hours. Railway service had already been discontinued, and soon the roads in the direction of the temporary rail-heads were filled with frightened people groping their way in the darkness and confusion, carrying as best they could their most necessary belongings. The sudden and peremptory dispersal of an entire population at Charleville, as elsewhere, inevitably caused many pitiful occurrences and dire distress.

The German advance-guard entering Mézières on the 25th found the bridges intact and unguarded and they hastened to occupy Charleville. They had no sooner crossed the river than the bridges were blown up by bombs which had been placed in position by the French before they retreated. At the same time a French detachment concealed in the apparently deserted houses opened fire upon them with machine-guns with deadly effect. But the Germans, who were likewise provided with machine-guns, also sought the cover of the houses, returned the fire of the French, and finally drove them from the town.

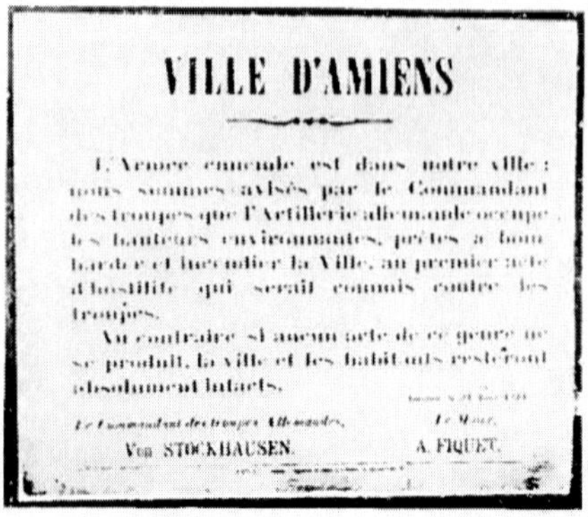

VILLE D'AMIENS

L'Armée ennemie est dans notre ville : nous sommes avisés par le Commandant des troupes que l'Artillerie allemande occupe les hauteurs environnantes, prêtes à bombarder et incendier la Ville, au premier acte d'hostilité qui serait commis contre les troupes.

Au contraire si aucun acte de ce genre ne se produit, la ville et les habitants resteront absolument intacts.

Le Commandant des troupes Allemandes, *Le Maire,*
Von STOCKHAUSEN. A. FIQUET.

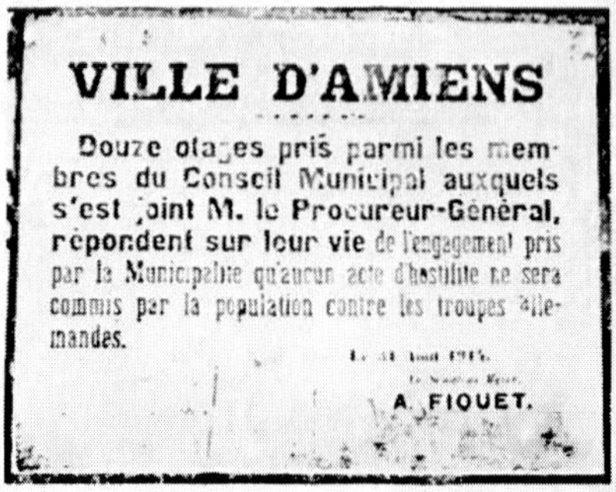

VILLE D'AMIENS

Douze otages pris parmi les membres du Conseil Municipal auxquels s'est joint M. le Procureur-Général, répondent sur leur vie de l'engagement pris par la Municipalité qu'aucun acte d'hostilité ne sera commis par la population contre les troupes allemandes.

A. FIQUET.

Notices posted in the city of Amiens.

CITY OF AMIENS

The enemy's army is in our city: we are notified by the Commander of the troops that the German Artillery occupy the neighboring heights, ready to bombard the city and set it on fire at the first hostile act committed against the troops.

On the other hand, if no act of such a nature takes place the city and its inhabitants will remain absolutely intact.

Amiens, August 31, 1914.

The Commander of the German forces, *The Mayor,*
Von STOCKHAUSEN. A. FIQUET.

CITY OF AMIENS

Twelve hostages selected from the members of the Municipal Council, together with the attorney-general, answer with their lives for the engagement made by the municipality that no hostile act shall be made by the inhabitants against the German troops.

August 31, 1914.

The Senator-Mayor,
A. FIQUET.

German infantry marching through Amiens, August 31, 1914. French women at a street hydrant handing water to the soldiers.

The French artillery on the heights commanding Mézières-Charleville occasioned much annoyance to the main body of the German forces until the latter, after throwing pontoon bridges across the river, brought up their heavier pieces and drove it away.

The principal part of the German Third Army under von Hausen, after taking the fortress Les Ayvelles on the Meuse, advanced to Rethel on the Aisne, where it defeated the French on August 30th. Givet, the frontier fortress at the northern extremity of a narrow projection of French territory extending far down the valley of the Meuse in the direction of Dinant, held out against the Germans until the 31st. Its prolonged resistance, like that of Longwy in the northeast, was made possible by the fact that the Germans concentrated their heavy siege-artillery against Namur, and later transported it to the positions before Maubeuge, leaving the lesser strongholds to be bombarded by the howitzers which accompanied the field armies, formidable pieces, it is true, but far less disruptive in their effect.

After the capture of Longwy the German Crown Prince advanced resolutely towards the section of the Meuse below Verdun. The old citadel of Montmédy on an eminence commanding an important passage of the Chiers was taken without much difficulty. The Germans claim to have found hundreds of packages of dum-dum bullets in it. The French attempted to blow up the railway tunnel at this point, but the Germans employed the services of their prisoners in clearing away the wreckage very quickly, and even laid an extra track around the elevation which the tunnel perforates.

There was some stubborn fighting before the Germans effected the passage of the Meuse. But the French who contested the crossing were finally driven from the heights on the left bank by the German artillery with its superior range. The French destroyed the bridge at Stenay, but in two days the German engineers had erected a temporary structure to replace it.

German concentration in the West had scarcely been completed when the civilian population in Germany was gladdened by reports of the fortunate progress of the offensive movement inaugurated in Lorraine on August 20th. Rumors of a great victory were greeted with rapturous expressions of delight in Berlin on the 21st. Buildings were decorated along the leading thoroughfares. The visit of the Kaiserin to the Crown Princess was the occasion for a public demonstration before the palace where the latter resided. In Unter den Linden crowds waited until after midnight for more definite news.

Very keen satisfaction was felt in Bavaria because the victorious army had been commanded by the Bavarian Crown Prince and was largely composed of Bavarian troops. The king addressed the throng before the Wittelsbach Palace in the following words:

"I am proud to see my son win such splendid success at the head of his intrepid soldiers. But this is only the beginning; great victories still lie before us. I have confidence in the quality of the German army, which will remain the victor, however great the numerical superiority of its enemies may be."

Tidings of a series of German victories followed with bewildering regularity confirming or surpassing the most sanguine expectations. The strongest fortresses had crumbled in a few days beneath the fire of the German siege-artillery. The French invasion of the Reichsland had been brought to a sudden and inglorious termination. The announcement that the British force, which had been joined by three divisions of French Territorials, had been decisively defeated north of St. Quentin with the loss of seventeen field-batteries, a heavy battery, and thousands of prisoners, was received throughout Germany in a spirit of exultation as the righteous retribution for British perfidy.

By the 28th the Allies had apparently been everywhere defeated and were in full retreat along the whole front from Cambrai to the Vosges. The Germans were pressing forward towards Paris with seemingly irresistible force, and after the startling performance of the heavy siege-artillery at Liége and Namur the reduction of the forts about the French capital was scarcely regarded as a serious undertaking. The vigor of the French army was apparently being hopelessly impaired by the lack of concord in the higher counsels of the nation. The fatal results of the democratic political control of military policy in France, as so often predicted in Germany, were manifesting themselves with unmistakable clearness. The Germans were convinced that a consistent, progressive policy was impossible in a country where there had been forty-two ministers of war in forty-three years. Joffre's strategical talent, whatever its value might be, was evidently hampered and embarrassed by the control of the president and ministry, who in turn were subject to the caprice of parliament and amenable to the fluctuating influence of public opinion.

Divergence of views might at times exist in Germany in the war councils, but when the decision had once been made, it received the unconditional sanction of a final authority.

Germany was animated with absolute assurance of victory and a boundless spirit of elation. There was just one shadow of apprehension. The Russians had burst into East Prussia with unexpected promptness and thus menaced the Kingdom of Prussia in a very susceptible part, a region in which the sentiment of the royal family and the material prosperity of an influential part of the nobility were very intimately affected. But when, after a few days, this danger also was suddenly removed, it seemed as if the impossible had been accomplished. The *Vossische Zeitung* expressed this feeling in the following words:

"The mind is almost unable to conceive what is told the German people about their victories from the East and West. It is, as it were, a judgment of God which condemns our antagonists as the criminal originators of this fearful war."

The French troops released by the abandonment of the offensive in Upper Alsace on August 25th were hurriedly transported westward. To the Seventh Army Corps, regiments of which had been at Mülhausen, four reserve divisions and General Sordêt's cavalry corps were added to form the Sixth French Army, which got into position to the left of the British on August 29th, thus relieving the pressure on the Expeditionary Force.

Yielding to custom and the convenience of a definite geographical indication we have treated Paris as the obvious goal for the German invasion of France. Paris is a unique center of communications; it is the nerve center of the country. Three times during the nineteenth century German armies had marched to Paris as their objective. The occupation of Paris in 1914 would have been a moral victory that would have created a profound impression in consequence of the common habit of associating such an achievement with the collapse of French resistance.

But the attribution of a definite point as the German objective is correct in only a restricted and conditional sense. For the supreme purpose of the Germans could have been no other than to destroy the enemy's field armies. The French capital was only incidentally the goal because, by directing their armies in the general direction of Paris, the Germans hoped to encounter and destroy the field armies of their adversary on the way. Without the destruction of the enemy's main armies, the possession of Paris itself would have had no final importance. But after the destruction of the field forces the resistance of no fortress would have availed to prevent a German victory.

Five of the German armies in the West were now sweeping forward into northern France like a great arm moving on the German position before Verdun as a pivot stretched out to envelop all the French forces in its course. The aim was not merely to outflank the Allies, but to encircle and roll them in together, and clasp them tight in a deadly embrace. The Germans proposed to repeat the victory of Sedan by a maneuver embracing all northern France. It was the most imposing military movement ever attempted.

By the 29th, the front of the Allied forces, beginning on the course of the Oise, extended on a line indicated by the position of the fortresses La Fère and Laon and continued eastward along the upper reaches of the Aisne and thence across to Verdun. The forces on both sides southeast of Verdun had become almost stationary.

The five German armies striking at the Allied front west of Verdun were advancing on lines that tended gradually to converge. Their common front presented a somewhat hollow contour, the western extremity distinctly protruding. The left flank of the Allies was correspondingly pressed back, foreshadowing as it were the ultimate success of the German turning maneuver.

The Germans probably hoped to bring the campaign in the West to a victorious culmination in a gigantic battle on the broad plains about Reims where the deployment of their tremendous forces would be unrestricted and where the Allies would be exposed on every side. A vigorous attack along the whole front would have served as prelude to the envelopment of one or both of the wings of the Allies and the gathering of all their field armies west of Verdun into the fatal snare.

This, we may assume, was the nature of the German plan as it appeared to General Joffre.

On the 29th the First and Third Corps in the Fifth French Army on the Oise executed a brilliant counter-attack in the direction of the Somme, driving back the Prussian Guard with its reserve and the Tenth German Army Corps and inflicting severe losses upon them. The German official *communiqués* which exultingly announced the "total defeat" of the British near St. Quentin a day or two before passed over in silence this exploit of the French. The resulting protuberance of the French front near La Fère must have accentuated the angle already created in this part of the Allied line by the recession of the British front towards the left in conformity with the course of the Oise.

But the Allies did not endeavor to make this partial success a turning point in the general course of their operations. It is a self-evident maxim of good strategy to do exactly the opposite of what the enemy wants. The Germans wished to bring the campaign in the West to a speedy issue while their vigor and numerical superiority were unimpaired, and before the progress of the Russian invasion in the East necessitated a redistribution of their forces. The Allies had thus far evaded by their systematic retirement the efforts of the Germans to accomplish their purpose. The German turning movement, at one moment on the point of execution, had not yet been achieved.

General Joffre subjected the whole situation to a careful scrutiny and visited Field-marshal General French on the 29th to confer with him on the plan for the immediate future. The situation was still fraught with imminent peril for the Allies. The Germans were pressing forward with undiminished vigor and greatly superior forces. The advance of the Russians in East Prussia had not abated the energy with which the offensive in the West was being pushed. The left flank of the Allied armies remained exposed. Apparently the conditions would have been advantageous to the Germans in a great battle fought in the actual position of the contending armies. All the material circumstances argued in favor of a further retreat for drawing the Germans on until the situation became favorable for the Allies to resume the offensive. Every day that intervened before the supreme trial of strength taxed the endurance of the invaders, increased the distance that separated them from their bases of supply, added to the French reinforcements concentrating in the rear, and intensified the Russian peril in the East.

But a further retreat was beset with the greatest difficulties of a moral nature. The inevitable abandonment of the wounded and much of the equipment in a long-continued retrograde movement tends to depress the soldiers, to undermine their *morale*. The temporary sacrifice of a considerable portion of the national territory and the consequent privations and suffering of millions of Frenchmen and disruption of the intricate network of industry and all human relations would react with tremendous force upon public opinion. It was doubtful whether the patience of an impetuous people, already sorely tested, would endure this additional trial. It required an implacable resolution to face such difficulties.

The evacuation of French territory had probably occasioned a conflict from the first between the political and military leadership of the nation. It is safe to assume that the reorganization of the ministry on August 26th was not without a very close connection with this fundamental disagreement. When the haze which still obscures the more intimate course of events at that time will some day be dispelled, a great moral victory of the French nation will probably be revealed, an achievement fit to rank with the defeat of the Boulanger conspiracy and the vindication of Captain Dreyfus as the third essential step for the perpetuation of the republic. In the two former crises the French people rallied to the defense of liberty against the artful intrigues of despotism. But in the late summer days of 1914, as will probably become increasingly apparent, this headstrong people, jealous of its freedom, abdicated voluntarily, for the course of the war, its privilege of vacillation, an essential attribute of liberty, and accepted the firm, dictatorial control which was indispensable for unity of plan and purpose.

General Joffre decided to continue the retreat, sacrificing all other considerations to the single object of victory. La Fère was evacuated after a sharp engagement.

About this time the Ninth French Army, made up of three corps from the south, took its position at the front between the Fifth and Fourth Armies to steady them before the German onslaught.

The official German communication of General Quartermaster von Stein on September 2d announcing the defeat of about ten French army corps between Reims and Verdun the day before, while the Kaiser was actually present with the army of the Crown Prince, was very clearly an endeavor to elevate the daily rear-guard engagements incidental to the systematic retreat of the French to the imposing proportions of a decisive victory for the Germans worthy of the anniversary of Sedan Day.

The French evacuated Reims, September 4th. The Germans fired about sixty shells into the city because there

The effect of the heavy German artillery on one of the forts at Liége.

Steel cupola of one of the forts at Maubeuge cracked by German high-explosive shells. Maubeuge was besieged by the Germans on August 27, 1914, and held out under severe bombardment until September 7th.

had been no formal capitulation after the departure of the French troops. On the 5th the Saxon troops of the German Third Army marched into Reims singing their national hymns and occupied the city.

The British withdrew on August 29th to a position a little north of Compiègne and Soissons, and continued their retreat on the following days in conformity with the general movement of the French armies, often by forced marches of extreme length. Their route lay through the great Forest of Compiègne which contains 35,000 acres. The rear-guard was attacked with special violence as it was emerging from the woods to the south of Compiègne on September 1st. The 1st British cavalry brigade and 4th Prussian Guards brigade were chiefly engaged in the ensuing encounter. After a spirited contest the British repulsed the Germans, capturing ten of their guns.

The British Second Corps followed the main highway from Compiègne through Senlis and the Forest of Chantilly in the direction of Paris, while the First Corps proceeded by a route further east. The British were closely followed by the Germans and many other rear-guard actions occurred.

By the 27th the railways were bringing great crowds of refugees to Paris from the north and east. This influx was just beginning to tax the resources of the authorities when it was more than offset by the exodus towards the south and west of vast throngs who were terrified at the prospect of another siege of Paris with all its privations and danger. This migration had gained considerable volume by August 30th, it was probably stimulated by the appearance of the first German aircraft over Paris, and it reached its culmination when the government's own example became known on September 3d. The railways, subjected at this time to an excessive strain in effecting the redistribution of the French forces, were unequal to the additional burden thus laid upon them. Huge crowds in a tumult of apprehension choked the railway stations and bivouacked before the termini of the principal lines. Numbered tickets indicating the order of admission to the trains had to be procured forty-eight hours in advance of the intended hour of departure. People gladly paid 250 francs ($48.25) for passage to Havre by river, and as much as 5,000 francs ($965) is said to have been paid for the hire of an automobile for conveyance to the same destination. There was a veritable stampede of motor-vehicles whose owners wished to leave the city before their machines were requisitioned or the highways were barricaded. Endless processions of vehicles of every description and of pedestrians filled the roads. A vast torrent of humanity representing every station of life poured forth from the metropolis, while Paris itself remained strangely tranquil.

At length the authorities, recognizing the advantage of reducing as far as possible the number of mouths to be fed in Paris, offered free transportation by rail to those who wished to depart for the provinces, making public announcement of the hours and places of departure and the destination of the trains which were made available for this purpose.

Paris is the center of an immense intrenched camp formed by a girdle of detached fortresses, erected for the most part in the decade following 1871, with a perimeter of ninety miles. It was estimated that 150,000 troops would be necessary to defend, and 350,000 to invest, the fortified area. But the lessons of Liége and Namur had shaken confidence in its impregnability. The open spaces between the forts would have offered the same problem of defense. Besides, it was evident that the Germans would not have invested the whole circuit of the intrenched camp. They would have massed their forces against a section of the perimeter, shattered two or three of the forts by the concentrated fire of their heavy siege-artillery, and rushed through the wide breach thus created.

The rapid approach of the enemy bringing Paris itself within the area of the war zone induced the government to decide to transfer its seat temporarily to Bordeaux, so as to remain in unrestricted contact with the country as a whole. To justify this decision, the following proclamation was issued on September 2d, signed by President Poincaré and countersigned by the members of the ministry:

"People of France!

"For several weeks relentless battles have engaged our heroic troops and the army of the enemy. The valor of our soldiers has won for them marked advantage at several points. But in the North the pressure of the German forces has compelled us to fall back.

"This situation has compelled the President of the Republic and the government to take a painful decision. In order to watch over the national welfare, it is the duty of the public powers to depart temporarily from the city of Paris.

"Under the command of an eminent chief, a French army, full of courage and zeal, will defend the capital and its patriotic population against the invader. But the war must be carried on at the same time on the rest of the territory.

"Without peace or truce, without cessation or faltering, the sacred struggle for the honor of the nation and the reparation of violated right must continue.

"None of our armies is impaired, If some of them have sustained very considerable losses, the gaps have immediately been filled up from the reserves and the levy of recruits assures us new reserves in men and energy for to-morrow.

"Endure and fight! Such must be the motto of the Allied British, Russian, Belgian, and French armies.

"Endure and fight, while at sea the British aid us in cutting the communications of our enemy with the world.

"Endure and fight, while the Russians continue to advance to strike the decisive blow at the heart of the German Empire.

"It is the duty of the Government of the Republic to direct this stubborn resistance.

"Everywhere Frenchmen will rise for their independence. But to insure the utmost spirit and efficacy in the formidable contest, it is indispensable that the government shall remain free to act.

"At the request of the military authorities the government is therefore transferring its headquarters to a place where it can remain in constant touch with the whole of the country.

"It requests members of Parliament not to remain away from it, in order that they may form with the government and their colleagues a bond of national unity in the face of the enemy.

"The government leaves Paris only after having assured the defense of the city and of the intrenched camps by every means in its power.

"It knows that it does not need to recommend calm, resolution, and coolness to the admirable population of Paris which is showing every day that it is on a level with its highest traditions.

"People of France!

"Let us all be worthy of these tragic circumstances. We shall gain the final victory. We shall gain it by unflagging will, endurance, and tenacity.

"A nation which refuses to perish, and which, in order to live, does not flinch either from suffering or sacrifice, is sure of victory."

A special train left the station of Auteuil with President and Madame Poincaré and the ministry at eleven o'clock on the evening of the 2d and arrived at noon in Bordeaux, which had been selected as the temporary seat of government. Other trains during the 3d brought the diplomatic corps and Council of State. The offices of government were distributed among the public buildings. The prefecture became the president's residence and the faculty of letters of the university was transformed into the Ministry of War.

The eminent chief, to whom the defense of the capital had been entrusted, General Galliéni, issued the following proclamation:

"To the Army of Paris and the inhabitants of Paris!

"The members of the Government of the Republic have left Paris to give a fresh impulse to national defense. I have been entrusted with the task of defending Paris against the invader. This task I will fulfil to the end."

This was the laconic utterance of a man of energy and quiet determination. It contrasts favorably with the pretentious rhetoric of General Trochu to whom the same mission was confided in 1870.

Joseph Galliéni, who was born at Saint-Béat, in the south of France, April 24, 1849, entered the military school at St. Cyr in 1868, and like General Joffre served as sub-lieutentant in 1870. He was a member of the group under Commander Lambert who surrendered after the exhaustion of all their ammunition at Balan, the village of the "Dernières Cartouches." Later Galliéni was engaged in service in the French dependencies, where he showed himself to be a capable organizer and intelligent administrator as well as an energetic soldier. He organized the French Soudan as lieutenant-colonel. After his elevation to the rank of colonel in 1891 he executed important missions in Indo-China and Madagascar. He was made brigadier-general in 1896 and appointed Governor-general of Madagascar, a post which he occupied until 1905. He wrote a number of works on French colonial subjects, particularly on the geography and administration of Madagascar. He was the youngest general of divisional rank in the army when raised to that grade in 1899. He was appointed Military Governor of Lyons and Commander of the Fourteenth Army Corps in 1906, and two years later he was admitted to the Conseil Supérieur de la Guerre.

The German armies in the extreme West were still advancing in the direction of Paris at a speed that is only attainable with the extensive use of motor transport on such roads as the incomparable highways of France, and in a level region. Both von Kluck's army and the British whom they were pursuing covered as much as thirty miles in a single day.

Suddenly an armed motor-car conveying about twenty soldiers would appear on the German line of march. Then groups of from five to twenty Uhlans would spring up as if by magic at different points in the zone to be traversed by the German columns. Larger masses of cavalry supported by rapid-fire artillery mounted on automobiles would next appear; and then came the main bodies of infantry, regiments, divisions, army corps. The compact columns were followed by heavy artillery drawn by motor-tractors. Air-scouts projected in every direction the vision of the intellectual management guiding the movement of the invading hosts with almost unfailing accuracy. The whole was impelled by a relentless determination and iron discipline.

It seemed almost a question of hours whether the Germans would succeed in their great venture or be balked of their prize before the very gates of Paris.

Just at the crucial moment, when the future of France, and perhaps of democratic institutions in Western Europe, hung in the balance, the dark war-clouds rose rapidly again upon Germany's eastern horizon, and this time with the menace of such immediate peril as to demand forthwith very earnest attention. After a fortnight's continual battles on a scale hitherto unknown, Lemberg was evacuated by the Austrians on September 3d. The Russian deluge was rolling westward across Galicia with irrepressible volume bearing the sorely battered Austrian armies before it. The gateway to Vienna and Berlin was in imminent danger.

CHAPTER VI

The Russian Offensive

Early operations on the Russo-German border. Exaggerated impression of Russian might. Natural objective for the Russian offensive. Why the Russians did not set out for Berlin from the nearest point on their frontier. Lack of natural boundaries. Important rivers: Niemen (Memel), Vistula, San, Bug, Narev, Bobr. Strategic consequences of the position of Russian Poland as a great salient. Teutonic fortresses and Russian fortresses. Russian invasion of East Prussia; the German forces, the Russian Vilna and Warsaw Armies. Encounters at Stallupönen and Gumbinnen. The Masurian lakes. Von Hindenburg and the Battle of Tannenberg: von Hindenburg, the man and his hobby; his appointment to the command in the East, August 22d; his concentration of troops and chances of success; his plan of battle resembling that of Hannibal at Cannae; Battle of Tannenberg: the contest at Hohenstein, August 26-28; the German occupation of Soldau and its results, August 26-27; operations by the German left wing, August 26-29; the consummation and extent of the victory. Operations against Rennenkampf. The situation in Galicia. The Austrians invade Poland while the Russians invade Galicia from the east. The operations of Russky and Brussiloff and the evacuation of Lemberg by the Austrians on September 3-4. Collapse of Dankl's offensive in Poland. The defeat of the Austrians on the Grodek-Rawaruska line and their withdrawal from most of Galicia.

Minor operations on the Russo-German border were reported as soon as war between the two countries was declared. The Germans claimed that the Russians inaugurated hostilities during the night of August 1-2 by acts of aggression committed by their boundary patrols, presumably before the result of the German ultimatum at St. Petersburg had become known on the frontier.

On the 3d a battalion of the 155th German infantry regiment and the 1st Uhlan regiment occupied Kalisz, the frontier town in Russian Poland on the railway line leading from Lodz into Prussia. It was at this place that Alexander I in 1813 summoned the Germans to rise against

Field Marshal Paul von Beneckendorff und von Hindenburg, commanding the German armies operating against the Russians.

Napoleon's tyranny. On the same day German frontier-guards occupied Czenstochowa, which possesses important coal mines and a religious shrine of great sanctity among the Poles, frequented annually by thousands of pilgrims. Bendzin, likewise important for its coal mines, was taken at the same time.

Allusion has already been made to the German method of employing cavalry as a screen to hide artillery and infantry. These tactics were successful in several minor frontier engagements. On the 5th the German cavalry near Soldau in East Prussia enticed a body of Russian cavalry to a point within range of concealed infantry and machine-guns, where they were mowed down or put to speedy flight. The German covering troops with artillery repulsed Russian cavalry near Eydtkuhnen on the 10th.

A profound impression of Russia's inexhaustible strength was generally prevalent at the commencement of the Great War. In spite of their exaggerated opinion of Russian inefficiency and unwieldiness, the Germans shared in the excessive feeling of awe produced by Russia's tremendous magnitude. The danger of living next-door to such a prodigy had lately become an obsession with them. Even the German General Staff formed its plan of campaign with Russia's potential might as the fundamental consideration. The mad dash to the very heart of France, disregarding the neutrality of Belgium, was prompted by the supposed necessity of concentrating most of the German forces later on the eastern frontier to repel the tremendous hordes which the Tsar could muster in time from all parts of his vast dominions. In the opinion of the General Staff, the completion of the concentration of the Russian field armies, which was supposed to require about six weeks, set the inevitable limit for the accomplishment of the necessary decisive action in the West. To the

Germans, in the wild exultation over the early victories, the thought of Russia's millions came like the dark specter of a threatening doom; to the western Allies in the depression of their first defeats, it was the token of ultimate victory. Russia's armies were likened to a ponderous steam-roller that would flatten out every obstacle in its course. The feeling was very common in neutral nations also that Russia's numbers would eventually be the decisive factor.

All parties failed to discount the fundamental difference, which all recent development in the art and practice of war has accentuated, between a mere recruit and an efficient soldier trained and equipped. The mobilization of Russia's armies of the first line was accomplished with a rapidity which astonished all observers and lent some color to the German insinuation that the Russians had been making surreptitious preparations for several weeks. But the process of concentration scarcely assembled a tithe of the men of military age in the Russian Empire. The countless millions that had fired the imagination of the West were not available. Equipment was utterly lacking for mobilization on a scale commensurate with the bigness of the country. The supply of material of war was the crucial problem for Russia from the beginning. The development of Russian industry, satisfactory as had been its progress in recent years, had not attained the capacity of supplying unaided the material required for warfare on such a vast scale, particularly the necessary munitions.

It was soon evident that the Russian army had undergone a far-reaching reorganization since the disasters in Manchuria. The assumed degree of inefficiency and the supposed phenomenal size of the Russian army were alike fictitious. In the numerical strength of her mobile forces, Russia did not stand in a class by herself. The aggregate strength of her field armies in 1914 was considerably less than the combined strength of the German armies on both fronts. But the rapidity and vigor of the Russian offensive did much to embarrass, if it did not itself effectively upset, the execution of the great German plan of campaign in the West.

The natural objective for the Russian offensive was assumed to be Berlin, just as that of the Germans was thought to be Paris. The old-time tradition of the Seven Years' War suggested this, and the supposition that the German capital was exposed to attack from the east. Some surprise was therefore felt when the Russians commenced their invasion of Germany at the most remote extremity of East Prussia, more than 400 miles from Berlin, rather than from the western confines of Russian Poland, whence the distance to be traversed was considerably less than half as great. The point where the Warta coming from Russian Poland penetrates Prussian territory, only 180 miles east of Berlin, almost exactly in the direct line between Warsaw and the German capital, is an apparently suitable spot from which to set out on such an enterprise.

Yet never, throughout the whole fluctuating course of operations in the eastern theater of war, were the Russian aggressive movements directed along an approximately straight line from Warsaw to Berlin. The tradition of operations in the past may have contributed somewhat to this. But the main cause must be sought in a consideration of the physical features and political geography of the whole eastern area of hostilities.

The lack of natural barriers to define political boundaries in this part of Europe was emphasized in the first volume. Between the Carpathians and the Baltic Sea a vast plain extends from the heart of Russia across northern Germany. The Russian boundary of the Teutonic empires taken as a whole resembles the figure of a capital S turned backwards, with East Prussia filling the upper recess, Russian Poland the lower, and Galicia falling just below the tail.

The physical character of East Prussia is scarcely distinguishable from that of the adjacent Russian provinces, and the boundaries of Russian Poland in the direction of Posen and Galicia are likewise, for the most part, marked by no physical distinctions.

The noteworthy features of this great plain are its watercourses. The general direction of the rivers is northwards, since they mostly rise in the Carpathians or their foothills and empty into the Baltic Sea. Beginning in the east we encounter the Niemen, which flows past the Russian fortresses Grodno and Kovno, in a course generally parallel with the eastern frontier of Prussia and about fifty miles east of it. Bending to the west after passing Kovno, the Niemen penetrates the Prussian boundary and reaches the Baltic Sea as the German Memel. The Vistula, by far the largest and most important river of this region, rising in the Carpathians, flows by Cracow and reaches Warsaw by a long, sweeping curve to the left, after forming the northern boundary of Galicia throughout a third of this section of its course. Just below Warsaw the Vistula turns westward and reaches the German boundary, after flowing in a generally northwestern direction, about twelve miles above the fortress of Thorn. Then traversing West Prussia it empties into the Gulf of Danzig. The Vistula receives two very important tributaries: the San, approaching it from the right near the point where it leaves the Galician boundary; and the Bug, also from the right, eighteen miles below Warsaw, after this latter tributary has been swelled by the waters of the Narev, which empties into it on the right. The Bug rises in eastern Galicia and flows through Brest Litovsk, 150 miles east of Warsaw, where

G
23°
H
24°
J
1
2
3
4
55
54
53
52
Kovno
Suwalki
Augustof
Grodno
Bialystok
Marjampol
Kalvarja
Bjelsk
FOREST OF
BJELOVJESH
Brest
Litovsk
Siedlce
Biala
Wegrof
Drusgeniki

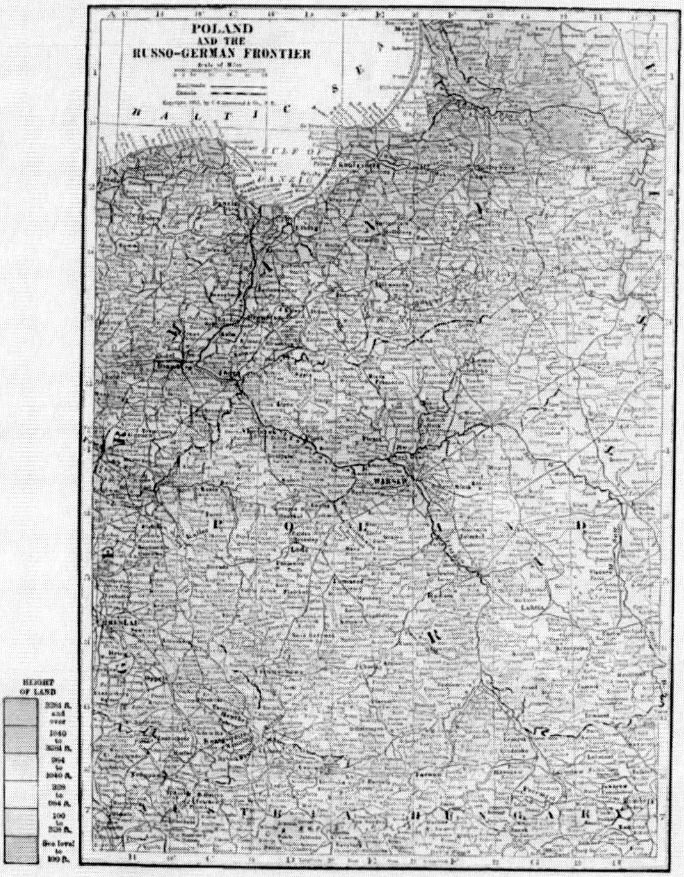

POLAND
AND THE
RUSSO-GERMAN FRONTIER
Scale of Miles
B A L T I C
S E A
GULF OF
DANZIG
WARSAW
P O L A N D
HEIGHT
OF LAND
Sea level
to
100 ft.

there was an immense military depot protected by a ring of forts, the main base of supplies for all Russian operations in Poland. East of Brest Litovsk there is a large tract of almost impenetrable country, the Pripet marshes, one of a number of such swampy areas in Russia, which are a consequence and characteristic of the ill-defined water-partings. The Narev with its tributary the Bobr and the frequent marshes along their course are an important defensive feature of the region opposite the southeastern frontier of East Prussia.

Russian Poland, which, for the sake of convenience, will be alluded to henceforth simply as Poland, occupies, as we have seen, a great salient in the western front of Russia, with the strategical attractions of a position protruding far into the enemy's territory, but at the same time beset with insidious perils. This westward extension not only brought Russia to a point only 180 miles from Berlin, but it bordered on Silesia, a hive of German industry, second only to Westphalia in the importance of its coal mining and manufactures of iron and steel.

But a Russian army advancing westward through the center of Poland would be threatened on both its flanks. The Teutonic allies could launch their blows from three sides at the heart of Poland, and the remarkable efficiency of their strategic railways enabled them to transfer their forces from one part of the frontier to another with such rapidity and secrecy that it was impossible for the Russians to foresee from what direction a deadly thrust was to be expected.

Poland is the most thickly populated part of Russia, but its railways were few as compared with the lines of Prussia and Galicia, and those which existed were partly single-track so that their capacity for military purposes was not to be compared with the German and Austrian systems. There were no essentially strategic lines in Poland, and the lack of a railway following the course of the frontier all around was a very serious deficiency, especially in the operations against East Prussia. Throughout the entire course of the German, and practically all of the Austrian, frontier one or more lines run parallel with the border, which rendered invaluable service for the Teutonic allies in shifting troops and material from point to point. A deliberate adaptation of the railway system to strategic requirements is evident even in Galicia, where at eleven points the railheads approached the Russian frontier with no communication beyond it.

While the Teutonic powers relied chiefly on their field-armies and splendid equipment of railways, they had not entirely neglected the construction of modern strongholds to support their defensive and serve as bases for offensive operations. Königsberg is a first class fortress, composed of a double enceinte and twelve detached forts, the only stronghold of much consequence in East Prussia. Danzig and Graudenz are both strong fortresses. But Thorn has been the most important German fortress for the operations in the eastern theater during the present war, although it has never once been attacked. Thorn was the pivot for von Hindenburg's railway strategy and one of the bases from which he delivered his repeated blows against Warsaw in the autumn of 1914. The fortified area at Thorn lies on both sides of the Vistula, eight of the detached forts being situated on the right bank and five on the left. While there is a very large intrenched camp at Posen, the rich province of Silesia was practically destitute of all natural or artificial defenses. Peremysl, sixty miles west of Lemberg, the capital, was the principal fortress in Galicia. But Cracow, the ancient capital of Poland, now in the western extremity of Galicia, was in one respect a far more significant position. Cracow was the constant goal of the Russians in their offensive in Galicia, although they never reached it. The fortress of Cracow guards the natural gateway from the valley of the Vistula to that of the Oder, a gateway that opens westward upon a natural vestibule from which corridors lead to Vienna and Berlin respectively. Cracow is defended by a girdle of six powerful forts on both sides of the river.

At the beginning of the war the Russians prudently abstained from any serious attempt to hold the western part of Poland. They even withdrew their frontier and custom house guards from some parts of the Austrian border, so that the boundary control was entirely relaxed and Poles were free to join the Polish regiments in Galicia. The concentration of the Russian forces was carried on in the eastern part of Poland, behind the Vistula, in the area covered by the fortresses and more important rivers.

Besides Brest Litovsk, the Russians had two very strongly fortified positions in Poland, Ivangorod, sixty-four miles southeast of Warsaw, with twelve forts, nine on the right bank of the Vistula and three on the left, and Novo Georgievsk at the confluence of the Bug and the Vistula.

The conditions were such that the Russians could not hope to advance into the heart of Germany before they had cleared their adversaries from one or both flanks of their position in Poland. The Russians undertook both these operations at practically the same time. For several reasons it is more convenient to consider first the movement on their right wing. The invasion of East Prussia advanced at first more rapidly than that of Galicia and was watched with keener interest. But very soon it collapsed entirely leaving very little impression on the subsequent course of events, except that it afforded the opportunity for the discovery of the greatest talent and the making of the most distinguished reputation on the German side throughout the war.

As we have observed, it was the intention of the Germanic powers to hold the Russian armies in check as far as possible until the fate of France had been decided. After presumably sufficient forces had been directed against the Serbian army and an army corps had been sent to help the Germans in Alsace, the remainder of the Austro-Hungarian army, nearly 1,000,000 men, was concentrated on the Russian border. But the Germans limited their forces in the East at the beginning of the campaign almost to the point of recklessness in their eagerness to give the fullest effect to their initial blow in the West. Twenty of their army corps were concentrated on the western front, so that only five were left to guard the more extensive eastern frontier, the First, Fifth, Sixth, Seventeenth, and Twentieth. Only the First and Twentieth Army Corps were in East Prussia, their headquarters being at Königsberg and Allenstein respectively. The Sixth was sent to reinforce the Austro-Hungarian forces in Galicia. The Fifth was probably distributed along the boundary of Posen. The Seventeenth may have been held at its headquarters at Danzig in readiness to support the forces in East Prussia. The First was brought up to the eastern frontier of East Prussia. Probably an unusually large proportion of reserve and Landwehr formations were assembled in this part of the country to compensate somewhat for the very great disparity in forces of the first line.

The Russian army corps ordinarily stationed near the boundaries of the Teutonic Empires with their respective headquarters were the following: Third, Vilna; Fourth, Minsk; Second, Grodno; Sixth, Bialystok; Fifteenth, Nineteenth, and Twenty-third, Warsaw; Fourteenth, Lublin; Eleventh, Rovno; and Twelfth, Vinnitza.

Field-marshal von Potiorek, commander of the Austrian army which invaded Serbia.

General Victor Dankl, commander of the First Austrian Army, directed to invade Poland.

General Moritz von Auffenberg, commander of the Second Austrian Army, intended to protect the flank and rear of the First Army.

The rapidity of the Russian concentration must have been a very disconcerting factor in the plan of the German General Staff. The general Russian advance into East Prussia commenced about August 16th, just as the German deluge was sweeping away the final barriers at Liége, and at least a week before the systematic invasion of France was begun.

Three railways which cross the frontier from Russia into East Prussia determined the general course of the Russian invasion on converging lines. The Vilna Army, as it was called, which had been concentrated on the Niemen, composed of the Second, Third, Fourth, and Twelfth Army Corps of the active army, and the Third and Fourth Reserve Divisions, and five cavalry divisions, under General Rennenkampf, was ordered to advance westward along the main railway line connecting St. Petersburg and Berlin which penetrates the East Prussian boundary at Eydtkuhnen. General Rennenkampf was one of the few Russian commanders who earned favorable distinction in the Russo-Japanese War, in which he commanded a division.

The task of the Narev, or Warsaw, Army, under General Samsonoff, probably made up of the First, Sixth, Eighth, Fifteenth, and Twenty-third Army Corps of the active army, and the Twelfth Reserve Army Corps, with three cavalry divisions, was to assail East Prussia from the southeast, from the region of the Narev, advancing mainly along the railway lines leading from Bialystok to Königsberg and from Warsaw *via* Mlava to Soldau and Danzig. Altogether more than 500,000 Russians had thus been massed as early as August 16th for the invasion of East Prussia.

The large complement of cavalry was a very useful adjunct to the Vilna Army in its advance across the northern part of East Prussia, a comparatively open, level expanse of territory, well-developed and with good roads.

A long-established popular tradition of the ruthless cruelty of the Cossacks was one of the chief causes for the general panic and exodus of a large number of the inhabitants at the approach of the Russian cavalry, who were soon scouring the country far and wide. The consternation and flight of the inhabitants was not unlike the havoc occasioned by the inroad of the Germans into Belgium at precisely the same time.

The rounding contour of East Prussia and its excellent network of railways might have provided the Germans with an opportunity of keeping their own forces together and striking the different masses of the invading forces separately in quick succession. But at first the Germans did not operate in accordance with such a plan, either from lack of a single commander, or because they underestimated the enemy's strength.

General von François confronted the Vilna Army with the First and probably a part of the Seventeenth Army Corps and their accompanying second line formations. The Russians outnumbered his forces in about the proportion of two to one. The Germans tried to delay the Russian vanguard at Stallupönen where there was a stubborn contest on August 17th. The German reports of the fighting in East Prussia at this time resemble in one respect the Belgian notices of the operations of the Belgian army at precisely the same period. Each series presents the seeming incongruity of a succession of victories on an ever-receding front. The public was informed, for instance, that the German army took 3,000 prisoners at Stallupönen and 8,000 three days later at Gumbinnen, and that its eventual retirement, leaving the way open to Insterburg, an important railway center, was solely due to strategical considerations. The advance of the Narev Army from the south was menacing its line of communications. But the immediate cause of the hasty retreat of the troops under General von François was undoubtedly their inferiority in number to the Russian army in front of them. The Germans were hopelessly outnumbered and overpowered; but this, in the circumstances, was no disgrace. The Russians advanced on a broad front. By concentrating the bulk of his available forces at special points so as to gain a temporary local superiority, General von François was probably able to delay and embarrass the Russian forces for a few hours at a time.

The principal encounter in this period was fought before Gumbinnen, twenty-two miles west of the Russo-German frontier at Eydtkuhnen, on the 20th. The Russians attacked the Germans in front in a succession of bayonet charges. After a courageous resistance of fourteen hours the Germans withdrew. In the broad sweep of their advance the Russians repulsed the Germans on the left at Goldap and occupied Tilsit on the right. The Germans, threatened with envelopment, retired in considerable disorder, abandoning much of their equipment along the road. The Russians entered Insterburg on the 24th. They did not at any time invest Königsberg or isolate it, although their cavalry advanced beyond Tapiau on the railway line from Eydtkuhnen.

Meanwhile, the Narev Army had invaded East Prussia from the southeast. Its left wing put to flight a German detachment which had occupied Mlava in Poland and drove the Germans from Soldau, an important junction point on the Prussian side of the border. At the extreme right a secondary mass, detached from the main body of the army, was advancing on Lyck by way of Osovietz. The Russians had to traverse a dreary portion of the province of East Prussia, but a region of exceptional strategic importance, as the approaching great events were soon to prove. This is the region of the now famous Masurian Lakes, a monotonous sandy waste, covered in large part by forests of stunted pines and birches. Innumerable shallow depressions retain the water in lakes, pools, and morasses, the latter often concealed beneath a treacherous layer of soil and vegetable growth.

This labyrinth can only be traversed in a few places by narrow passages and causeways between the lakes. The intricacy of these lines of communication is very baffling except to those who are thoroughly familiar with the region. The different parts of an army advancing in separate columns through these various defiles are necessarily out of touch one with another and afford an excellent opportunity for the enemy to attack them separately. The region presents the gravest perils for armies in retreat by reason of the difficulty of distinguishing the trustworthy from the deceptive openings between the marshes and sheets of water. The railway system of East Prussia was admirably designed for making the most of the defensive possibilities of this part of the province. A main line with double track facilitated the distribution and shifting of troops and supplies behind the cover of the lakes.

The Russian column arriving at Lyck separated, one part passing south of the lakes through Johannisburg, the other part proceeding to the northward of Lake Spirding. Thus the army of General Samsonoff was spread out over a very broad front where communication and coöperation were rendered difficult.

The Twentieth Army Corps and some Landwehr formations opposed the march of the Narev Army, particularly its right wing. The Germans made a stand for two days, August 23-24, at Frankenau in a prepared position, and then were forced to retire, and the Russians occupied the headquarters of the Twentieth Corps at Allenstein, an

Burial of Austrian dead after repulse of a Russian attack on Peremysl.

Neidenburg in East Prussia. The destruction is said to have been wrought by Russian gunfire.

important junction point on the main railway line from Berlin *via* Thorn to Insterburg and Eydtkuhnen. General Rennenkampf's front was now on the line Friedland-Angerburg, and the cavalry of the two Russian armies had nearly established contact.

The progress of the Russian offensive, particularly the advance of the Narev Army, unless immediately arrested, would in a few days have cut off the army of von François, isolated Königsberg, swept the Germans from all the rest of their territory east of the Vistula, and so inflicted upon them a national calamity involving even more serious moral effects than the formidable physical losses.

By August 22d the German General Staff had become convinced that drastic measures were required to cope with the alarming situation in East Prussia. A dispatch from the Kaiser summoned General von Hindenburg to take command in the East, a man sixty-seven years of age, who was as little known outside purely military circles in his own country as many other commanders who have won great distinction in the Great War.

Paul von Hindenburg (Paul Ludwig Hans Anton von Beneckendorff und von Hindenburg) belonged to a family of stalwart Prussian Junkers who had served the state with loyalty for more than two hundred years. His career began as a lieutenant at Sadowa, where he received a slight wound. He took part in the battles of St. Privat and Sedan and in the operations before Paris in 1870. Later he attended the War Academy. While serving on the staff of a division at Königsberg during the years 1881-1883 his attention was enthralled by the strategical possibilities of the Masurian Lakes for the defense of East Prussia. This subject became at once his occupation and his pastime. He was obsessed by it. He explored every nook and corner of this bewildering region. Later, when called to the General Staff and to the professorship of applied tactics in the War Academy, von Hindenburg had an excellent opportunity to develop his favorite theme in the course of his lectures. Some of his colleagues regarded his apparent infatuation with good-natured ridicule. But von Hindenburg's military career gave him a comprehensive experience. He rose through the various grades of troop commander until he became commanding general in 1903. He resigned his post at the age of sixty-four in 1911. At the commencement of the war he offered his services, but he had almost given up hope of an opportunity of actually conducting the campaigns which he had so long anticipated in imagination when the call finally came.

On the night of the 22d a special train conveyed von Hindenburg from Hanover, where he had been living in retirement, in company with his Chief of Staff, von Ludendorff, towards the eastern theater of hostilities. They arrived at Marienburg, the temporary headquarters, the next afternoon. Three days later the battle began which made von Hindenburg famous throughout all the world.

Von Hindenburg is a man of relentless energy and prodigious capacity. The massive cast of his features suggests an indomitable resolution. If genius consists in clear and searching comprehension, unfailing adaptability to circumstances, and indefatigable patience, then von Hindenburg is a genius. He used the means which were at hand, his methods involved no startling innovations; but all his operations were characterized by a marvellous grasp, rapidity, and thoroughness.

He proceeded without delay to concentrate with the utmost rapidity all the available German forces scattered throughout this part of the country. An uninterrupted procession of troop trains day and night taxed the capacity of the main line from Thorn to Osterode, as far as the railway could be safely operated under German control. No feature of von Hindenburg's generalship is more significant than his extensive and effective employment of the railways. To the Twentieth Corps and its Landwehr auxiliaries, which were already in the vicinity, were added the First Corps and Landwehr formations, part or all of the Seventeenth Corps from Danzig, and a reserve corps. Thanks to the faultless service of the railways the delicate operation of withdrawing the troops of von François from the front of the Russian Vilna Army was successfully performed. But in all not more than the equivalent of nine divisions was concentrated.

The aggregate Russian forces in East Prussia outnumbered this German army more than two to one. Von Hindenburg's only hope of victory depended upon his ability to deal with the Russian masses in detail. The armies of Samsonoff and Rennenkampf were only two or three days' marches apart. Von Hindenburg naturally directed his attention first to the army of Samsonoff, the nearer of the two. Every moment was precious; for the situation involved imminent peril. If Rennenkampf had come up in time and united his forces with those of his colleague, the Germans might have suffered an appalling disaster. Just at this time in the West the sensational dash towards Paris was entering upon its final stage. The withdrawal of large forces from the western theater, which would have been absolutely indispensable if the Germans had suffered such an overwhelming defeat in the East, would have been fatal to their entire plan of campaign. Everything depended upon the chances for victory in East Prussia and upon the skill of a single man.

In his calculations, von Hindenburg, who had been an attentive observer of Russian methods during the Manchurian campaign, assumed a degree of hesitancy and lack of initiative on the part of Russian commanders that justified the venture which he was about to undertake.

Four different railways form a convenient framework for the region in which the decisive encounter between von Hindenburg and Samsonoff took place: a section of the strategic line which follows in Prussian territory the sinuous course of the Russian boundary and a section of the main line from Thorn to Insterburg, together with portions of two other lines, as intercepted by these two first mentioned, one between Soldau and Deutsch Eylau and another between Ortelsburg and Allenstein. The four junction points here mentioned may be regarded as the corners of an irregular quadrilateral having a length of about fifty miles from southwest to northeast. It contains other railway lines of lesser importance. In fact, the remarkably convenient communications for a German army confronting an invader in this area was a factor of fundamental importance in the situation. On the other hand, the eastern part of the quadrilateral, as defined above, consists in large part of lakes and marshes, with few traversable openings between them and infrequent highways. It is part of the general region of the Masurian Lakes.

All the means by which an elastic but closely-knit political organism, with its resources well in hand, controlled by leaders of vigilance and discernment, can repel the attacks of a vastly larger, but cumbersome, neighbor were revealed in the rapid series of operations by which von Hindenburg swept the Russians from the soil of Prussia in the first part of his campaign of 1914.

Von Hindenburg's defeat of Samsonoff recalls very plainly the crushing of the Roman army by Hannibal on the plain of Cannae in 216 B.C. which afterwards became proverbial. To Professor Hans Delbrück is due the clearest, most consistent analysis of this very famous battle of antiquity, and on the basis of his able interpretation Count von Schlieffen, formerly Chief of the German General Staff, once declared that Cannae was the prototype of the kind of plan which should be the ideal for the modern commander. We may assume, therefore, that Professor Delbrück's explanation of the Battle of Cannae was current in higher military circles in Germany.

Both the Battle of Cannae and the notable contest which will presently be described were victories of nimbleness and dexterity against stolid, unintelligent force. In each the more talented commander enveloped and hopelessly ensnared his opponent. Hannibal led 50,000 men against the 70,000 Romans. The Germans claim to have won the no less complete victory in East Prussia against similar odds, while the Russians assert that the Germans were numerically superior to themselves. Both statements may possibly be true, each in a particular sense. The entire army of General Samsonoff was undoubtedly more numerous than that of von Hindenburg, but it was scattered, as we have remarked, and the Russians declare that only seven divisions of their troops were actually engaged in the decisive conflict against nine divisions of the Germans. The difficult character of the country and the distance separating the different routes by which the Russian army was endeavoring to advance make this declaration seem not entirely improbable. But even if it were admitted, the circumstance that through quickness of perception and skill in maneuvering von Hindenburg secured a local superiority at crucial points no more detracts from his fame than do such other favorable factors as his superior knowledge of the country, a closer contact with his base, better transportation facilities, and a more intelligent staff. Military renown that could not endure analysis would rest upon a very unstable basis.

Hannibal at Cannae drew up his less effective troops in the center of his battle-line in rather attenuated formation against the solid mass of the Roman infantry, and placed the more steadfast elements of his army in compact array on the wings which overlapped the Roman front. The weaker Carthaginian center received the shock of the Roman onslaught and yielded ground before it, but without breaking, until the wings were in position on the Roman flanks. Then, when the squadrons of Numidian cavalry had enclosed the Roman rear, the essential maneuvers were accomplished. The Romans were completely surrounded, compressed into a congested position, where they were unable to deploy and make their numbers count. The closer they were crowded together the more unwieldy became their efforts and the more deadly the action of the enemy. What followed was not a battle; it was simply wholesale slaughter.

Leading authorities on the art of war unanimously insist that the paramount aim in military operations should be the destruction of the enemy's field forces, not the occupation of his cities or territory. Von Hindenburg perceived at once that the successful prosecution of the campaign in East Prussia required the immediate destruction of the Narev Army and with iron consistency he directed every resource to the attainment of this single purpose. He adopted the plan of the victor at Cannae, the double turning movement for the envelopment of his opponent, the method inculcated by German doctrine and confirmed by a victorious experience for compassing the enemy's destruction. It is significant, furthermore, that this method produced at Cannae the most conspicuous example in ancient times of the destruction of an adversary's army, and in von Hindenburg's first great victory the only instance in the present war where the same conclusive result may be said to have been accomplished.

The natural conditions and means at his disposal permitted von Hindenburg to economize in the use of his forces, as compared with Hannibal, in two very important respects. The almost impenetrable character of the territory in the rear of the Russians served as effectually as the Numidian cavalry to complete the hostile ring, while the excellent railway communications across the rear of his own position enabled von Hindenburg to shift his forces quickly and secretly and thus use the same troops successively for the culminating operations of the conflict in different parts of the field. By rapid intrenching he compensated for the withdrawal of these troops from the positions which they had been chiefly instrumental in securing.

The Russians were advancing northwestward on a broad front, evidently with the intention of coöperating with Rennenkampf's army for completing the conquest of East Prussia as the necessary prelude to an advance beyond the Vistula.

The Russian intelligence department must have been quite defective. Perhaps the Russians were thrown off their guard by the inadequacy of the German forces which they had encountered and dispersed thus far. The movement of troops directed by von Hindenburg may have been largely concealed by the forests which cover much of this region. For Samsonoff was apparently not aware that considerable forces were being concentrated against him until the 26th, when he encountered the German army posted along the line Gilgenburg-Lautern.

In his initial dispositions, von Hindenburg, like Hannibal, concentrated his chief strength on his wings, reducing the forces at the center to the lowest degree at all compatible with safety. The German right wing was composed of the First Corps and a Landwehr division; the left was formed of the Seventeenth Corps near Lautern and a

Reserve Corps opposite Allenstein, which had been occupied by the Russians. There remained only six Landwehr regiments for the position in the center opposite Hohenstein.

The Russians advancing from Hohenstein attacked the Germans who fell back at first without seriously engaging themselves. The fighting near Hohenstein, which was very severe, lasted from the 26th to the 28th. The function of the German center was to hold firm before the Russian attack until the double flanking maneuver had been executed by the wings. That the German center did not break before the repeated assaults of the Russians is largely due to the superiority of the German artillery. The Russian gunners were said to have handled their pieces with exceptional efficiency and precision. But the Russian shrapnels in exploding did not diffuse their charge of bullets as extensively as the German. Besides, the movement of the Russian artillery was rendered difficult by the character of the country; whereas the Germans had control of a good provincial highway along which field-pieces and munitions could easily be conveyed. Later, when the Twentieth Corps, half of the Reserve Corps, and other parts of the Landwehr Corps had come to the support of the center, the Germans turned to the offensive. They concentrated their heaviest artillery, subjected the Russians to a pitiless shower of shrapnel, destroyed their imperfect trenches, and fairly blew Hohenstein to pieces with explosive shells. The Russians fled as best they could, abandoning most of their guns. Hohenstein and its vicinity presented a terrifying appearance for days afterwards. The once cheerful town in the midst of the dark pine forests was reduced to a blackened, distorted shell. Tall trees, shorn of their branches, twisted and scorched, stood like grim specters of an awful tragedy. The dead covered the

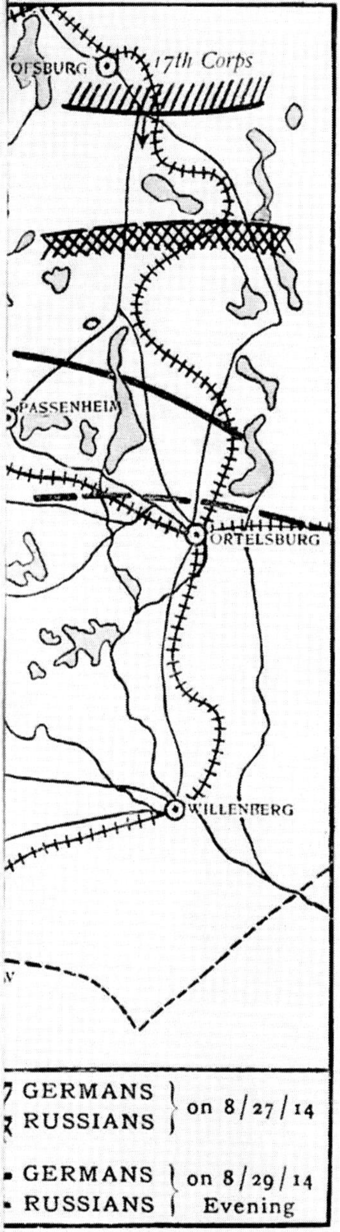

OFSBURG
17th Corps
PASSENHEIM
ORTELSBURG
WILLENBERG
GERMANS
RUSSIANS
on 8/27/14
GERMANS
RUSSIANS
on 8/29/14
Evening

nnenberg.

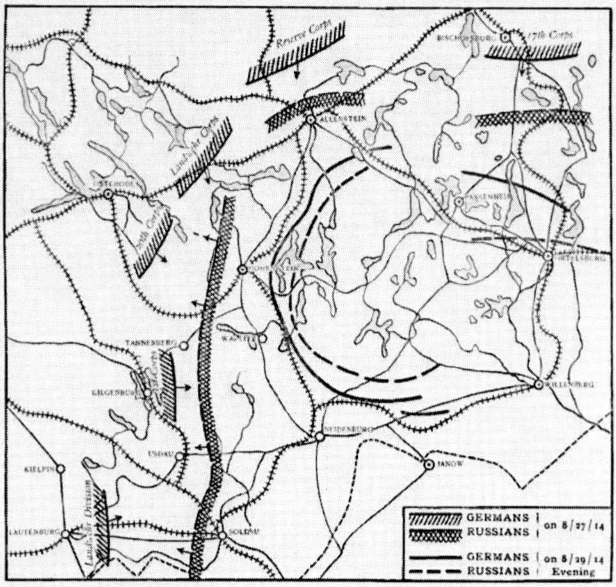

Map showing the disposition of the German and the Russian forces at the Battle of Tannenberg highway and filled the ditches along the sides, some of them disfigured beyond recognition or literally torn to fragments by the explosion of the larger shells. A slight layer of gray dust that settled like a pall over all the hideous wastage of war made the spectacle more revolting by its mockery of concealment.

In the meantime, on the 27th, von Hindenburg concentrated strong forces on his extreme right and pressed the Russians back from Soldau. This was part of the double flanking maneuver. But its immediate purpose was two-fold. It secured for the Germans possession of an important junction point, severing the only direct line of communication or retreat for the Russians in the direction of Poland, and it deceived the Russian commander as to the point from which von Hindenburg's heaviest blow was to be delivered. The Russians had committed the fatal error of strengthening their center at the expense of their wings. Samsonoff saw his mistake when it was too late. He tried to collect sufficient forces to drive the Germans out of Soldau, but the movements of his troops and artillery towards this quarter of the battlefield were impeded by the poor roads. Besides, the Germans had already intrenched themselves in a position covered by marshes. Far from recovering Soldau the Russians were driven back in the direction of Neidenburg.

As early as the 26th there had been fighting on the extreme German left, where the Seventeenth Army Corps repulsed a Russian army corps which had been advancing towards Lautern, driving it back in the direction of Ortelsburg and capturing many cannon.

The field intrenchments protecting Soldau made it possible to dispense with a considerable part of the German forces on that wing. These were rapidly transferred by railway to Allenstein, which the Russians evacuated, whence von Hindenburg was planning to deliver his culminating blow. The Germans advanced, driving back the Russians, as far as Passenheim on the 27th. Von Hindenburg now controlled the main railway line as far as Allenstein and the branch line over to Passenheim. His army was in position on three sides of the Russians, and on each side it had command of a good highway and all the motor-vehicles in the countryside had been requisitioned to supplement the railways in the conveyance of troops and supplies.

Von Hindenburg had only to draw the fatal noose closer and closer about the entangled Russians. Eastward the series of lakes offered an effective barrier to their escape. It was a comparatively easy matter for von Hindenburg, thoroughly acquainted as he was with the whole region, to occupy most of the solid intervals between these lakes, where small detachments operating in the necessarily restricted spaces could hold their own against vastly superior forces.

The uninterrupted pressure of the German wings on the 28th and 29th bent back the Russian front into a semicircular outline. In proportion as the perimeter of the Russian position was contracted in this way, the concentric fire of the German artillery became more effective. Distributing their heavy guns as they chose the Germans poured their shells into the congested masses of Russians floundering hopelessly in the swamps or staggering confusedly through the forests.

The Russian army was finally separated into two parts, one escaping eastward by the only available defile, along the railway in the direction of Johannisburg, leaving behind their wounded and most of their heavier equipment, the other surrounded and forced for the most part to surrender.

Whatever may be the degree of his responsibility for this appalling defeat of the Russians by reason of his carelessness, Samsonoff died a hero's death, struck down together with his chief of staff in a last vain effort to rally his men on the 31st.

The Kaiser conferred the rank of colonel-general and the Iron Cross of the first class upon von Hindenburg as a reward for this auspicious victory.

An official German communication on the 29th, with the customary terseness which these documents exhibit when they relate victories as well as when they condescend to report reverses, announced von Hindenburg's glorious exploit in the following terms:

"Our troops in Prussia under the command of Colonel-general von Hindenburg have defeated, after three days' fighting in the region of Gilgenburg and Ortelsburg, the Russian Narev Army consisting of five army corps and three cavalry divisions, and are now pursuing it across the frontier."

At first it was reported that the German army had taken 30,000 prisoners. But on August 31st Quartermaster-general von Stein made the following announcement:

"The victory of Colonel-general von Hindenburg reported in the East is of far greater importance than it was possible at first to recognize. Although fresh troops coming by way of Neidenburg made an attack, the defeat of the enemy was complete. Three army corps were annihilated, and 60,000 prisoners, including the commanding generals, and many cannon and colors fell into our hands. The Russian troops who are still in the northern part of East Prussia have commenced a retreat." On September 1st came the news that the number of prisoners was 70,000 including 300 officers; and finally, on the 3d, the following official communication was made public:

"The troops of Colonel-general von Hindenburg in the East are garnering further fruits of their victory. The number of prisoners is growing daily; it has already reached 90,000. It is impossible to determine how many cannon and other trophies are still concealed in the Prussian forests and swamps. Apparently not two, but three, Russian commanding generals have been captured. According to Russian reports, the Russian army commander (Schilinsky) fell." The reference to Schilinsky, who was Samsonoff's colleague, was simply due to a confusion of identity.

If the Germans took 30,000 wounded prisoners, as has subsequently been reported, in addition to 90,000 unwounded, the number of the Russians who were slain in battle or perished miserably in the lakes and swamps was probably about 30,000. The entire German loss in fighting strength was from 10,000 to 15,000 at most.

The Germans have chosen to call this victorious contest the Battle of Tannenberg, although no fighting of any importance occurred in the vicinity of the village of this name. But Tannenberg recalls a crushing defeat inflicted by the Poles upon the Teutonic Knights, the pioneers of German civilization in East Prussia, on July 15, 1410, a sort of Prussian Kossovo. Thus the ignominy of a great defeat inflicted by Slavs upon Germans was now after five centuries erased by a glorious victory won by Germans over Slavs.

It was commonly supposed that the prompt invasion of East Prussia by the Russians contributed materially to the failure of the Germans to consummate their design in the West by compelling them to detach large bodies of troops from their forces in France. This view has undoubtedly been exaggerated. Any great shifting of forces at a critical period in the operations in the West would assuredly have been made before the destruction of Samsonoff's army. The earliest notice of an eastward movement of German troops is the statement that on Friday night, August 28th, 160 German troop-trains passed through Belgium travelling from southwest to northeast, presumably withdrawing at least an army corps from the western front to reinforce the army in the East. But these troops, reserves or Landwehr of course, would undoubtedly have been entrained after the climax of danger in East Prussia had already passed. If the notice is authentic, the troops which were seen returning through Belgium may have been intended for service on the Galician frontier. It is not the least part of von Hindenburg's title to glory that he won the Battle of Tannenberg solely with the troops at hand in the provinces immediately threatened, most of whom had already suffered the moral depression of defeat by the Russians, without making any extra demand upon the resources of the Fatherland at a period of such extreme suspense.

As soon as von Hindenburg had disposed of the army under Samsonoff he struck out towards the northeast, scarcely allowing his soldiers any respite, for the purpose of dealing with the Vilna Army. Instead of continuing to push vigorously westward, or hastening to unite forces with Samsonoff, Rennenkampf, grown suddenly cautious, had begun to intrench himself on a line running from Lake Mauer to Tapiau, where his army faced in the direction from which von Hindenburg's attack was naturally to be expected.

The Vilna Army, drawn up in a generally concentric position covering the entire northeastern part of the province, presented far too extensive a front for the immediate application of the supremely effective double turning maneuver. But von Hindenburg hoped that the envelopment of the enemy's left wing by a turning movement through Lötzen, between Lakes Spirding and Mauer, in the direction of Goldap would lead Rennenkampf to shift the bulk of his forces southward, compressing them on the narrow front Gerdauen-Nordenburg-Angerburg, where a repetition of the tactics of Tannenberg would probably be successful.

In the interval following the Battle of Tannenberg the Eleventh Corps, a Reserve Guard division, and a Saxon cavalry division arrived as reinforcements for von Hindenburg. The Landwehr corps was left behind to deal with the scattered remnants of the Russian Narev Army. For the operations against Rennenkampf there were thus available four active corps, one and one-half corps of the Reserve, and the Saxon cavalry division, besides some minor auxiliary forces posted before Königsberg.

The First Corps with the Saxon cavalry division was ordered to turn the Russian left wing by marching northeastward in the direction of Goldap, while the other corps advanced against the Russian front, their attack converging towards Insterburg. General von Morgen with a Reserve division, ordered to parry any blow from the southeast, successively repulsed on September 7th, 8th, and 9th, with severe Russian losses, a Siberian army corps which was trying to attack the German right wing in the rear from the direction of Lyck and Marggrabova.

The Germans commenced their main attack on the 9th. The enveloping movement succeeded, and by the 10th the Russian left wing, its front contracted into a narrow arc, was retreating in the general direction of Insterburg and Gumbinnen. But instead of leading his remaining troops to the threatened wing and engaging his whole army in a restricted position facing southward, thus exposing them to a fatal snare, Rennenkampf abandoned the imperilled left wing to its fate and retreated eastward as rapidly as possible with the main part of his forces. Insterburg was abandoned on the 11th. In some parts the retreat became a rout, so that between thirty and forty thousand prisoners and 150 guns were left in the hands of the

Germans. In a very few days East Prussia was entirely cleared of the enemy and the Germans advanced into Russian territory, occupying Suwalki, where they set up a German administration.

The splendor of von Hindenburg's rapid achievements was somewhat clouded by the failure of his invasion of the Russian territory lying eastward of East Prussia undertaken in the latter part of September. There was continuous fighting in the Forest of Augustovo east of the boundary from September 25th to October 3d. The Germans attacked Osovietz, a fortress guarding the crossing of the Bobr. Situated in the midst of a marshy tract, the Germans could approach it only by a narrow defile. They reached a point where the fortress was within range of their heavy artillery. But the Russians made a sortie by night, attacking and outflanking the Germans, who were on difficult ground, and put them to flight after a conflict lasting thirty-six hours. After the failure of repeated attempts to cross the Niemen, von Hindenburg gave up the offensive and fell back into East Prussia.

To the professional enthusiasm and impetuous temperament of the Supreme Commander-in-chief of the Russian armies may be attributed in no small degree the initial rapidity and dash of the Russian offensive, but probably not the lamentable blunders which frustrated the imposing invasion of East Prussia. The Grand-duke Nicholas (Nikolai Nikolaievitch), second cousin of the present Tsar, was born in 1856, and began his active military career in the Russo-Turkish War of 1877-1878, where his father, a brother of Tsar Alexander II, was commander-in-chief of the Russian forces operating in Europe. Grand-duke Nicholas became lieutenant-general in 1893, and held the position of Inspector-general of Cavalry at the time of the Russo-Japanese War. He was relieved of the post of President of the Council of Defense, to which he was elevated in 1905, in consequence of opposition in the Duma. The command of the military district of St. Petersburg, conferred upon him in 1906, involved the assumption of responsibility for the Tsar's person at a time when the disturbed condition created by the revolutionary agitation was still serious.

The grand-duke's marriage in 1907 with Princess Anastasia, daughter of the King of Montenegro, the pledge of a fervent attachment, is one of the forces that made him a resolute leader of the Pan-Slavist movement. This circumstance and his thorough devotion to the military profession marked the Grand-duke Nicholas as the appropriate supreme commander in the national struggle.

At the outbreak of the Great War he was the only member of the imperial family who had dedicated his time and energy unreservedly to the serious cultivation of the art of war. Though a cavalryman by early training and inclination, his experience extended to the command of the other arms. Having a towering stature, graceful carriage, and a keen, penetrating expression, and accustomed to deal in a direct, open, compelling manner, his appearance and qualities are such as impress and inspire his associates and the vast armies ready to encounter any difficulties and hardships with patriotic enthusiasm. The operations inaugurated under his auspices were as signally successful in Galicia as they were disastrous in East Prussia.

To the Austro-Hungarian forces had been assigned in the general plan of the Teutonic powers the principal task of embarrassing and breaking up the concentration of the Russian forces, so as to prevent as long as possible the commencement of their expected invasion of Germany. This function would of course be performed by an aggressive action launched from Galicia against the strategic

General Russky, commander of the Russian army operating against Lemberg.

General Ivanoff, commander of the Russian army which repelled the first Austrian invasion of Poland.

General Rennenkampf, commander of the Russian Vilna Army.

positions in Poland. The forces assembled by the Dual Monarchy in Galicia were divided into two armies intended for the more active field operations and a third which was held chiefly in reserve. The first of these armies resting on Jaroslaw and Peremysl faced northward towards the heart of Poland. This army was to undertake the vigorous offensive movement in the direction of Lublin and Chelm. It was commanded by General Dankl and consisted of seven army corps with some minor reserve formations. It is said that in general the more reliable elements of the Austro-Hungarian military establishment were incorporated in this army.

The second army, with Lemberg as its base, faced to the northeast. Its task was to protect the flank and rear of the first army during the latter's advance into Poland. Its commander was General von Auffenberg.

The reserve army lay westward of the two others. Its active rôle in the plan for the earliest operations was limited to the invasion of the parts of Poland lying west of the Vistula, where no considerable masses of the enemy were to be expected. Each of the active armies numbered about 300,000, and the reserve army about 200,000 effectives.

The Austrians evidently believed that while their second army warded off invasion from the south of Russia, where the concentration of the Russian forces might presumably proceed more rapidly, the first army could strike a blow with paralyzing effect at a sensitive point in Poland, where the assembling of the hostile forces would still be in its early stages.

The invasion of Poland by the first army commenced as early as August 11th. General Dankl advanced rapidly, encountering very little opposition at first, with the intention of cutting the railway between Lublin and Chelm and threatening the communications to the rear of Warsaw. His army repulsed two Russian corps near Krasnik on August 23d. Then it moved forward to within eleven miles of Lublin, and with such comparative ease as to make it seem probable that the Russians were deliberately drawing General Dankl as far as possible from his bases and from contact with General von Auffenberg, and into a position from which he would find it difficult to extricate himself.

The advance of this northern army was suddenly checked by alarming news of a formidable Russian invasion of Galicia from the east and southeast. The Austrians had committed the fatal mistake of greatly overestimating the relative tardiness of Russian concentration. They paid a heavy penalty for their unfounded assurance. The Russians displayed, particularly throughout this campaign, a surprising facility of movement independently of railways. The broader gauge of the Russian railways, originally adopted no doubt from a motive of self-protection, should have embarrassed the Russians in an offensive movement against Austria or Germany in consequence of the impossibility of employing their own rolling stock beyond the border. But the invasion of Galicia was pushed forward with a remarkable degree of expedition in spite of this serious drawback.

General Russky, to whom was entrusted the Russian army operating directly against Lemberg, had been commander of the military district of Kieff, where he brought the organization to a high degree of efficiency. He was thoroughly familiar with all parts of the country where the early operations in Galicia were destined to take place. The startling success of the Russian offensive was due to General Russky's thorough administrative work in the district of Kieff no less than to his intelligence and skill in the field.

His invasion of Galicia from the northeast with eight army corps commenced in earnest on August 17th. Russky's army was directed against von Auffenberg's center and left wing. Very soon another Russian army, composed of five army corps and three divisions of cavalry, made its appearance in the southeast, advancing against von Auffenberg's right wing. This army was under command of General Brussiloff. To oppose these two Russian armies von Auffenberg seems to have had six army corps, the Third, Seventh, Eleventh, Twelfth, Thirteenth, and Fourteenth, together with five divisions of cavalry and possibly some minor reserve formations. Thus the Russians were at first vastly superior in numbers to the forces which confronted them. In this situation a part of the reserve army already mentioned, which was commanded by the Archduke Joseph Ferdinand, was hastily moved eastward to the support of von Auffenberg. In the ensuing tremendous conflicts in Galicia, conducted on a scale hitherto unparalleled in the annals of warfare, as many as 1,200,000 men were probably engaged altogether in the armies on both sides.

Russky occupied Brody on the 22d and on the same day Brussiloff crossed the Galician border at Voloczysk, the frontier station on the railway line from Odessa to Lemberg and Cracow. Russky advanced on a broad front. With his right he struck out straight westward to outflank the Austrian forces covering Lemberg and drive a wedge between Dankl and von Auffenberg. The center and left wing advancing directly against the army of von Auffenberg were soon engaged in desperate encounters, attacking the Austrians in front in reckless onslaughts. It was the task of Brussiloff to advance upon the right flank of the Austrian armies. Thus the Russians proposed to envelop the Austrians on both flanks and roll them together or else force them to seek safety by evacuating Lemberg.

After two days' delay before the Zlota Lipa, where he finally forced a passage, Brussiloff established contact with Russky's army. Von Auffenberg fell back to a position of great natural strength on a line more than seventy miles in length across the front of Lemberg, where trenches and barbed-wire entanglements had been carefully prepared in advance.

For days the furious attacks of the Russians,—their bayonet charges,—made no impression on the Austrian center. An official communication on September 1st declared that in a battle lasting a week the army of von Auffenberg had won a complete victory, capturing 160 guns, and that the Russians were retreating across the Bug. Whatever the extent of this rather shadowy victory may have been, it had no appreciable effect in impeding the progress of the Russian offensive, which was being most actively carried out on the flanks.

The left wing of Brussiloff's army particularly executed an extensive outflanking movement. Sweeping far to the south it attacked the Austrian forces defending Halicz, broke their line by the evening of the 31st, and crossed the Dniester on pontoon bridges after the Austrians had destroyed the permanent bridge to cover their retreat. Consequently, the Austrian retirement became a panic. In the meantime, the Russian wedge was successfully driven forward on the right as far as Tomasof, where the Austrians were also defeated.

By September 3d the wings of the Austrian army before Lemberg were pushed back like the extremities of a horseshoe, and the prompt abandonment of the city and systematic withdrawal from the neighboring region became imperative to avert a catastrophe.

Lemberg was evacuated on the same day and the Austrians were in full retreat by the 4th. The official organs

Lieutenant-general von Heeringen and staff, commander of the German Seventh Army.

German dead on the field of battle after a charge.

were at some pains to emphasize the point that Lemberg was not taken by force, but abandoned voluntarily by the Austrians themselves, since its defense was not part of their plan and would only have involved the city in useless destruction, and that their retreat was due solely to strategic considerations. The distinctions which the Austro-Hungarian authorities endeavored to establish in this way may have satisfied national pride, but they were without any practical significance. It is very doubtful whether the Russians experienced any great chagrin at being deprived of the supreme, but empty, glory of taking Lemberg by assault, or at accomplishing their aim by strategical, rather than tactical, combinations.

Von Auffenberg's army fell back by parallel routes in the direction of Peremysl, harassed on the march by the tireless Cossacks. Besides the gain in prestige, the acquisition of Lemberg was a very valuable advantage for the Russians in a material sense. It is the most important railway center in Galicia, where seven lines converge. The capture of thirty locomotives and many cars of the standard Austrian gauge was a matter of no inconsiderable importance at just that moment. Vast stores had been accumulated at Lemberg as a military base, and these fell into the hands of the invaders.

While these very significant events were occurring in Galicia, the Russians were collecting a large army on the line Lublin-Chelm under General Ivanoff. In consequence, a general battle developed with the army of General Dankl. The situation of this Austrian army very soon became precarious and the turning point in the first invasion of Poland was the failure to pierce the Russian line, in a final effort on September 2d, when the Tenth Austro-Hungarian Army Corps bore the brunt of the fighting.

The offensive in Poland collapsed and the initiative passed to the Russians under Ivanoff. As the Austrians fell back there was desperate fighting at Krasnik in which two German divisions were engaged. Only skilful management saved the Austrian army from an overwhelming calamity. It fell back with its left wing on the Vistula, gradually contracting the breadth of its front from about eighty to forty miles.

Besides the army of General Ivanoff closely pursuing the retreating Austrians, a Russian force was likewise advancing along the left bank of the Vistula, keeping pace with their march. Dankl arrived at the San on September 12th, hoping to be able to restore and reorganize his exhausted troops beyond this barrier. The passage of the river was effected only with the greatest difficulty and frightful losses. The Russians occupied some elevations commanding the bridges, shelled the Austrian forces while they were crossing, and are said to have taken no less than 30,000 prisoners. They captured the bridge at Krzeszof before the Austrians could destroy it, and were thus able to surmount the supposed obstacle in their pursuit without very much difficulty or delay.

Von Auffenberg resolved to make a determined stand with the forces retreating from Lemberg on a line passing through Grodek, in positions prepared in advance. Reinforcements had in the meantime been hurried forward to strengthen and extend the left flank of von Auffenberg's army, in an effort, probably, to fill the gap that separated his forces from the army of Dankl. Eventually an entire new Austrian army was formed from parts of the 3d army, two corps hastily transferred from the Serbian frontier, and some German contingents.

The new Austrian position was about sixty miles in length, but the critical points were Grodek and Rawaruska, on the right and left wings respectively. For eight days there was continuous, terrific fighting before Rawaruska, becoming more and more intense as the Austrians were gradually forced back from one trench to another and the crucial moment approached. The defense was carried on with the utmost valor, but the north wing of the Austrians was finally dislodged. The attack on Grodek began on the 6th, and the fighting continued for five days without interruption. During the greater part of this time the Austrians fought with the greatest stubbornness in trenches choked with the decomposing bodies of their fallen comrades, and without regular supplies of food. By the 13th, the official dispatches admitted that in consequence of the superior forces of the enemy and the danger that the left wing would be enveloped, the Austrians had been forced to retire from this position.

The Austrian armies, outnumbered and repeatedly worsted in battle, pursued and harried without mercy, threatened with complete disorganization, retreated westward seeking safety or a respite. The western part of Galicia offers superior opportunities for the defensive. The Carpathians and the Vistula, which furnish the requisite protection for the flanks of a hard-pressed army, converge towards the west, so that the front to be defended becomes continually shorter. This province is traversed by the San, Visloka, and Dunajec Rivers, which flow from the Carpathians to the Vistula, forming successive natural barriers. The Austrians attempted to rally at each of these rivers. The failure to hold out at Grodek and Rawaruska compelled them to retire behind the San, about seventy miles west of Lemberg. But the crossing of the San near its mouth by the Russians and the fall of Jaroslaw, an important fortress on the same river about twenty miles below Peremysl, commanding the railway to Cracow, on September 21st, destroyed their hope of defending this line.

Peremysl alone remained on the San, garrisoned by about 100,000 men under General Kusmanek, like another Maubeuge, an island in the midst of the spreading deluge of invasion.

A brief stand on the Visloka was followed by a retreat to the line of the Dunajec and Tarnow, eighty miles west of Jaroslaw. But even this position was threatened by the Russians who pressed relentlessly upon the heels of the retreating Austrians.

The heterogeneous character of their forces was probably a serious drawback to the Austro-Hungarian armies. The Slavic sympathy of an important part of the population doubtless contributed to the numerous desertions and the vast numbers who surrendered. Thus in the short period, September 11-14, alone, the Russians reported the capture of 83,531 prisoners. But in spite of this factor, and of the anticipated numerical advantage of the Russians, the Austrians had confidently expected, in view of the superior organization and quality of their troops, to carry on a successful campaign. Their conspicuous failure and the rapid progress of their disasters seemed to portend the approaching dissolution of the monarchy.

By the end of September the Austrian armies were apparently on the point of complete demoralization. They had seemingly contracted again their old-time "habit of defeat." The results were manifesting themselves vaguely but unmistakably in the attitude of neutral states, especially Italy and the Balkan Kingdoms.

CHAPTER VII

The Battle of the Marne
(*September 6-10, 1914*)

Germans in the neighborhood of Paris. The deviation in von Kluck's march and the reasons for it; the fundamental change in the German plan, the design of crushing the center of the Allies. Von Kluck's oblique movement and passage of the Marne. Joffre's tremendous responsibility and his plan of battle. Relative strength of the combatants and the positions of the forces. The assumption of the offensive by the Allies on the 6th. The concentric attack on von Kluck's army. The plight of the French Sixth Army on the 8th; the Paris "taxis" to the rescue. The climax on the 9th and the German retreat. Violent German attack on the Anglo-French center. General Foch's successful tactics. Discomfiture of the Prussian Guard. The Crown Prince's attack on Troyon. The retirement of the Duke of Württemberg and the Crown Prince. General reflections upon the character and consequences of the Battle of the Marne. The fall of Maubeuge.

It was commonly regarded as more important for the Germans than for their opponents to bring the contest as quickly as possible to a decisive issue. The Germans themselves appear to have acted upon this conviction in the most resolute manner, as we have already observed. They did not scruple to spend their men and ammunition lavishly whenever the saving in time justified the sacrifice. They did not hesitate to employ all their available reserves from the first so as to augment very greatly the striking force of their regular army corps. They inaugurated the great turning movement on a scale that far exceeded the expectation of their opponents and promised to confound all the latter's previsions and preparations. The flying wing of the German armies had thrown aside or driven before it all resistance in its path.

At the beginning of September von Kluck was advancing with giant strides straight towards Paris. The German advance-guards descended the Oise from Compiègne towards Creil. The British cavalry executed the brilliant rear-guard action on September 1st which has already been recorded, but in general they retired before their adversaries. The Germans applied the torch to Senlis, one of the most picturesque towns in France, situated only twenty miles from the outer defenses of Paris. They raided Chantilly, dispersing the army of English trainers, jockeys, and stable-boys at the famous racing center. German advance patrols penetrated as far as Pontoise; they were even encountered near the bank of the Seine. They arrived almost within gunshot of the outer forts of Paris. The world realized with a convulsive shudder that Paris might be foredoomed to fall again a prey to the invader.

On August 30th the new military governor of Paris issued a decree commanding all proprietors of premises in the regions of fire of the forts and defensive works of the capital to demolish and remove completely all buildings from such areas within four days, failing which the authorities themselves would carry out the measure. This was a stern reminder of the impending danger of a siege. The population of Paris, temporarily reduced by more than a third, awaited its fate with a stoical composure that belied the predictions of the enemy, though hourly expecting the commencement of a bombardment.

But at the very moment when Paris seemed in most imminent peril, the tide of danger had been deflected. The final stage in von Kluck's frenzied pursuit of the British had been only a screen to cover the alteration in the direction of the movement of his main forces. By September 4th British and French air-scouts reported that from the neighborhood of Compiègne the German army had begun to move in a southeastern direction instead of continuing southwest on the capital, and that von Kluck was marching towards Meaux and Coulommiers.

Two theories have been advanced in explanation of this sudden change. One view assumes that von Kluck was entirely misled in regard to the strength and position of his immediate opponents, believing that the British had been crushed and disorganized and that no French field forces worth considering remained beyond his grasp in the west. According to this theory, von Kluck was convinced that the culminating moment had arrived and he swung to the left in expectation of performing his indispensable function in the decisive climax by gathering the French armies into the fatal snare.

Another theory explains the alteration in von Kluck's line of march as the consequence of a deliberate change in the fundamental method of German strategy and tactics inaugurated with adequate knowledge of the actual circumstances.

The second view may be unreservedly adopted. True, von Kluck was probably ignorant of the extent and exact disposition of the forces on the extreme left of the Allies. If it is true that British and French aircraft had established a tactical superiority over that of the enemy, this would doubtless have limited von Kluck's opportunity for obtaining information. Furthermore, the buildings of Paris concealed to a considerable extent the passage and concentration of troops in the metropolis itself. Yet it is scarcely conceivable that von Kluck, in swerving to the left, was ignorant of the fact that considerable available forces of his adversaries would remain beyond the reach of his turning movement. The British Expeditionary Force had certainly exhibited a vigorous sign of life in the spirited action near Compiègne on September 1st when ten of the German guns were captured.

The Germans had hoped to repeat the exploit of Sedan on a mammoth scale by encircling the Anglo-French armies in an open region with no insurmountable obstacles to impede the progress of their evolutions. On August 29th the advancing German armies extending across northern France had presented a somewhat concave front, the right flank reaching forward as if to clutch the prey. But the Allies still eluded the grasp of their pursuers. The German front continued to move forward until its right wing was about to encounter the defenses of Paris and the left wing was already in contact with those of Verdun. The Allies had now retreated to a position where both their flanks were covered, the left by the intrenched camp of Paris and the Seine, the right by the eastern barrier fortresses. They had thwarted the German hope of enveloping them where their flanks were exposed. In the circumstances the Germans could not think of investing Paris. Even if they could have silenced one or two of the forts by the concentrated fire of their heavy artillery and rushed through the breach thus made, the achievement would have been a sterile, or even a dangerous victory, as long as the Allies had large, unconquered armies in the field. The Allied lines were now inseparably bound to powerful fortresses which it was inexpedient, if not absolutely impossible, to beset. The Germans discovered that the morsel which they had coveted involved a larger mouthful than they were able to swallow at this time. It would have been perilous to strain the front any further in an effort to envelop Paris.

But the German armies could not safely remain stationary in the existing situation. It was necessary for them either to withdraw very quickly or to strike southward with all their available energy so as to break through the enemy's front and drive a wedge between the French

Paris motor-buses transformed into army transports.

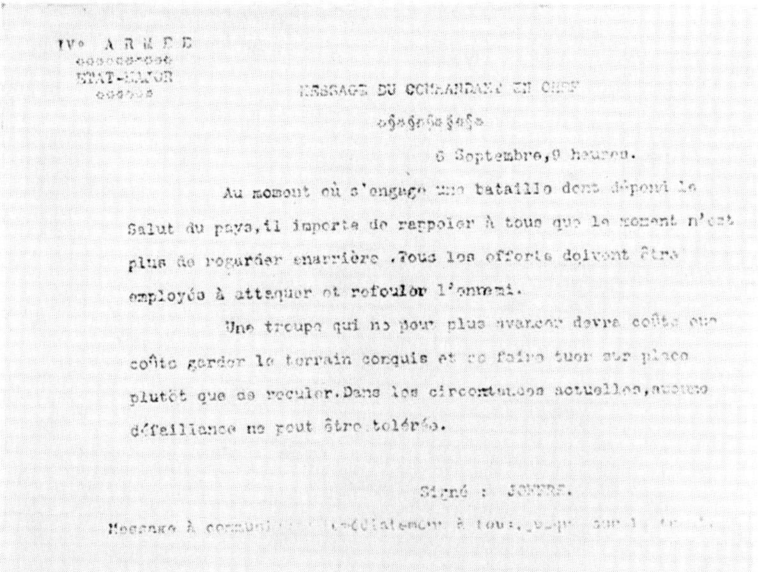

IVe ARMÉE
ÉTAT-MAJOR

MESSAGE DU COMMANDANT EN CHEF

6 Septembre, 9 heures.

Au moment où s'engage une bataille dont dépend le Salut du pays, il importe de rappeler à tous que le moment n'est plus de regarder en arrière. Tous les efforts doivent être employés à attaquer et refouler l'ennemi.

Une troupe qui ne peut plus avancer devra coûte que coûte garder le terrain conquis et se faire tuer sur place plutôt que de reculer. Dans les circonstances actuelles, aucune défaillance ne peut être tolérée.

Signé : JOFFRE.

Message à communiquer immédiatement à tous [illegible]

Facsimile showing in what form General Joffre's order of the day, issued on the morning of the first day of the Battle of the Marne, was communicated to the troops. *(For translation see page 170.)*

masses east and west. The first course would have been a palpable admission of failure; the second, which was actually adopted, offered the possibility of winning an important, perhaps a decisive, victory. The general situation impelled the Germans to strike at once and to strike hard so as to escape the fatal contingency pressing so closely upon them, which threatened to arrest abruptly their great effort in the West. The Austrians had evacuated Lemberg on September 3d and were staggering back under the Tsar's tremendous blows. Austria-Hungary's military power was crumbling and so the days of the German offensive in the West were numbered.

Other considerations contributed also to determine von Kluck's change of front. We are apt to think of the German invasion as a broad curtain drawn down across the north of France. But in reality this figure, though convenient, is not strictly pertinent, for the invading armies advanced in separate columns and not with an uninterrupted front. The columns of a single army were in intimate communication; but the intervals separating adjacent armies might naturally be greater, and the headlong rush of the western armies in pursuit of their opponents expanded the field of invasion to such a degree that the communication between the different armies was undoubtedly threatened and the front as a whole was exposed to the danger of penetration by a counter-attack of the enemy. The situation required that the dangerous intervals should be closed by the deflection of the western armies towards the southeast.

The opinion has been expressed that the awkward position of the Germans here described, and in a measure the miscarriage of their whole plan for the offensive, was due to the tardiness of the German Crown Prince, his failure to keep up with the progress of von Kluck. The latter masked the fortress of Maubeuge and hurried on, but the Crown Prince allowed himself to be delayed by the resistance of Longwy. Later, according to this view, the Crown Prince was drawn aside and hopelessly entangled in contact with the French barrier fortresses. The proverbially impetuous Crown Prince erred, in other words, by reason of his excessive caution! By September 3d, however, we may assume that the Crown Prince at one extremity of the German front was on approximately the same line east and west as von Kluck at the other. The solution of the question of possible responsibility for the German failure rests upon the exact determination of the German objective. If the German plan required the convergence of all the armies on Paris, the Crown Prince, who was still 160 miles from this goal when von Kluck was only thirty, was hopelessly outdistanced. But so were all the other German commanders, though in lesser and varying degrees.

Such an assumption, however, is opposed to the obvious facts. The lines of march of the different German armies had at no time converged on Paris. In general, their line of advance bore southward, and in relation to this direction the Crown Prince seems to have progressed as far as von Kluck, in spite of the latter's tremendous rapidity of movement. The Crown Prince's movement, since he was near the pivot, was necessarily much slower. Moreover, it would probably have been perilous for the Crown Prince to disregard Verdun. The position of the Germans with respect to the forwarding of supplies was still somewhat precarious, since Antwerp, Maubeuge, and Verdun threatened the most serviceable lines of communication running into northern France.

In defiance of the most elementary laws of prudence, von Kluck marched diagonally across the front of his adversaries, exposing his right flank to attack by the Anglo-French forces concentrated near Paris, whose strength he probably underestimated. The British army which had been falling back before von Kluck reached a position south of the Marne between Lagny and Signy-Signets by September 3d. In accordance with instructions by General Joffre, the British destroyed the bridges crossing the Marne on their front and then fell back about twelve miles further south.

Von Kluck was still moving with the frenzied speed of his dash towards Paris. His main forces were probably in the vicinity of Compiègne on September 3d. Moving southeastward he passed the Ourcq and the Marne. His forces and those of von Bülow were seen crossing the Marne at Changis, La Ferté-sous-Jouarre, Nogent, Château Thierry, and Mézy on the 5th. By the evening of the same day the head of von Kluck's column had reached the neighborhood of Coulommiers.

No greater responsibility had ever rested on any individual than that borne by General Joffre in the dramatic situation now before us. The future condition and fortunes,—perhaps even the existence,—of France were at stake. The awful significance of the moment bewilders the imagination. Through twenty centuries swept the imposing spectacle of French history, teeming with great examples of virtue and baseness, lofty aspirations and dark passions, a thrilling, sensational epitome of the life of all humanity. France had been the leader in revolution, the pioneer in political innovation, the guide and standard in art and literature, the model in elegance and taste. The experience of the French people had become an organic and necessary part of the cultural equipment of all civilized nations. But suddenly this brilliant course of development was threatened with an immediate and sudden eclipse. A nation of tougher fiber had arisen, a people who knew how to curb and utilize its individual energies for collective effort, whose uneasy ambition sought joy in the consciousness of difficulties surmounted through toil. With unbounded assurance in their well-disciplined vigor, the Germans challenged the traditional ascendancy of the French, indifferent to the havoc involved in the world of spiritual relationships. All the resolution, endurance, and energy which the French nation could muster would hardly avail to sustain it in this supreme and final trial.

But General Joffre faced the crisis with coolness and determination. He had probably discerned the nature of the coming great encounter as soon as the Allies were forced to relinquish the line of the Sambre. He had probably forecast the Battle of the Marne and planned its chief details as early as his interview with Field-marshal French on August 29th. The German armies were now gliding into the positions which General Joffre had deliberately selected, as molten metal flows into the prepared molds. By the 5th he perceived that the strategical situation for which he had been waiting had arrived. The time had come for turning to the offensive. He visited Field-marshal French and explained his plans to him. He issued his orders in accordance with the dispositions for a general assumption of the offensive on the morning of the 6th. The Sixth Army on the extreme left under General Maunoury, which had been drawn up like a screen protecting Paris from the northeast, was directed to advance eastward in the direction of Château Thierry, so as to cross the Ourcq between May-en-Multien and Lizy-sur-Ourcq northeast of Meaux, with its front extending at an angle of 135% with the general line of the Anglo-French forces. The purpose was to strike at the flank and rear of von Kluck's army and threaten his communications. The British were instructed to execute a change of front by

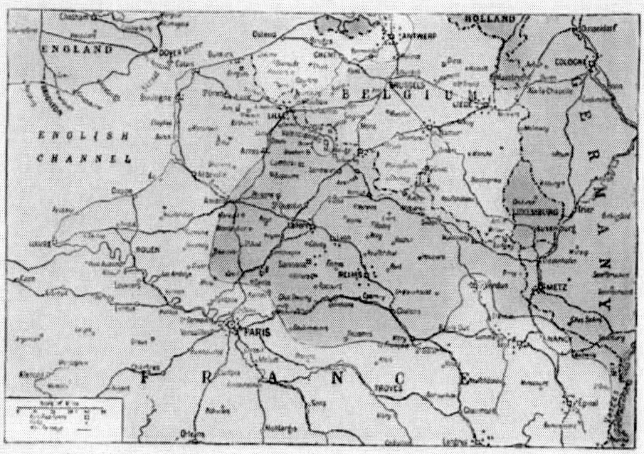

Map showing by the shaded portion the extent of the German advance into Belgium and France just prior to the Battle of the Marne, September 5, 1914.

wheeling to the right so as to form a sector of the battle-line between the Sixth and Fifth French Armies. The Fifth Army on the line Courtaçon-Esternay was ordered to attack towards the north, and the Ninth Army to hold the line south of the marshes of St. Gond. The Fourth and Third Armies also received directions in conformity with the general plan.

In making the turning maneuver of his left wing the significant feature of his offensive, General Joffre borrowed the essential principle of his adversaries' tactics, just as the Germans, in delivering a terrific frontal attack for the purpose of penetrating the hostile battle-line, had adopted precisely the characteristic tactical method of the French. The main part of the Anglo-French line was expected to hold firm against the anticipated violent onslaught of the enemy, while the left wing turned the German flank, rolled back the hostile line, and destroyed the enemy's communications.

The strength of the position of the Allies depended upon the security of their flanks. In consequence of the deflection in von Kluck's line of march the Allies were relieved from apprehension regarding the safety of their left wing. On the right the barrier fortresses shielded the French from the tide that might have swept down upon them from the east. The tenacity of this defensive line was an essential condition of success in the Battle of the Marne.

An element of German superiority throughout all the early part of the war,—an advantage more striking even than their superiority in numbers,—was predominance in artillery. We have already considered the sensational effect of the German siege-artillery, which had been developed to such a size and degree of adaptability that it overcame the resistance of the strongest fortifications with startling rapidity. The existence of the powerful barrier fortresses on the French border had doubtless stimulated progress in the heaviest types of mortars and howitzers. But the ascendancy of the Germans in the various forms of artillery which accompanied their field-armies was no less significant. With 6.4 guns for every 1,000 rifles the Germans surpassed all nations in the relative strength of their artillery, France followed with about 5.5, while the United States, by way of comparison, with 3.1 guns per 1,000 rifles was the weakest nation in this respect.

The French in their standard 7.5-centimeter field-piece, the famous "seventy-five" (that is, 75 millimeters), had developed an instrument of unequalled precision and efficiency. But with too implicit confidence in the advantages afforded by the superior merit of this single piece, the French had allowed the Germans to surpass them very greatly in all the other forms of mobile artillery.

The Germans had apparently foreseen that in consequence of the more stationary character of battles and the employment of motor-transport there would be a far greater opportunity of using heavier artillery on the battlefield. They had been engaged very diligently for four years in supplying their army with heavy field-pieces, so that about one-fourth of their field-artillery consisted of 10.5-centimeter field-howitzers, which deliver explosive shells capable of destroying the smaller field-batteries of the enemy. Moreover, a section of four batteries, each of four pieces, of 15-centimeter howitzers was allotted to each German army corps, artillery matching in caliber the hitherto standard large fortress and siege guns.

The "seventy-fives" with their average effective range of about 5,000 yards were helpless when exposed to the fire of the heavier German field-artillery from points beyond their reach. The range of even the 10.5-centimeter howitzer was more than twice as great as that of the "seventy-five." But the French had been loath to introduce heavy artillery into their field formations. They were almost without field-howitzers at the beginning of the Great War.

In general, the practice of estimating the relative numerical strength of the belligerents on the sole basis of their permanent army corps is apt to lead to erroneous conclusions. For this method either ignores entirely the reserve contingents or assumes that their strength is uniformly proportionate to the number of the army corps. But a comparison of the conditions in the German and French military establishments will illustrate the fallacy of such an assumption. At the beginning of the Great War Germany had twenty-five army corps and France twenty-one. But Germany's superiority in available trained men was very much greater than this relationship would indicate. The peace establishment, upon which the division into army corps is based, did not stand in the same ratio to the available war-strength in the two countries. For France kept a much larger proportion of her trained men in the active army in time of peace. The standing army of France numbered 742,000 men, that of Germany, 870,000, according to the modifications introduced in the two countries in 1913. But the effective numerical strength in time of war is based upon the size of the year-classes which are available for mobilization, and it required three year-classes to provide the 742,000 men of the French army, and only two year-classes (except in the cavalry and horse-artillery) to make up the 870,000 men of the German army.

In the absence of precise information regarding the number of reserves, including troops of the second or third line, actually called to active service at the beginning of hostilities, the proper basis for comparison of the available forces of the belligerent nations in which compulsory service prevails is obviously the size of the annual classes of recruits as incorporated in the standing armies of the respective powers and therefore provided with the necessary training at the time of mobilization for war. For a number of years preceding the enlargement of the German levy in 1913, the annual class of recruits accepted for active service in France had been roughly equivalent to two-thirds of the corresponding class in Germany. We shall not be very far wrong, therefore, in adopting the ratio 8:12 to represent the numerical relationship of the forces of France and Germany respectively as available for active service in the early stages of the war, the aggregate strength, in other words, of the corresponding serviceable year-classes in each country. Assuming that one-sixth of the German forces were concentrated on the eastern frontier, we still have a ratio of 8:10 to represent the relationship in numerical strength of the French and German forces in the western theater. Even the addition of the Belgian and British armies would scarcely suffice to bring the aggregate strength of the forces of the western Allies as compared with that of the Germans confronting them to the ratio 9:10. In other words, at the beginning of the contest the German forces in the West had a superiority of 10-15% over the combined strength of all their adversaries. While at first the Germans do not seem to have called out as many year-classes as the French, their more rapid mobilization and concentration probably compensated for this difference in producing the approximate initial superiority in numbers here indicated.

The whole course of events in the early part of the campaign in the West, especially the ability of the Germans to drive back a formidable French invasion of Lorraine at the same time as they were moving overwhelming forces through Belgium, is incomprehensible on any other

Scenes of destruction in and around Senlis, which was set on fire by the Germans on their retreat. The views show a totally destroyed dwelling, a piano looted by the Germans and left in a field, French Zouaves and German dead, and a destroyed bridge.

assumption than that the Germans were considerably superior in number to all their opponents in the western theater.

But the relative strength of the German forces actually available for fighting in the West must have diminished continually in consequence of the detachment of second-line troops for guarding the lines of communication, the transference of reserve formations and possibly troops of the first line to reinforce the armies on the eastern frontier, and the progress of the mobilization of the older year-classes in France, until finally in the Battle of the Marne the Germans were probably outnumbered by their opponents.

The consideration of the numbers engaged must not be allowed to obscure our impression of the formidable advantage possessed by the Germans in the bewildering rapidity and impetuousness of their offensive, their unanimity of will and purpose, the numerical superiority and more varied character of their artillery, and the phenomenal thoroughness of their preparation in every detail.

The general region of the great encounter was the basin of the Marne and of its tributaries, which is encircled by somewhat elevated plateaus. The positions chosen by General Joffre, generally across the southern part of this depression or along the low ridges bounding it on the south, offered several distinct advantages for the Allies. The Marne receives the Ourcq from the north and the Petit Morin and the Grand Morin from the southeast in the neighborhood of Meaux. All these streams flow in a westerly direction throughout the greater part of their course. The Germans had to fight, therefore, with several parallel water-courses in their rear hindering the transportation of supplies, ammunition, and the heavy artillery. The Germans were separated from their most convenient lines of communication by rail. The railways with which they were in immediate contact south of the Marne were comparatively few and of secondary importance. The Allies, on the other hand, enjoyed excellent communications by a network of railways, including several trunk lines in the rear of their positions. Before the close of the battle the Germans suffered severely in some parts of their line from a shortage of food and ammunition.

When, shortly before the Battle of the Marne, the German wings encountered the obstacles which arrested or deflected their movement, the intervening masses were still borne forward by their own momentum and by the impulse of the new design of bursting the enemy's line, and engaged themselves in the hollow produced by the recession of the French center, giving the German front a bulging, bow-shaped outline. Thus, in a general way, on the eve of the battle, the contending forces occupied concentric, semi-elliptical positions nearly 150 miles in length, with the extremities resting on a line from Paris to Verdun as the common base.

It is very important to obtain as clear a notion as the available evidence will permit of the composition and distribution of the opposing armies on the morning of the memorable Sunday, September 6, 1914, when the greatest battle of all history commenced.

The French Sixth Army, consisting of the Seventh Army Corps, a reserve corps, three divisions of Territorials, and Sordêt's cavalry corps, under General Maunoury, was advancing eastward on a front extending from near Betz to the vicinity of Meaux. The five divisions of the British Expeditionary Force faced the northeast with a front stretching from the vicinity of the Marne to the neighborhood of Vaudoy, the center being near Mauperthuis. Conneau's cavalry corps occupied the interval between the British and the French Fifth Army. The latter, composed of four regular army corps under General Franchet d'Espérey, held a section of the front about twenty miles in length from Courtaçon to Sézanne, and faced northward. The Ninth Army, three regular corps and two reserve divisions under General Foch, continued the French front for a distance of twenty miles to Sommesous. The Fourth Army, four regular army corps under General de Langle de Cary, occupied an extensive sector of the French front from Sompuis to Sermaize, twenty-five miles, its center opposite Vitry-le-François. General Sarrail's Third Army, probably three army corps, held the line from Revigny to Souilly, facing nearly westward.

The interval of ten miles between Foch's right and de Langle de Cary's left and probably the gap between the latter and Sarrail were filled provisionally as well as possible with artillery and cavalry, while a division from Verdun was stretched across the opening between Sarrail's extreme right and this great barrier fortress.

General von Kluck's First German Army, consisting of the Second, Third, Fourth, and Seventh Corps, and the Fourth (and possibly the Seventh) Reserve Corps and two cavalry divisions was thus confronted by the Sixth French Army, the British Expeditionary Force, and the left wing of the Fifth French Army. The Second Army, the next eastward in the German line, composed of the Ninth and Tenth Army Corps, the Guard, and the Tenth Reserve Army Corps under General von Bülow, was opposed by the right wing of the Fifth French Army and by the left of the Ninth. The Third German Army, made up of the Twelfth and the Nineteenth Army Corps and the Twelfth Reserve Corps, led by General von Hausen, was faced by the right wing of the Ninth French Army and by the left wing of the Fourth. The Fourth German Army, probably the Eighth and Thirteenth Army Corps and reserve corps, was directed against the Fourth French Army. Finally, the Fifth German Army, probably consisting of the Sixteenth and Eighteenth Army Corps, with probably one of the Bavarian Corps and reserve formations, was confronted by the Third French Army. We may safely assume that reserve formations of considerable strength were included in all the armies, as is indicated in the case of the First, Second, and Third German, and Sixth and Ninth French Armies.

In consequence of the extreme reticence of the official communications it is impossible in most instances to determine the apportionment of the individual French army corps to the different armies at this time.

In the early days of the campaign the First, Second, Third, Sixth, and Tenth French Army Corps were probably concentrated on the Meuse below Verdun, while all the others had either actually taken positions along the border to the southeast of Verdun or were being transported in that direction. The Eighth, Twelfth, Fourteenth, Fifteenth, Twentieth, and Twenty-first seem to have remained on the eastern frontier after the redistribution of the French forces in the latter part of August, while the Fourth, Fifth, Seventh, Ninth, Eleventh, Thirteenth, Sixteenth, Seventeenth, and Eighteenth were transferred to positions westward of Verdun and the Nineteenth was brought from Algeria to this portion of the front. The Fifteenth and Twenty-first were transferred to the French line south of the Marne during the course of the battle.

Accordingly, in the Battle of the Marne, conceived in the broadest sense, including all the field operations west of Verdun from the sixth to the tenth of September, the Anglo-French armies contained at most nineteen and a half regular army corps, the British being equivalent to two and a half army corps, while the Germans numbered fourteen and a half army corps, the three divisions of the Guard being reckoned as an army corps and a half. Assuming, as seems not unlikely, that the percentage of second-line troops or reserve formations at the front was at this time about the same on both sides, and that they were about half as numerous as the troops of the first line represented by the regular army corps, we may estimate the aggregate forces of all classes as somewhat more than 1,100,000 for the Allies and rather less than 900,000 for the Germans. Thus the Allies had now for the first time, as it would seem, a distinct numerical advantage.

But of greater actual significance was the relative numerical strength at the critical position, and in this respect the situation had been completely changed since the Germans shattered the resistance on the Sambre and the Meuse. The Germans had surprised and disconcerted the French by concentrating overwhelming masses of troops on the right wing in Belgium so as to sweep around and destroy their adversaries' left. This preponderance continued as the tide of battle rolled across northern France. But all the time the new concentration of French troops was proceeding in the rear. With unfailing discernment General Joffre distributed the reinforcements in such a way that they would offset the superiority of the enemy's right flank. The formation of the Sixth Army reduced the disparity. The introduction of the Ninth Army under General Foch between the Fifth and Fourth Armies was like inserting the keystone in the arch. Finally, the movements just preceding the Battle of the Marne completed the process whereby the tables were turned. All the way to Compiègne the Germans had been following the main lines of communication leading to Paris, which afforded convenient means for the conveyance of reinforcements, munitions, and supplies. But by turning to the left the

Germans forsook these lines, marched obliquely across the routes to Paris, and shifted the relative position of their right wing in such a way as to expose it to a converging attack, so that the number of units which could profitably assail it was very greatly increased. At the beginning of the Battle of the Marne, instead of outnumbering its opponents nearly two to one, the German right wing found itself in its turn outnumbered in about the same proportion.

General Joffre made his headquarters with General Franchet d'Espérey and the Fifth Army, practically at the center of the battle-front. On the 6th he issued the following memorable order of the day to the soldiers:

"At the moment of engaging a battle on which the fate of the country depends, it is my duty to remind every one that the time has passed for looking backward. Every effort must be made to attack and to drive back the enemy. Troops that can no longer advance must at all costs stand their ground and let themselves be killed on the spot rather than retreat. In the present circumstances no hesitation can be tolerated."

The westernmost armies assumed the offensive on the morning of the 6th. The Sixth French Army under General Maunoury advanced towards the Ourcq in the direction already indicated. The Fourth Reserve Army Corps forming von Kluck's rear-guard which was descending the valley of the Ourcq east of the river faced about to meet this attack. The Second German Army Corps was either in the same general position at the time or was immediately sent back to support the Fourth Reserve. Upon learning that the latter was hard pressed, von Kluck ordered a Landwehr corps in Compiègne to advance to its assistance. Compiègne was nearly two days' march from the scene of battle. But the measures immediately put into execution by the Germans seem to have been sufficient to maintain an equilibrium in the fighting in the valley of the Ourcq.

By advancing through Coulommiers to attack the French Fifth Army, von Kluck very rashly exposed his right flank to the British. This was probably due not so much to a misconception arising in consequence of the gap between the Fifth and Sixth French Armies, which existed as late as September 5th, and was later filled by the British, as to another, a more general cause.

Stupidly inflexible national prejudices or preconceptions are often a most formidable obstacle to successful generalship. In spite of their intellectual keenness and scientific detachment, the Germans are peculiarly subject to certain prepossessions which, for the most part, have been either designedly propagated for political reasons or produced by the abstract process of academic speculation.

Misleading impressions of this kind had acquired enormous influence by the force of their insinuating appeal to national antipathy or conceit. Such delusive impressions might expose an army to the gravest danger from the subtle strategems of a wily antagonist. Prominent among them was the conviction that the British forces were defective in military qualities. It would be too much to affirm that General Joffre had deliberately planned to exploit this particular vagary of the German imagination. However, von Kluck exposed himself to the great peril of a British flank attack in a manner most satisfactory to his opponents, and he probably did so in contemptuous disregard of any loss that the Expeditionary Force might inflict upon him.

The British were now organized as three army corps, Lieutenant-general W. P. Pulteney having taken command of the recently formed Third. In conformity with the instructions of General Joffre, the British, who had retired to a line running east and west about twelve miles south of the Marne, wheeled to the right, at the same time advancing somewhat to the northeast, so as to swing into place between the Sixth and Fifth French Armies. In this position the British attacked the army of von Kluck on its exposed right flank while the Fifth French Army assailed it in front. The Germans were forced back, evacuating Coulommiers, and by the evening of the 6th the British occupied the left bank of the Grand Morin, while the Fifth Army continued the line through Esternay.

Resuming the battle on the 7th the British continued to advance while the Fifth French Army forced the Germans back in the direction of the Petit Morin. Again on the 8th, the British, coöperating with the French, assailed the Germans with the same success on the Petit Morin. The French occupied Montmirail after a fierce attack.

In the meantime, Maunoury was not obtaining the measure of success upon which the hope of a decisive victory depended. Attacked with much violence on the 7th, his troops were forced to yield ground. The difficulties which had probably impeded the conveyance of the heavier German field-pieces to the principal battle-front south of the Marne favored the overwhelming concentration of artillery against the French Sixth Army north of the Marne. From the hills overlooking the Ourcq the long-range guns and howitzers swept the French position with galling effect.

Maunoury's situation was very difficult on the 8th, when the Germans launched a turning movement against his left flank and succeeded by attacks of great violence in occupying Betz and Nanteuil-le-Haudouin. The Germans in the valley of the Ourcq were fighting with their faces set towards Paris, with the possibility before them of a speedy

A corner of the garden of the Château de Mondemont, which changed hands four times in a few hours during the Battle of the Marne.

Bridge at Lagny destroyed by the British forces in their retreat before the Germans and pontoon bridge built to pursue the latter on their retreat during the Battle of the Marne.

and glorious termination of the campaign to stimulate their exertions, and the thought of the long weary routes behind them, the peril involved in defeat far from their bases, and the ignominy of failure to strengthen their resolution. The French contested every foot with stubborn determination. The fate of Paris, and perhaps of the cause of the Allies, depended upon the constancy with which they sustained the onslaught of their opponents. To Maunoury's appeals for assistance, General Joffre could only reply urging the Sixth Army to hold out a few hours longer. By the evening of the 8th it was apparent that the following day must decide the issue of the entire battle.

The situation of General Maunoury's army had become alarming by the morning of the 9th. The Germans, supported by their powerful artillery, were hammering it unmercifully in front and were driving home the turning movement to envelop its left or northern wing, where the arrival of the Landwehr troops from Compiègne would soon increase still further the odds against the French. The fighting had attained a terrible degree of ferocity. That the Germans would win on the Ourcq by the very tactics Joffre had chosen, that they would turn the tables by outflanking the French flanking movement, seemed to be almost inevitable. But Paris was free from the danger of attack, for the time, at least, and a portion of the garrison could safely be spared to reinforce the field army engaged in a life and death struggle so near. General Galliéni had decided to play the final card. On the evening of the 8th the notice was quietly distributed to the drivers of about 5,000 taxi-cabs and motor-vans in Paris summoning them to assemble early the next morning at certain indicated places to transport the garrison troops to the support of the threatened left wing of the Sixth Army. That soldiers should be conveyed to battle in taxi-cabs is only one of the many seemingly paradoxical innovations of the Great War. In the early hours of the 9th interminable lines of motor-cars roused the sleeping towns of the *banlieue*, and stretched out like continuous, flexible bands along the highways. The reverberation of distant artillery gained volume and distinctness as they advanced. It was another "Sheridan's Ride" on a mammoth scale, and in a modern setting. The hour of destiny had sounded; the fate of France was swaying in the balance. Every available element was being thrown upon the scales on this supremely eventful 9th of September. It might have seemed that the future of democracy, with its inherent disorder, but with all its sublime possibilities, depended upon the speed of this amazing procession that rolled across the *Ile de France*.

General Galliéni succeeded in his brilliant, decisive stroke of transferring possibly 50,000 men a distance of forty miles in six hours and delivering a decisive blow at the enemy's flanking movement. The danger of the complete failure of the French maneuvers at the critical point was averted.

On the same day the British reached the Marne. The First and Second Corps forced a passage after violent hand to hand encounters in which they drove the Germans in some places into the river. The Third encountered the desperate resistance of the part of the German forces which had retired into La Ferté-sous-Jouarre on the north bank of the Marne. The bridges had been destroyed, but after several ineffectual attempts to throw pontoon bridges across during the day, the British finally forced a passage at this point also after nightfall. On the same day the French Fifth Army made its way to Château Thierry, driving the Germans back across the Marne with heavy losses.

Von Kluck's situation rapidly became serious with the changes in the situation produced on the 9th. He suffered prodigious losses in vain attempts to repel the renewed offensive in the valley of the Ourcq. The climax came in the evening. Realizing that further hesitation might involve the destruction of his whole army, he gave the order for the general retreat. The decision had a solemn significance; it was a palpable indication that the tide had turned. For the first time since Jena, since the establishment of the principle of universal, compulsory service, a Prussian army had been defeated in a pitched battle. The splendid prestige, the tradition of invincibility which had flourished for more than a century, suffered a first and ominous disfigurement.

The Germans had felt absolute confidence in the supposed infallibility of their General Staff. In all their sacrifices for military supremacy the resolution of the German people had been sustained by the proud conviction that their armies could at any time march to Paris. This belief was generally held, however much individual opinions might differ in other respects about the prospects of a general conflict. To doubt it was regarded as heresy. Even the most unwarlike persons were cheered by the comforting reflection that Germany could at any time take Paris, if she wished, a source of satisfaction recalling a human trait observed by Juvenal; "even those who want the will, pant for the dreadful privilege to kill."

But now the first cloud of perplexity appeared on the superb horizon of national assurance. The German armies had strained every faculty to accomplish a definite purpose and had failed, and in spite of the efforts made to palliate or disguise this failure, the fact is evident that although they had advanced almost to the outer defenses of Paris, they had been obliged to retreat to a considerable distance and had accomplished nothing.

The thrilling incidents and impressive circumstances of the contest in the western region, such as the imminent peril to the capital of France so barely averted, the most striking features of General Joffre's inimitable tactics, the palpitating suspense which attended the critical moment, and von Kluck's precipitate retreat dramatically contrasted with his sensational advance,—significant as all these are,—must not monopolize the attention and interest at the expense of the operations at the center, where the contest raged with even greater fury and the immediate danger to France was more vital. The Germans hurled themselves against the French center with all the force at their command. The Ninth Army commanded by General Foch bore the brunt of this formidable onslaught. The most violent attacks were delivered in the neighborhood of Sézanne and especially of Fère Champenoise. The pressure became severe on the 7th. The Germans were impelled by a frantic determination to achieve their purpose of breaking through the French lines, which quivered beneath the terrific impact. The right of the Ninth Army was pushed back as far as Gourgançon on the 8th and the left wing, in spite of its obstinate resistance, was forced to yield ground.

The 9th was the critical day in this section also. The Germans were pushing forward in eager expectation of winning a decisive victory. With the wings of both the Ninth and Fourth French Armies bent back, the Germans were pressing through the gaps at the extremities as a flood pours through crevices in a retaining dam. The Twelfth (Saxon) Reserve Army Corps after penetrating the cavalry screen that covered the opening on Foch's right flank was entering Mailly. The Twelfth Army Corps had driven the French from Sommesous, and further west the Guard was forcing Foch's right center to retire before it south of Fère Champenoise. The whole Allied center was tottering. The impetuous invaders seemed about to cleave the barrier and break France into two parts. The situation was apparently almost hopeless when Foch sent his heroic reply to the commander-in-chief: "My left has been forced back, my right is routed; I shall attack with the center."

But factors had been silently preparing which were soon to be instrumental in reversing the tide of battle in this very vital section. Foch is said to have discovered through his air-scouts on the evening of the 8th that the Saxon troops in their advance against his own right wing had impaired their contact with von Bülow's army towards the west. Furthermore, the Fifteenth French Army Corps, hastily transferred from the front in Lorraine to reinforce the right wing of the Fourth Army, had arrived in Bar-le-Duc and repulsed the German Eighteenth Corps at Robert Espagne on the 8th. This enabled de Langle de Cary to shift the equivalent of an army corps to the support of his hard-pressed left wing. After the capture of Montmirail in the course of the victorious advance of the Fifth French Army on the 8th, Franchet d'Espérey was able to direct the Tenth Army Corps against von Bülow's right flank on the 9th so as to relieve the pressure on Foch's left wing. Finally, during the 9th, the Twenty-first Army Corps brought from the front in Lorraine, came into action in the gap between the Ninth and Fourth Armies, attacked the German troops who had penetrated the cavalry screen, and restored the cohesion of the French front.

With these more favorable circumstances both Foch and de Langle de Cary took the offensive on the 9th, the former attacking Fère Champenoise and the latter Sompuis with great energy. The struggle raged throughout the day with the utmost violence but without producing very definite results. With what fierceness the possession of every point of vantage was contested is illustrated by the desperate combats engaged at the Château de Mondemont, which is situated about six miles east of Sézanne on a slight elevation overlooking the marshes of St. Gond from the south. For three days the French had held this position against repeated assaults. At the climax of the battle it changed hands four times within a few hours. The French, first driven out by the Germans, brought up some of their "seventy-fives," opened a breach in the garden wall and rushed the château. But the Germans returned in greater force and occupied it again; and they were so sure of undisputed possession this time, that preparations were in progress for the officers' midday meal, when the French returned to the attack with very much greater fury.

Crowding into the interior of the château, where the restricted space barred the use of their rifles, the French soldiers grappled their opponents of the Prussian Guard, fighting like demons with whatever weapons came to hand. An adjoining outhouse yielded a store of short iron bars which the French employed with murderous effect. All the passions which had provoked this war seemed to have concentrated their fury to inflame the writhing mass of combatants with a delirium of rage. The Germans were finally ejected from the château, where the floors were a welter of carnage and the ceilings were dripping blood.

By the evening of the 9th the Tenth French Army Corps belonging to the Fifth Army had penetrated eastward as far as Baye, north of Sézanne, threatening von Bülow's rear. At the same time General Foch executed the daring maneuver which turned out to be decisive. After a heated contest on his left, in which a division from Morocco under General Humbert had conducted itself with special distinction, repelling many furious attacks, the Germans had abandoned their offensive. The commander of the Ninth Army accordingly moved all, or, at least, the greater part of the troops composing his left wing eastward to the support of his center and right, assailing the Prussian Guards and the Saxon corps unexpectedly on their flanks and driving a wedge into the aperture between the armies of von Bülow and von Hausen. Thus the Prussian Guards were outflanked and driven into the marshes of St. Gond, which were unusually treacherous in consequence of the wet weather. There they were entangled and subjected to the deadly fire of the French field-artillery. They suffered terrible losses and many of their guns were engulfed. Some detachments of the Guards who refused to surrender were annihilated where they stood. Von Bülow broke contact with the French as best he could and retreated northeastward on the 10th. The Ninth Army pursuing him crossed the Marne on the 11th at Châlons-sur-Marne and in its neighborhood.

The progress of events in the fighting in the different parts of the Battle of the Marne was apparently determined by impulses starting from the west. Each important stage in the development of the conflict, inaugurated by the action of the most western armies, was reached by the others consecutively in the order of their succession eastward.

The action of the German Fourth Army scarcely began before the 8th and did not become severe before the 10th, when the battle further west was practically terminated. The attitude of the Germans in entering the Battle of the Marne is illustrated by an order of the day addressed by the lieutenant-general commanding the Eighth Army Corps from his headquarters at Vitry-le-François on the 7th, the eve of battle, to his soldiers. The text is as follows:

"The object of our long and arduous marches has been achieved. The principal French troops have been forced to accept battle after having been continually forced back. The great decision is undoubtedly at hand. To-morrow, therefore, the whole strength of the German army, as well as that of our army corps, is bound to be engaged all along the line from Paris to Verdun. To preserve the welfare and honor of Germany, I expect every officer and man, notwithstanding the hard and heroic fights of the last few days, to do his duty unswervingly and to the last breath. Everything depends on the result of to-morrow."

The retreat of the Fourth Army on the 11th was probably necessitated primarily by the exposure of its right flank in consequence of the flight of the German forces which had covered it.

The army of the German Crown Prince occupied St. Ménéhould on the 6th, but its action was mainly directed to the region eastward of the forest belt of Argonne towards the French barrier fortresses.

The retirement of the French armies before the Battle of the Marne left Verdun at the extremity of a salient projecting at least twenty-five miles to the north of the general front of the French forces. A rupture in the line of fortresses reaching to Verdun would open a very convenient avenue of communications for the German armies to the west and isolate one of the most important and formidable of the French strongholds. Accordingly, the German Crown Prince commenced the bombardment of some of the barrier forts on the 9th directing his chief efforts against Troyon.

The inevitable necessity of falling back as a result of the outcome of events to the west arrested the Crown Prince's offensive on the 13th, just in time to save Troyon, which was being battered to pieces. The fact that fluctuations in the course of the Battle of the Marne started in the west and the circumstance that the Crown Prince did not

German guns captured near La Ferté-Milon.

Limber-wagons left by the Germans during their retreat in the battle of the Marne.

withdraw until three days after von Kluck's retirement are additional evidence disposing of the theory that the modification in von Kluck's movements just before the battle was determined by the necessity of extricating the Third Army from a perilous situation.

Coincidently with the great struggle in the region of the Marne a very violent attack against the defensive positions before Nancy was launched by the Germans in the presence of the Kaiser and the General Staff. The offensive in this quarter, commenced on the 5th, was probably intended to support the efforts of the Crown Prince in blasting an opening through the French fortified barrier. But the Germans abandoned the offensive and retreated very suddenly on the 12th, probably because the impending retirement of the Crown Prince would have released a large part of the French Third Army which could have been turned to the side of Nancy. The French immediately recovered Lunéville.

In the official reports of the progress of the Great War the deliberate tendency to deception has generally manifested itself in the suppression, disguise, or concealment of disagreeable information. In other words, misrepresentation in the official bulletins usually consists in implication rather than specific falsehood. Many interesting examples of this tendency might be cited; but perhaps the most noteworthy was the effort of the German authorities to disguise the fact that there was a general pitched battle on a grand scale in the neighborhood of the Marne with the results which have just been described.

The German authorities emphatically denied the truth of the report from hostile sources indicating that they had suffered a defeat. They themselves furnished the information about the operations and their outcome which we shall consider in the words of the original bulletins.

Quartermaster-general von Stein announced on September 3d:

"The cavalry of Colonel-general von Kluck is pushing its forays as far as Paris. The western army has crossed the line of the Aisne and is continuing its march towards the Marne, which some of its patrols have already crossed.

"The enemy is retreating before the armies of von Kluck, von Bülow, von Hausen, and the Duke of Württemberg to the Marne and behind it."

Thus in the evident expectation that decisive results were at hand, and that there would be no retrogression, the military authorities made public the forward movement to the Marne.

Another bulletin from the same source, dated the 10th, announced:

"The units which had pressed forward to the Marne and penetrated beyond it in pursuit of the foe east of Paris have been attacked by superior forces from the direction of Paris and between Meaux and Montmirail. In a severe two days' engagement they arrested the enemy's offensive movement and even advanced themselves. At the report of the approach of strong reinforcements for the enemy the wing was withdrawn. The enemy made no attempt anywhere to follow. Fifty guns and several thousand prisoners have thus far been taken as spoils of victory."

This was a very skilful attempt to represent von Kluck's defeat in a favorable light. The first sentence is a concise and accurate statement of the situation of von Kluck's forces in the initial stages of the battle both south of the Marne and north of the Marne in the valley of the Ourcq. But the following statements are true only of the course of operations north of the Marne, which was favorable until the 9th. The statements in the last two sentences it is difficult to reconcile with conditions in any part of von Kluck's battle-ground.

After the Germans had retreated to the line of the Aisne, forty miles to the rear, with a precipitation which in some instances resembled a rout, it was too late to conceal the previous advance to the Marne and beyond it and very difficult to account for such a conspicuous retrograde movement when the goal of the campaign had seemed to be in sight, without admitting that they had suffered a serious defeat in a great battle. The General Staff tried to evade the difficulty by suppressing every significant indication of locality in their reports of the renewed fighting on the Aisne. Thus the following announcement was issued at the Great Headquarters on September 17th:

"The battle *between the Oise and the Meuse* has not yet been brought to a final issue, but there are indications that the resistance of the enemy is beginning to weaken."

The apparent indication of the western and eastern limits of the battlefield has no practical significance. What is really important about the situation of the battle-lines, their position north and south, is conspicuously lacking. Any great battle in Northern France at this time would naturally have been fought "between the Oise and the Meuse." For many weeks the German authorities continued to announce progress between these discreetly irrelevant longitudinal limits, without once mentioning the Aisne, although it was the most conspicuous physical feature in the vicinity of the new front.

German commentators have tried to disguise the defeat on the Marne by denying the existence of such a battle and by claiming that the German armies retreated at their own discretion. According to their view the presence of the Germans on the Marne was a casual circumstance and the fighting there did not constitute a single, coherent pitched battle, but a series of local encounters with varying results passing over by an uninterrupted transition into the operations on the Aisne. The withdrawal of von Kluck and von Bülow, as they assert, was a voluntary measure of prudence for the elimination of a salient rendered awkward and unprofitable in consequence of the extension of the German front towards the northwest and the greater attention henceforth paid to the French barrier fortresses in the east.

As for the contention that the German armies were merely maneuvering for position, or without a distinctive purpose involving definite success or failure, we need only remark that if pushing forward almost a million men with so much energy and vehemence was a comparatively aimless performance, the world has still to discover the tremendous measure of exertion which the Germans would put forth in a supremely significant undertaking. The assertion that a single Battle of the Marne did not take place is a claim that might equally be made in regard to any other great battle of the war. For the vast territorial extension of the great battles of the present necessarily deprives them of the unity and compactness which characterized the pitched battles of the past. But it makes no practical difference whether the Germans were defeated in a single battle of unprecedented magnitude or in a series of smaller encounters. The argument, moreover, that they were not defeated on the Marne because the fighting there was prolonged uninterruptedly by the operations on the Aisne, which produced a draw, is mere idle jugglery with words and definitions. The assertion that von Kluck and von Bülow did not suffer a tactical defeat because they retreated deliberately to avoid destruction is not unlike attempting to prove that a victim of theft has not actually been robbed because he voluntarily relinquished his valuables to escape

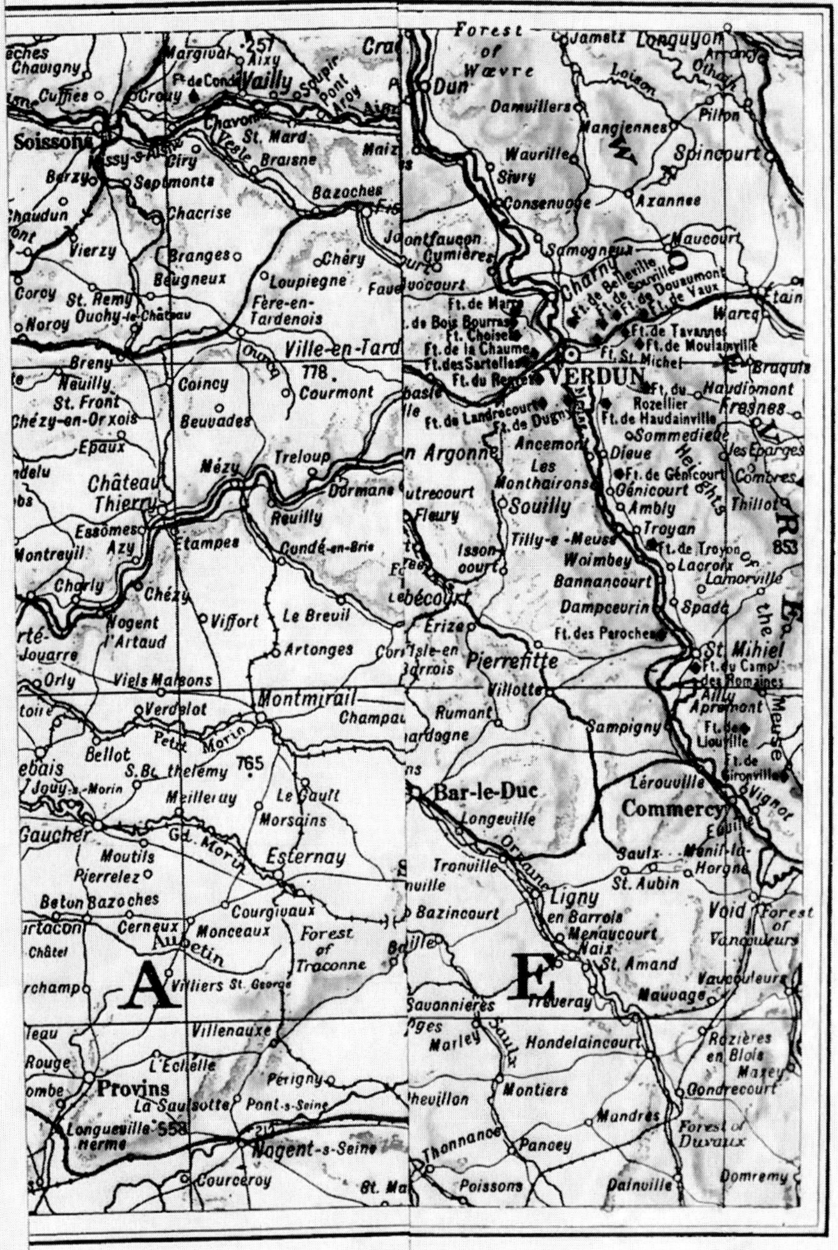
Forest of Wævre
Jametz
Longuyon
Dun
Damvillers
Mangiennes
Pillon
Spincourt
Loison
Othain
Wavrille
Sivry
Consenvoye
Azannes
Samogneux
Maucourt
Charny
Ft. de Belleville
Ft. de Souville
Ft. de Douaumont
Ft. de Vaux
Warcq
Etain
Ft. de Tavannes
Ft. de Moulainville
Ft. St. Michel
Braquis
Ft. de Marre
Ft. de Bois Bourrus
Ft. Choisel
Ft. de la Chaume
Ft. des Sartelles
Ft. du Regret
VERDUN
Haudiomont
Fresnes
Ft. du Rozellier
Ft. de Haudainville
Ft. de Landrecourt
Ft. de Dugny
Sommedieue
Ancemont
Dieue
Les Eparges
Les Monthairons
Ft. de Génicourt
Génicourt
Combres
Souilly
Ambly
Thillot
Tilly-s-Meuse
Troyan
Ft. de Troyon
Woimbey
Lacroix
Lamorville
Bannancourt
Dampcevrin
Spada
Ft. des Paroches
St. Mihiel
Ft. du Camp des Romaines
Ailly
Apremont
Ft. de Liouville
Ft. de Gironville
Meuse
Heights of the Meuse
Pierrefitte
Villotte
Rumont
Sampigny
Lérouville
Commercy
Vignot
Euville
Bar-le-Duc
Longeville
Ornain
Tronville
Bazincourt
Ligny en Barrois
Menaucourt
Naix
St. Amand
Void
Forest of Vaucouleurs
Vaucouleurs
Mauvage
Sauix
Ménil-la-Horgne
St. Aubin
Savonnières
Treveray
Marley
Saulx
Hondelaincourt
Rozières en Blois
Mazey
Gondrecourt
Montiers
Mandres
Forest of Duvaux
Thonnance
Pancey
Poissons
Dainville
Domremy
Margival
Aixy
Vailly
Soupir
Pont Arcy
Ft. de Condé
Croûy
Cuffies
Chavigny
Soissons
Chavonne
St. Mard
Missy-s-Aisne
Ciry
Vesle
Braisne
Berzy
Septmonts
Bazoches
Chaudun
Chacrise
Vierzy
Branges
Chéry
Coroy
Beugneux
Loupiegne
St. Remy
Ouchy-le-Château
Fère-en-Tardenois
Noroy
Ville-en-Tardenois
778
Breny
Neuilly
St. Front
Coincy
Courmont
Chézy-en-Orxois
Beuvardes
Epaux
Mézy
Treloup
Château Thierry
Dormans
Reuilly
Essômes
Azy
Étampes
Condé-en-Brie
Montreuil
Charly
Chézy
Nogent l'Artaud
Viffort
Le Breuil
Jouarre
Artonges
Orly
Viels Maisons
Verdelot
Montmirail
Petit Morin
Bellot
S. Barthelemy
765
Jouy-s-Morin
Meilleray
Le Gault
Morsains
Gd. Morin
Esternay
Moutils
Pierrelez
Beton Bazoches
Courgivaux
Cerneux
Monceaux
Forest of Traconne
Aubetin
Villiers St. George
Villenauxe
L'Echelle
Provins
Perigny
La Saussotte
Pont-s-Seine
Longueville
553
Nogent-s-Seine
Courceroy
A
E

Map shov

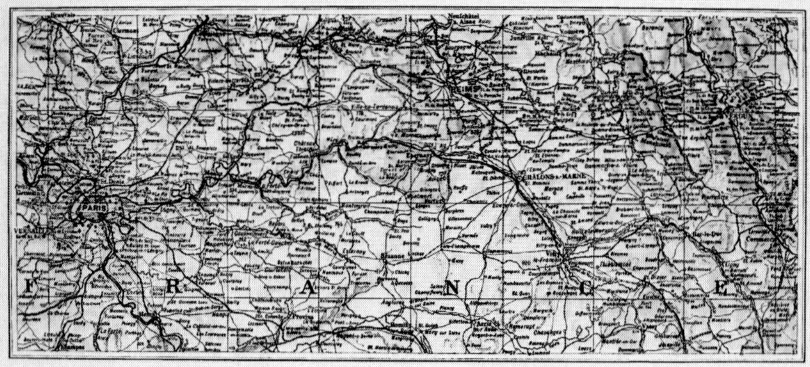

Map showing the Paris-Verdun district.

the extreme violence of the malefactor. To account for the retreat of von Kluck's army on the assumption that its position had become an inconvenient salient is to disregard the fact that it had stood in general alignment with the German front and that its retirement preceded that of the other armies farther east. It may be added that the western German armies were withdrawn to the Aisne before the prolongation of the German position, which is alleged as the motive, had actually been undertaken, probably before it was even contemplated. The resort to arguments based on pretended considerations of strategic expediency has become a very trite method of palliating disagreeable facts. Even Napoleon's disastrous retreat from Moscow might, by a conceivable exaggeration of this method, be interpreted as a well-calculated maneuver for the purpose of removing an awkward salient.

All these attempts to shield or assuage German pride hinge upon subtle distinctions in the definition of victory and defeat and the factors which produce them, conceptions which are necessarily relative and subject to variation. But no amount of casuistry can conceal the essential fact that the Allies won an important victory, not conclusive in the sense that a German army was destroyed or that the end of the war was brought in sight, but decisive in the sense that it was the turning point in the campaign. The Germans had launched a tremendous invasion of France with the professed intention of striking a decisive blow as soon as possible. In consequence of the prodigious struggle on the Marne they failed to accomplish this purpose. But, on the other hand, the German armies did escape from the snare which Joffre had devised for them.

The dexterity and promptness with which the German armies, after weeks of excessive exertion and a disconcerting defeat, extricated themselves from a position of imminent peril and retreated to the Aisne are no less deserving of admiration than their astonishing performances in the great rush towards Paris. The tenacity displayed by the German military organization under the terrific strain is almost incredible.

It is impossible to determine even approximately the losses in the Battle of the Marne. Some authorities estimate the aggregate dead and wounded at about 10% of the numbers engaged.

The German armies retired to positions already prepared for them on the Aisne. The capture of the fortress Maubeuge while the Battle of the Marne was in progress contributed materially to the security and consolidation of the German position in northern France.

We saw in a previous chapter how Maubeuge was left like an island in the inundating flood of invasion. When the Allies retired behind the Marne, Maubeuge, far away on the Belgian frontier, was the only French fortress in the north which still held out against the Germans. The heavy siege-train was transferred from Namur to the positions around Maubeuge. The Austrian motor-batteries, which had been employed at Namur, were conveyed thither after the bombardment of Givet, and rendered the same effective service.

The bombardment of the forts composing the ring around Maubeuge began on August 27th and continued until September 7th, when the white flag was displayed at 11.50 in the morning. The Germans took 40,000 prisoners including four generals, 400 guns, and much material of all kinds. It was jubilantly reported that a single German corps had reduced an important stronghold with a garrison equal to its own number. But of course the great superiority of the German siege-artillery more than counterbalanced the advantage of the defensive position of the French behind fortifications.

CHAPTER VIII

Operations on the Line of the Aisne
(*September 10-23, 1914*)

Significance of the operations on the Aisne. The new German front. Natural features of the valley of the Aisne. The position of the Anglo-French forces. The action on September 12th. The passage of the Aisne by the Allies, on the 13th, and their subsequent penetration of the heights to the north of the valley. Great difficulties and hardships suffered by the Allies in their exposed trenches. Their abandonment of frontal attacks. Great buoyancy shown by the German military organization.

Although the Germans undoubtedly still clung to the hope of pressing the war in the West to a decisive issue in the spirit of their original intentions, it became ever more apparent that their first great design of shattering the resistance of France once and for all by a sudden, irresistible blow had received a definite, conclusive check in the desperate combats near the Marne. The war of rapid, sensational maneuvers was now transformed into a war of position. While we may assume that the course of German operations in the West continued to be influenced, although to a somewhat diminishing extent, by the original strategic conceptions throughout the remainder of the campaign of 1914, the fact that the Germans were the first to dig themselves in, to sacrifice mobility for a definite line of front and the assured possession of a limited amount of the invaded territory, is a certain indication that the irrepressible vehemence of their initial operations had been impaired. The development of the international situation, as well as the military events themselves, was exercising a sobering influence on the mental attitude of the German authorities. In this connection we may only note the conclusion at London on September 5th, on the eve of the momentous Battle of the Marne, at the most critical moment in the campaign in the West, of the following fundamental agreement between the great powers that composed the Triple Entente:

"The British, French, and Russian Governments mutually engage not to conclude peace separately during the present war. The three governments agree that when the terms of peace come to be discussed, no one of the allies will demand conditions of peace without the previous agreement of each of the other allies."

The German government had undoubtedly been encouraged by the confident hope that the coalition against them would not endure to the end. That the association of their enemies was an incongruous and unnatural, as well as a monstrous, combination, seemed to the Germans to be a perfectly self-evident proposition. By comparison with it the alliance of the central empires presented a compact, united front. The disparity of tradition, attitude, and interests of Germany's enemies, especially of Great Britain and Russia, would, as it seemed, necessarily prevent a sincere, lasting coöperation. While the agreement of September 5th did not destroy this German expectation, it was one of many disappointing signs indicating that its fulfilment was very much more remote than had been supposed.

The German General Staff must now have realized the terrible extent and gravity of the task before them and that the war would be a long, hard struggle. Consequently, the desire to hold the enemy as far as possible from the frontier of the German Empire began to assert itself, coördinately with the aim of terminating the war by vigorous offensive strokes, as a motive influencing strategy and military policy.

General Gallieni, Military Governor of Paris, who sent an army in motor-vehicles to support the French Sixth Army at a critical moment during the Battle of the Marne. *After the painting by L. Rybet.*

General de Maud'huy, commander of the French Tenth Army, a new army formed by General Joffre in the latter part of September, 1914.

It can be said with much more truth of the fighting on the general line of the Aisne than of the Battle of the Marne that it resolved itself into a number of partial, detached engagements. The so-called Battle of the Aisne was a transition battle, which could scarcely have been decisive in any conceivable circumstances. The Germans contested desperately the further relinquishment of every yard of the French territory which they had occupied. Their immediate aim was to delay the decisive stage of the operations until the recuperation of their forces, a favorable turn of events in the East, and the arrival of reinforcements permitted them to resume an energetic offensive. We ought to note as significant features of this stage of the campaign in the West that for the first time since the completion of their mobilization the Germans had to accept the defensive rôle, which is so hateful to all their strategists; that the struggle on the Aisne inaugurated the comparatively stationary, siege-like operations of trench-warfare; and that the outcome destroyed the eager hope of the Allies that the repulse near the Marne would result in the immediate collapse of the German campaign in the West and that the Germans would be driven from France without delay.

The movement to the line of the Aisne, or in other words, the changes in the position of the opposing armies occasioned by the defeat of the Germans in the Battle of the Marne, was not fully carried out before the evening of the 13th. The German armies were compelled to take up one by one, in order of their succession eastward, the retrogressive movement inaugurated by von Kluck. In general we may say that the German front in receding pivoted on the left flank of the army of the Crown Prince east of the Argonne.

The new German front between the Meuse and the Oise, destined to mark the limit of German occupation for many months, left the Meuse near Consenvoye, traversed the rough, wooded region of the Argonne, and the plains of Champagne northward of Reims, followed the course of the Suippe northwestward to its confluence with the Aisne and crossed the latter just above Berry-au-Bac, gained the elevated plateau of Craonne, and continued westward along the summit of the heights on the north of the Aisne to the left bank of the Oise. The Germans also retained territory on the right of the Oise. The flanks of the new position were thus protected by natural barriers, and a considerable portion of the front, which was about 125 miles in length, offered exceptional advantages for defense. The work of intrenching this front had undoubtedly been undertaken as a precautional measure before the conclusion of the Battle of the Marne. The German armies retreating to this position were strengthened by the forces which had been released at the capitulation of Maubeuge and by numerous other reinforcements which probably restored the equilibrium completely.

The most important characteristics of the following sanguinary engagements are due to the special physiognomy of the western part of the region of hostilities, the valley of the Aisne from its confluence with the Suippe to its union with the Oise. This alone gives an unmistakable appearance of unity to the series. The Aisne flows in a winding course from east to west through a narrow, flat valley from about a half mile to two or three miles in width. An elevated ridge dominates this valley on the north at a distance varying from three to five miles from the river. A gradually sloping plateau, wooded in parts, extends southward from the ridge and is terminated at a distance of a mile or two from the river by an abrupt descent of 300 or 400 feet to the level floor of the valley. Deep ravines penetrate this hillside, scalloping the margin of the plateau above with many spurs and indentations. The slope on the south side of the river is essentially similar in character to that on the north. The valley through which the river flows is chiefly occupied by unenclosed meadows, interspersed with patches of woodland. It is dotted with villages and intersected with roads bordered by rows of poplars and fruit trees. The river itself is a formidable obstacle on account of its depth of fifteen feet, although it is less than 200 feet in breadth. Most of the bridges were destroyed by the Germans after they retreated to the northern bank of the stream. Villages situated at the crossing-points helped to make the bridges, or their temporary substitutes, conspicuous targets for the artillery posted on the heights to the north, from which the Germans had presumably measured all the ranges with the utmost care.

In advancing to the new line the armies of the Allies preserved the same relative position among themselves, with the French Sixth Army, the British Army, and the French Fifth, Fourth, and Third Armies in the order named, beginning in the west. The British experienced very little opposition on the 11th. They crossed the Ourcq where its course is east and west and bivouacked on the line between Oulchy-le-Château and Longpont. The cavalry arrived in the vicinity of the Aisne the same evening.

On the next day the Sixth and the British Armies advanced to positions overlooking the Aisne. The Sixth Army took up a position extending from Soissons towards the west, while the British drew up on a front about fifteen miles in length from Soissons eastward. The Germans were posted in strong positions on both sides of the river. They occupied Soissons.

There was a long-range artillery duel throughout most of the day on the 12th. The British artillery coöperated with the right wing of the Sixth Army in the attempt to expel the Germans from Soissons and the French gained possession of the southern half of the town during the night. Eastward on the same day the Fifth Army reached the line of the Vesle, which flows westward between the Marne and the Aisne and empties into the latter about eight miles east of Soissons.

There was general action throughout the valley on the 13th. The three British corps, which maintained the same consecutive order as before, with the First on the right wing, were ordered to cross the Aisne at the available points along their front, and the movement ushered in the most spectacular stage of the operations in this region. The First Division crossed on a viaduct by which a canal traverses the river. The Second utilized boats and a single girder of a half-demolished bridge at Pont-Arcy on which they passed in single file. A pontoon bridge was completed by five in the afternoon. Only a part of the Third succeeded in reaching the northern bank. The Fifth, finding the bridge at Condé intact but swept by the enemy's fire, were rafted across. The Fourth repaired the permanent bridge at Venizel and supplemented it by a pontoon bridge.

The Eighth Army Corps, forming part of the Sixth Army to the left of the British, crossed the river west of Soissons under cover of a furious cannonade.

The Allies everywhere encountered the most determined resistance. There were desperate hand-to-hand encounters in many of the villages, where the streets were filled with dead. Furious fighting continued all day, the terrific action of the artillery converting the valley into a hideous inferno. The lower part of Soissons burst into flames from the shower of projectiles poured into it. The meadows north of the Aisne were swept by such a fierce fire that it seemed impossible that living beings could exist in them. By

The Cathedral of Notre-Dame at Reims before the bombardment by the Germans. *Built in the thirteenth and fourteenth centuries, it is not only celebrated for its architectural beauties, the façade being one of the masterpieces of the Middle Ages, but also historically, as here used to be crowned the kings of France.*

nightfall throughout the greater part of the front, the Fifth, British, and Sixth Armies had established themselves on the northern bank. Before morning the British had placed eight pontoon bridges in position and had repaired several of the damaged bridges in spite of the almost incessant fire of the German heavy artillery.

The main part of the German forces withdrew to the positions on the heights north of the river, where their system of intrenchments had been very carefully prepared and heavy artillery commanding all parts of the valley had been mounted in cleverly concealed positions. At least as early as the 15th, the heavy artillery from the positions before Maubeuge was being employed by the Germans on the line of the Aisne. The most numerous, and probably, in the actual circumstances, most useful, of the heaviest pieces were the 21-centimeter howitzers with their effective range of 10,000 yards. Tremendous cannonading, calculated to unnerve and terrify the Allies, preceded the charges of the German infantry. The Allies had no means of replying to the heavy artillery of their opponents.

The Germans left strong detachments in advanced positions where they could sweep the approaches to the principal lines of trenches by the crossfire of field and machine-guns. These outposts were placed on projecting ridges or spurs of the plateau and even in the valley, as at Condé, an important bridge-head. Powerful searchlights disclosed any attempt of the Allies to approach the German lines by night.

The most striking performance on the 15th was the bold attempt of the British right wing to push northward up the ravines. After a day of very heavy fighting the First Corps under Sir Douglas Haig occupied a position on the line of the main highway, called the Chemin-des-Dames, which runs along a ridge on the plateau, and succeeded in intrenching themselves there. The great importance of this position is shown by the repeated attempts to drive the British from it, all of which were repulsed with great loss to the Germans. The Eighteenth Corps in the Fifth Army supported the British in this forward movement.

The chief efforts of the Allies throughout the front were directed primarily to securing positions on the plateau overlooking the valley on the north, where the enemy would be within range of their firearms, and the distance to be traversed in the contemplated final onslaught would be short. As they realized that the resistance of the Germans was no mere delaying action to cover a further retreat, but a determined stand in carefully fortified positions, the Allies gradually elaborated their system of trenches to secure the greatest possible protection against the shells of their opponent's heavy artillery.

In the early days on the Aisne the Germans by their furious attacks, repeated at frequent intervals, deprived their enemy of rest by day and night. After a very severe but unsuccessful attack on the 15th against the right wing of the Sixth Army and the left of the British, the Germans delivered no less than ten distinct attacks in the following night, all of which were repulsed.

The British Sixth Division arrived from England on the 16th completing the normal complement of the Third Corps. It was posted on the south side of the Aisne as a general reserve, making it possible to relieve regularly the detachments on service in the forward trenches. Four 6-inch howitzer batteries arrived from England at Sir John French's request on the 23d.

Very heavy rains continuing many days about the middle of September transformed the usually placid Aisne into a turbulent flood and impeded the restoration of adequate communication between the two banks, and exposed the

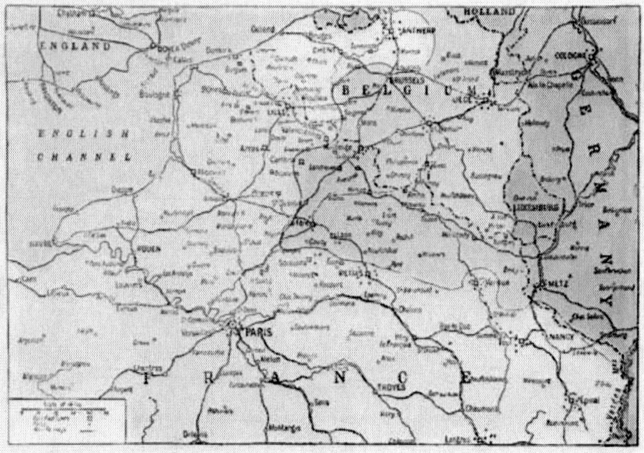

Map showing by the shaded portion the territory occupied by the Germans after the Battle of the Marne, September 13, 1914.

soldiers in the trenches to great discomfort, a forecast of the many coming months of privation and excessive hardship in consequence of the inclemency of the weather.

Service in the advanced trenches of the Allies, which approached in some places to within two hundred yards of the enemy's foremost line, required almost superhuman endurance. With the pouring rain for days the water stood knee-deep in the trenches. Many of the British had lost their great coats during the rapid retreat from Mons to the Marne. The soldiers in the trenches were covered with a stiff layer of clay from the saturated, chalky soil. The frequent bombardment allowed little rest or relaxation for limbs or nerves. In some sections of the trenches it was almost certain death for the soldiers to show their heads above the parapet during the day. Food and water were brought and the detachments on outpost duty relieved only by night, the men crawling on all fours. The trenches were repeatedly the target for the heavy artillery firing at high angles in positions behind the German lines. Falling on both sides of the trenches, the shells detonated with terrific violence, throwing up great quantities of earth and stones and perforating the surface with bowl-shaped craters. Whenever one of the larger shells exploded in a trench, there was no escape for the soldiers in position near the spot. The howitzer shells were called Jack Johnsons by the British soldiers, on account of the great columns of black smoke which they discharged.

The general nature of such trench-fighting involved some exceptionally terrible circumstances. With the lines drawn as closely as possible and every yard of ground or element of advantage either actually seized or jealously contested, the tension of a single battle was continued indefinitely without respite or truce for removing the dead and wounded left in the track of successive attacks and counter-charges in the zone of death between the opposing trenches. Soon the decomposing bodies of the slain filled the air with foul, contaminating odors. The inevitable suffering of the wounded was greatly augmented in the more exposed trenches, from which they could be removed only at night. The severely wounded in the trenches had frequently to lie many hours in awkward positions or on water-soaked straw. But the imagination is powerless to comprehend the horrible fate of the severely wounded left between the firing-lines, whom it was death for their comrades to approach. Many, unable to crawl to either side of the bullet-swept area, exposed continually to fresh wounds or lacerations, awaited hours or even days for the coming of death as a merciful deliverance from their agony.

During the early operations on the Aisne the Allies suffered far more than the Germans from the inadequacy of their trenches and their exposure to the fire of heavy artillery, while the Germans endured considerable hardship from the irregularity and insufficiency of their food supply. The Germans did not suffer the terrible bombardment to which they subjected their foes; but their losses in the trenches from the fire of small-arms must have been greater than that of their opponents, in consequence of the superiority of British marksmanship.

Upon recovering Reims in consequence of their victory in the Battle of the Marne, the French strengthened the defenses of the city and made it a prominent point on the line of their front. The Germans commenced a counter-offensive against Reims about the 20th and took the commanding heights of Craonelle near it. In accordance with the approved method they supported their infantry attacks by vigorous action of their cavalry, bombarding not only the lines of trenches but the city itself, where some of the French batteries had been posted, probably on account of

German Red Cross men at work under fire in the Argonne district.

German dead in a trench just captured by French troops.

the cover afforded by the buildings. Fires were caused in several places and the city suffered considerable damage from the German bombardment.

The Germans claim that their artillerists were ordered to respect the white flag which was raised on the famous cathedral, and that every effort was made to spare this beautiful monument of Gothic architecture, until an observation post was discovered in one of the towers directing the aim of the French artillery.

A German official communication stated that only shrapnel fired by the field artillery was employed for dislodging the observers in the tower, but admitted that the roof of the cathedral was afterwards seen to be in flames. It is scarcely conceivable that such a conflagration was due to shrapnel. A later communication added that a single projectile was fired by a mortar at the observation post, since the field artillery had proved inadequate for the purpose.

Unfortunately, in this and subsequent bombardments the cathedral suffered serious damage, but without being destroyed as an edifice. The circumstance was eagerly utilized for the purpose of inflaming the feeling of bitterness against the Germans, especially in neutral countries. But the supposition that the Germans were actuated by purely wanton maliciousness is preposterous. In many instances the rage and lawlessness of invividuals found vent in senseless acts of destruction. But the systematic action of artillery, as in defacing the cathedral of Reims, is not controlled by freaks of individual passion, and the German authorities were certainly not seeking gratuitous enmity.

It gradually became evident that without a far greater preponderance of forces than the Allies were able to muster they could not hope to surmount by frontal attacks the stronger positions, innumerable machine-guns, and powerful siege-artillery of the Germans. The offensive action against the front of the German position was therefore allowed to wane and General Joffre turned to the plan of threatening the German right flank by striking northward from the western section of the front on the Aisne, under cover of the forests. This movement was undertaken by the Fourth and Thirteenth Corps belonging to the Sixth Army. But General von Kluck anticipated precisely such a turning movement by extending his lines westward so as to intercept its line of advance. The French Corps were defeated south of Noyon, losing several batteries.

The German military organization exhibited its marvellous tenacity and resiliency in the crisis created by the defeat in the Battle of the Marne. At one moment the prodigious enterprise in the West collapsed and the German armies seemed to be entrapped or routed. In the next, the necessary adjustments had been accomplished, the armies in the West were in a firm position waiting an opportunity to resume the offensive, and soon we shall behold the Germans, while displaying an astonishing amount of activity in all sections where hostilities are in progress, pushing operations in the East with apparently as much vigor as we should have expected if their plans in the West had been successful.

CHAPTER IX

The Race to the Sea
(September 23-October 15, 1914)

The more complicated character of the second part of the first campaign in the West; the diversity of purposes. The Allies' offensive and the race to the sea; every effort of the Allies countered by the Germans. Transference of the British to Flanders. Lille taken by the Germans. Resumption of the German offensive. Intimate connection of Verdun and Antwerp in the deliberations of the Germans. The perforation of the French eastern barrier at St. Mihiel and its futility. Retrospect of the situation of the Belgian army. Destruction of Louvain. The German operations against Antwerp: the fortifications, the opposing forces, the beginning of the bombardment of the forts, September 28th, the removal of the Belgian base, the penetration of the outer girdle. The general outlook and departure of the Belgian field-army. The fall of Antwerp. The race from all sides towards the southwestern section of the Belgian coast. The arrival of the Belgian army on the line of the Yser and completion of the barrier from the Swiss boundary to the North Sea.

For reasons that have already been explained, we ought to regard the Battle of the Marne as the great turning point of the campaign of 1914 in the West, dividing it naturally into two parts. The interpretation of much of the action in the second part, the nature of the underlying purposes, is not so clear as in the first part. The confusion is inherent in the nature of the transition brought about by the above-mentioned preëminent event. For the immediate change in the purpose and method of the operations was not a complete revolution. Thus, although the earlier dashing maneuvers and lightning strokes gave way largely to the steady, comparatively stationary, grind of trench-warfare, a conspicuous amount of mobility was still displayed by the operations in some parts of the contested area for at least a month. With the turning of the tide in the Battle of the Marne the initiative passed from the Germans, who had held it quite exclusively since the completion of their concentration, to their opponents. But the latter were unable to retain it consistently very long. It fluctuated from time to time and from place to place and later returned quite unmistakably into the power of the Germans.

Consequently, the imagination is no longer enthralled by a single tremendous movement, beside which all else is insignificant, as in the earlier part of the campaign. Then the one great purpose of the invaders dominated the field and determined the action of their opponents, who strove for the time merely to hinder and delay their progress. In the later period this unity in the course of events is lost. Each contestant strove to carry out a distinctive, characteristic plan. The attention is claimed by important series of operations developing simultaneously in different parts of the western theater. The really significant exertions of both sides were actuated by their own positive, individual intentions. It is necessary then for us to distinguish clearly the respective designs of both contestants as the indispensable condition for comprehending the course of events, which is complicated by the fact that one of the contrasted plans was not simply the counterpart or reverse of the other.

The position of the initiative, resting with the Allies at the outset, later fluctuating, and finally passing over completely into the hands of the Germans, suggests the proper order of treatment. We shall consider, first, the purpose of the Allies and the continuation of their offensive movement which was started near the Marne on September 6th, and secondly, the intention of the Germans and the renewal of their offensive, which had only been suspended temporarily.

The Allies, foiled in their frontal attacks on the line of the Aisne, transferred their offensive to the left wing and resumed the effort to outflank the enemy's right. But every movement of the Allies for the purpose of circumventing their opponent's position and striking at his lines of communication was matched or surpassed by the efforts of the Germans, so that the net result of the struggle of each army to outflank the other was the rapid extension of the opposing fronts on parallel lines northward in a race which terminated at the sea about the middle of October. Meanwhile, the Germans, with renewed energy and greatly augmented strength, had extended the range of their aggressive action and consolidated their position in the invaded territory. Finally, they took up a vigorous offensive and hurled themselves in repeated attacks with fearful violence against the position of the Allies in Flanders. Thus the movement inaugurated by the French, which resulted in the extension of the fronts to the North Sea, and the determined resumption of the German offensive, which culminated in the tremendous effort to break through the barrier formed by the Allies, are the dominating features of the operations throughout the remainder of the campaign of 1914 in the West. The first is the main subject of the present chapter, the second will be reserved for treatment in the next.

A variety of considerations doubtless swayed the leaders of the Allies in their choice of a plan for the continuation of the offensive. In the first place it was natural for them to revert to a method which had been successful in the Battle of the Marne. Then the turning movement in the west doubtless seemed at the time to offer the largest prospect of success as well as the greatest advantages. Alsace was too narrow for aggressive operations on an extensive scale, and as long as the German lines were only sixty miles from Paris, a great offensive movement in Alsace would have produced a dangerous dissipation of forces. Another advance through Lorraine would have been foredoomed to failure without the previous reduction or investment of Metz or Strassburg, which would have involved insurmountable difficulties.

There was still the possibility of a flanking movement directed from Verdun towards the north. But in respect to facilities for the concentration of troops, Verdun was at a great disadvantage as compared with any corresponding base of operations in the western part of France. Besides, the prospective advantages were apparently very much greater in the case of a movement launched from the western end of the intrenched position between the Oise and the Meuse than in that of a similar maneuver started from the eastern extremity. A movement in the west would provide for the permanent safety of the Channel ports and secure the possession of the most important industrial region of France in the north. Lille, for example, a very wealthy city, on the border of Belgium, was the great center for the manufacture of locomotives, automobiles, and sundry appliances constantly required in warfare. The Germans had evacuated Lille in consequence of the Battle of the Marne. But the French held it by a precarious tenure. Furthermore, a northward movement in the west promised to bring the Anglo-French forces into actual contact with the Belgian army, so that the united efforts of all would become much more effective, and the advantages offered by the position of the fortress of Antwerp as a sort of projecting bastion would be realized.

If supremely successful, the turning movement of the Allies would sever the principal lines of communication of the German army running northeastward through Belgium, crush the army of von Kluck, and drive the invaders with great loss and confusion from French soil.

Strange as it may seem in comparison with later conditions, the vast area extending from the western extremity of the opposing lines, as they were at that time, near the confluence of the Aisne and the Oise, to the North Sea and the course of the Scheldt, was debatable territory, but only loosely guarded or patrolled in parts by Territorial troops or detachments of cavalry. But such a situation was no longer compatible with the close, intensive disposition along the Aisne, where progress was measured in yards, as in siege operations.

The armies on both sides were constantly being augmented by the incorporation of reserve troops of various categories whose state of training or preparation had been inadequate at the beginning of the war. These impatient masses, restrained by the baffling equilibrium of opposing factors in front, tended inevitably to spread out laterally and press eagerly into the unoccupied spaces. Since the Allies still possessed the initiative, they naturally took the lead in this movement. But the Germans had by no means renounced their fundamental design of crushing France as the preliminary step in the direction of an ultimate universal triumph. There was every reason to suppose, therefore, that von Kluck was merely awaiting a favorable opportunity for resuming the turning maneuver in the west which had so nearly succeeded before the Battle of the Marne. The extension of the French lines on the left might also be regarded, therefore, as an indispensable measure of self-protection.

In the feverish contest for expansion northward the Germans had a distinct advantage in the concentric form of their front at its western extremity, which shortened their transports. Both the French and the Germans were now bringing many new formations into the field. But the unexpected alacrity with which the Germans thwarted every attempt of the Allies to outflank them is an indication of their numerical superiority.

The German Ersatz Reserve is composed of those who, although they are physically fit and have arrived at military age, have never been enrolled for service in the regular army. It was a reservoir of potential military strength, a source of raw fighting material. The Ersatz Reserve was required to report for training at the regimental depots on the outbreak of war so as to supply the necessary drafts replenishing the active units in the field. But the accumulated margin between the actual annual classes of recruits and the whole number of men qualified for military service was so large in Germany that independent Ersatz divisions were formed about this time, when the members of the Ersatz Reserve were completing their minimum emergency training.

About September 20th a new French army, the Seventh, was concentrated between the Oise and the Somme, on the left of the Sixth Army, under the command of General de Castelnau, the capable leader who had saved Nancy from the German counter-offensive which followed the unsuccessful French invasion of German Lorraine. He was instructed to extend his left flank to the north of the Somme and to ascend the valleys of the Oise and Somme in the direction of St. Quentin.

At first the movement thus inaugurated must have been very perilous for von Kluck, whose right wing barely covered his principal lines of communication coming from Belgium. By the 21st, de Castelnau's right wing had reached the vicinity of Noyon, and on the left a detachment occupied Péronne on the 23d.

But the Germans concentrated their forces in the neighborhood of St. Quentin, the threatened point, with the

One of the principal streets in Lille after an engagement between German Uhlans and French infantry.

utmost energy and speed. Reinforcements were dispatched from different parts of the front, notably from Lorraine. It is even said that the whole, or at least the greater part, of the Sixth German Army was transferred to this section. On the 25th the French began to retire from the vicinity of Noyon under the formidable pressure of the Germans.

The almost uninterrupted fighting in this quarter from the 25th to the 27th was part of a general conflict which raged along the entire front in northern France from the Somme to the Meuse in connection with repeated assaults of the Germans. The 26th saw the renewal of activity on the heights to the north of the Aisne and the determined effort of the Germans to recover the advanced position held by the British right. Some authorities regard the waning of these attempts of the Germans to establish an ascendency in this section on the 27th as the end of the Battle of the Aisne.

The Germans again attacked with great fury the French positions before Reims. The bombardment did much damage to the city, setting it on fire in several places and killing a number of the inhabitants.

In the extreme west the Germans repulsed a French division advancing towards Bapaume, forced the French to retire from Péronne, and attacked Albert, an important crossing-point on the Ancre, north of the Somme, on a main highway leading westward to Amiens. The contest at Albert was prolonged for several days with great violence. In spite of the burning of the town the French held their position tenaciously. They took the offensive on the 30th, but their endeavor to advance eastward was checked by the Germans.

On the same day the Germans occupied the heights near Roye in the plain of the Somme about twenty miles north by west of Compiègne, where they threatened the center of de Castelnau's army and the communications between Compiègne and Amiens. But they were unable to penetrate further in this direction.

Thus while the converging movement of de Castelnau's army on St. Quentin was frustrated, the German counter-offensive was checked in return.

In the meantime, forces on both sides were pushing northward with the greatest energy. It is even reported that the French were compelled to march as many as twenty-five miles a day in order to meet the advance of the Germans, who were moving on interior lines. Everywhere the German cavalry appeared in advance of the infantry, overrunning the country, spreading terror and confusion, and seizing points of vantage.

To forestall the continued northward progress of the Germans, which threatened to open the way for a formidable offensive movement, General Joffre decided to form a new French army, the Tenth, about September 30th, in the neighborhood of Arras and Lens, the hilly country between the Somme and the valley of the Lys. As a preliminary step, two cavalry corps were sent northward as far as the Scarpe, where they were to coöperate with Territorial forces which had advanced from Dunkirk to Douai. But the French were immediately confronted in the district of Arras by a strong German army consisting of the Prussian Guard, four army corps, two reserve corps, and two cavalry corps.

The French Tenth Army was placed under the command of General Maud'huy. With its right flank resting on the Ancre, it was intended that it should secure the front Arras-Lens-Lille, and then, probably, advance in the direction of Valenciennes. The general aim of the Germans in this quarter was apparently to capture Lille and turn Maud'huy's left flank.

But a battle lasting many days commenced on October 1st east of Arras, a handsome city situated on the Scarpe, a tributary of the Scheldt, thirty-one miles north-northeast of Amiens, the birthplace of Robespierre. The once famous fortifications of Arras designed by Vauban have long since lost all practical significance. The Germans captured Douai and Lens and threatened to outflank Maud'huy's left. Most of the French forces were thus compelled to retire to the hills west of Arras. A large part of the population of the city fled from their homes. The rest took refuge for the most part in cellars during the bombardment, which lasted intermittently for three days, October 6-8. The venerable townhall, a beautiful example of secular Gothic, was wofully shattered, although the tower survived this first bombardment. The Germans forced their way into Arras but were subsequently ejected. But in this battle Maud'huy's offensive stroke was parried before it had been fairly launched.

General Joffre decided to concentrate still another army, the Eighth, to cover Maud'huy's left flank, establish the front between Lens and Dunkirk, and stem the vast returning tide of German invasion. The Germans had penetrated into the suburbs of Lille. They occupied Ypres, October 3d, their cavalry had taken Armentières and Bailleul and was pressing westward up the valley of the Lys. Their outposts had arrived in the neighborhood of Hazebrouck and Cassel. Dunkirk and Calais were menaced. The French offensive seemed on the point of complete submersion and the situation was rapidly becoming critical. The Eighth Army was entrusted to the command of General d'Urbal.

The British army which had commenced the campaign on the extreme left wing of the Allied forces, now found itself, in consequence of the general lateral shifting of the lines and the expansion of the fronts towards the northwest, practically in the middle of the Allied position. Obvious considerations of expediency prompted General French to urge that the British should resume their initial place on the left, in other words, that they should be transferred to the section of the theater of operations in the extreme northwest. Their position would thus be nearest England. Their lines of communication, which, as matters stood, crossed those of several French armies on the left, would be greatly shortened and the transport of supplies would be very much simplified. The conveyance and deployment of the expected reinforcements of British, Indian, and Colonial troops would also be facilitated.

The perilous situation in the north added weight to Sir John French's arguments, and it was decided that the British army should be transferred to a position on the left of the French in Flanders. The process of withdrawing the British from the foremost trenches in the region of the Aisne and replacing them with French troops had to be performed section by section in the night with the greatest caution. For in some places the opposing lines were less than a hundred yards apart. The operation of transferring the British army was begun on the 3d and completed on the 19th without loss and almost without any hitch. The soldiers were transported partly by rail and partly by motor-vehicles.

It was naturally hoped that the British would coöperate effectively in the defense or recovery of Lille, where the situation was precarious, since the garrison consisted of an inadequate force of Territorial troops. Maud'huy's progress had been checked at Arras and Lens. The arrival of the British in the northern section was not expected before the 11th. Many of the inhabitants of Lille departed by rail or on foot terrified at the prospect of a bombardment

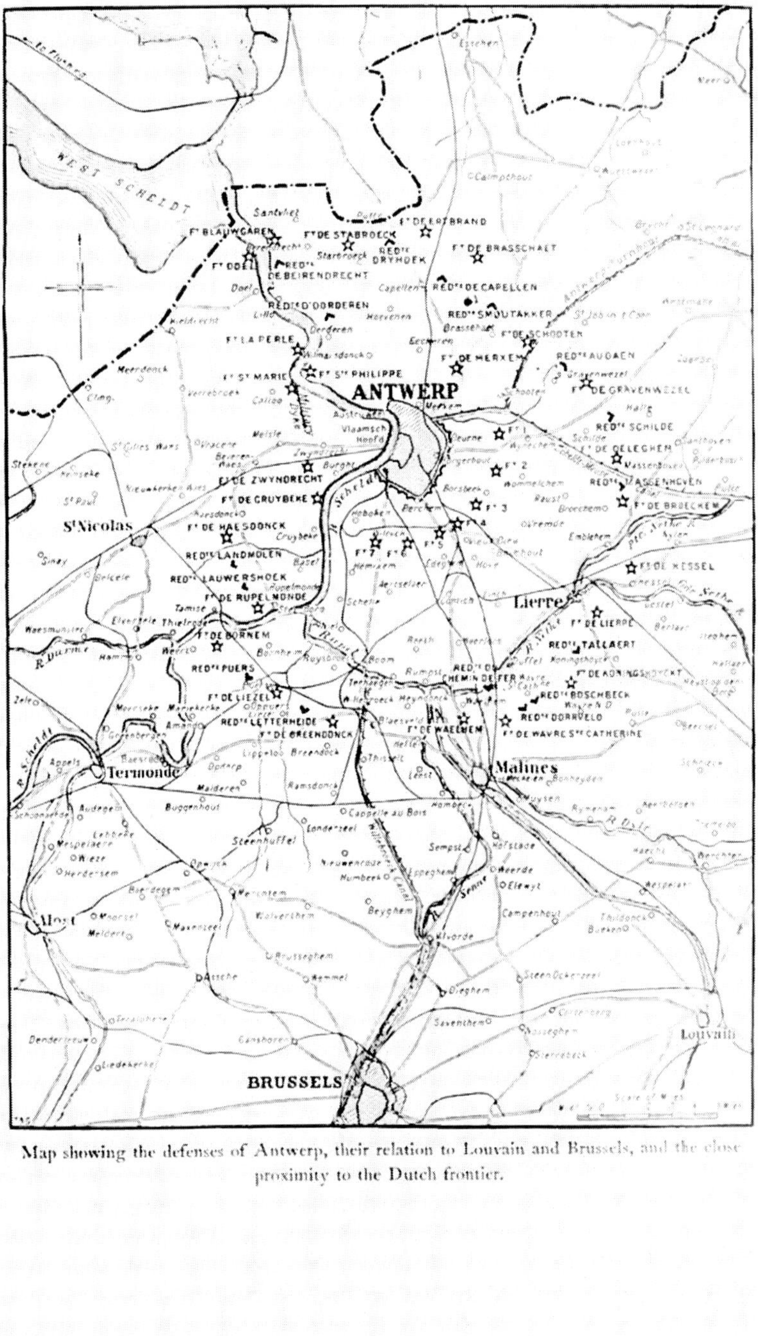

Map showing the defenses of Antwerp, their relation to Louvain and Brussels, and the close proximity to the Dutch frontier.

and of the capture of the city by the Germans. There were violent encounters in the vicinity on the 4th and 5th consummating in the repulse of the German advance-guard on the 6th.

But the Germans were constantly increasing their numbers. They took La Bassée and occupied the region between the Béthune-La Bassée-Lille Canal and the Lys, so as to threaten Maud'huy's left flank and intercept any attempt to relieve Lille.

General Foch, who deserves scarcely less than General Joffre the title Savior of France for his service in the Battle of the Marne, was appointed to correlate the efforts of de Castelnau, Maŭd'huy, French, and d'Urbal, to supervise, in other words, all the Anglo-French operations north of the Oise. In a conference of the military chiefs held on October 8th, the road from Béthune to Lille was adopted as a convenient line to separate the activity of the armies of French and Maud'huy operating to the north and south of it respectively. Maud'huy's Tenth Army was now established on a front extending from the Ancre across the hills to Béthune in the plain of the Lys, with its center at Arras.

It was decided that the right wing of the British army should pivot on the French at Béthune and strike northeastward at the Germans who had been threatening Maud'huy's left flank and that, if they forced them to retire, they should advance concurrently with the French in the direction of Lille.

The arrival of the British Second Corps, which would constitute the right wing of the British army in the new position was expected in the neighborhood of Aire and Béthune on the 11th. The detrainment of the Third Corps which would, for the present, form the left, was expected at St. Omer on the 12th. The First Corps, whose withdrawal from the advanced position on the right wing in the region of the Aisne was probably a more delicate operation, was not looked for before the 19th. The British left was to operate north of the Lys in conjunction with General d'Urbal's army which was then being formed.

The country between the Béthune-La Bassée-Lille Canal and the River Lys, west of Lille, is mainly an industrial region abounding in coal mines and factories. The numerous villages are frequently almost continuous. The serviceable cover thus afforded, together with the enclosures and other obstructions, made operations very tedious and difficult for an attacking force. Great caution had to be used in forestalling surprises and ambuscades. The Germans had carefully intrenched themselves in many of the villages and were well supplied with machine-guns which swept the exposed approaches. The British infantry was powerless, therefore, without the support of the field-artillery. The Second Corps went into action almost immediately after its detrainment in the vicinity of Béthune. But their progress was slow and costly. The British relieved the pressure on Maud'huy's left; but far from relieving Lille, the Allies were unable to eject the Germans from their advanced position at La Bassée.

The Germans transferred a part, if not all, of their First Army to the region of Lille and supplemented it by units drawn from other quarters. Thus the Nineteenth Corps, which had been part of the German Third Army, marched 108 miles in five days from the German front east of Reims to the vicinity of Lille.

The bombardment of Lille was begun on October 10th. A great panic ensued and a renewed exodus of thousands of civilians from the city, which was surrendered on the 14th to save it from destruction. Considerable destruction of property was caused, however, by the bombardment, but the loss of life was slight.

Meanwhile, the British Third Corps detrained at St. Omer and joined forces with the Eighty-seventh and Eighty-ninth French Territorial Divisions and the Fourth Cavalry Division under General d'Urbal. Together, they swept the Germans from the region west of the line Comines-Ypres-Dixmude and marched into Ypres on the 13th, thus altering the apparent situation in southern Flanders very quickly.

We must now turn our attention to parts of the western theater where other movements were developing simultaneously which united eventually in the same tremendous climax.

As long as the Germans expected to force a decisive turn of the conflict in the West by rapid, overwhelming strokes, they gave little heed to the symmetry or coherence of the occupied territory. While their armies were sweeping across Belgium or striking southward through northern France in eager expectation of a speedy triumphal entry into Paris, time permitted only the most elementary measures of precaution for covering their flanks and lines of communication. The astonishing rapidity of movement was therefore associated with a lack of definite demarcation which afterwards seemed incredible.

But the repulse on the Marne and the prospect that the war would be a long one, in which endurance would count as much as brilliant maneuvers, compelled the Germans to reflect upon their geographical situation and devote serious attention to rounding out and consolidating their position and securing consistent protection for their lines of communication.

In this connection Verdun and Antwerp were intimately linked in the motives actuating the conduct of the Germans, although these two places were situated near the opposite extremities of the vast semicircular rim of the chief mass of enemy territory which they occupied in the West. As long as the French held Verdun the chief lines of communication for the huge aggregation of German troops in northern France had to pass through Belgium, where the hostile army under cover of the forts of Antwerp was a constant menace. The reduction of one or both of these two strongholds must have been regarded by the German General Staff as a vital necessity. To illustrate this statement, the fact may be cited that the failure to surround and isolate Verdun was followed immediately by the attack on the forts of Antwerp.

The Germans were impressed from the first with the importance of the position of Verdun, which holds the railway from Metz to Paris at the northern extremity of the chain of fortresses extending forty miles to Toul. The design of capturing Verdun was part of the larger project of widening the zone of direct communication between Germany and the occupied portion of northern France, a dream which the Germans never abandoned. It was desirable to make their position in France self-sustaining in case a mishap should jeopardize communications through Belgium. The capture of the Verdun-Toul defenses, furthermore, would have secured an advanced base of supplies, one that was admirably situated on the shortest lines from the heart of Germany to the most vital objective points in France, capable of convenient connection with the railway system in Germany, and in German hands almost unassailable.

The principal aim of the German Crown Prince had been to mask, and if possible to capture, Verdun and the fortresses in alignment with it. During the Battle of the Marne the Germans made a desperate effort to pierce this fortified barrier and nearly succeeded. The garrison of Troyon on the east bank of the Meuse, subjected to simultaneous bombardment by the Fifth Army on the west and

From the Quay Van Dyck looking toward the cathedral, Antwerp.

The Bourse at Antwerp.

the Sixth Army on the east, after defending itself with unflinching heroism and repelling many formidable attacks by the German infantry, had finally been reduced to the last extremity when the Germans were compelled to withdraw on September 13th.

The disruption of the fortress barrier would probably have reversed the tide of victory in the Battle of the Marne. The ultimate purpose of the fierce attacks delivered before Nancy at the time of that great battle was probably to open a way by which the Germans could sweep around the southern extremity of this line at Toul, attack the French Third Army in the rear, and completely isolate all the fortresses northward to Verdun.

The prominent position of Verdun at the extremity of a wedge-shaped salient in the French lines was particularly galling to the Germans because it impaired the cohesion of their front. The German positions east and west were linked by only a narrow corridor of French territory, and the Germans scarcely made any progress at this supremely important point after the first few weeks of hostilities. The only railway line affording communication east and west controlled by the Germans in this section of France, between Verdun and the Belgian frontier, connecting Trier, Diedenhofen and Metz with Sedan and the west, passed through a tunnel only seventeen miles north of Verdun, the destruction of which would have gravely embarrassed the German transports. As matters stood, troops moving between the German fronts in Lorraine and northern France were largely conveyed by a long detour through Belgium.

In the middle of September the Sixth German Army, under the Crown Prince of Bavaria, reached from a point on the German front opposite Lunéville to Consenvoye on the Meuse, about eight miles north of Verdun, where it connected with the left wing of the Fifth Army, commanded by the German Crown Prince. The Germans naturally cherished the idea of breaking the line of fortresses and encircling Verdun by the combined action of these two armies.

The Crown Prince of Bavaria launched a movement of his right wing across the Woevre region in the direction of the Meuse and the fortified barrier about September 20th. This maneuver was executed mainly by the Fourteenth Army Corps and the Sixth Bavarian Division.

The Meuse is bordered on the east throughout almost all this part of its course by abrupt, wooded elevations constituting a rather formidable natural obstruction. The forts composing the barrier are situated, now on one side of the river, now on the other, according to the local conditions. The German forces advanced towards the Meuse, in part at least, along the Rupt de Mad, a stream which rises very near Commercy on the Meuse and flows into the Moselle in the vicinity of the German border, passing through a defile which forms a natural thoroughfare from one river to the other across the Woevre.

The wings of the German forces deploying in the Woevre were checked and pressed back by the attack of troops from the two great fortresses, Toul and Verdun; but the center, encountering only weak resistance, pushed forward to the Meuse at St. Mihiel midway between the extremities of the fortress barrier. The resulting contracted situation was preserved for many months in the curiously-pointed salient of the German lines, touching the Meuse with its sharp stiletto tip.

Close to St. Mihiel was the modern fortress, Camp des Romains. After thirty hours of preparatory conflict and bombardment, the 12th brigade of the Sixth Bavarian Division took this fortress by storm in a desperate struggle at close range on the 25th, while the 11th brigade warded off attempts to relieve the garrison. Of the latter 508 were taken prisoners; the rest perished in the fortress, which was reduced to a heap of ruins.

The Germans had now opened a breach in the fortified barrier. But it remained to be seen whether they could make profitable use of their partial success. They crossed the Meuse at St. Mihiel and occupied the suburb on the left bank on the evening of the 25th, and by the morning of the 26th their heavy guns were in position on the bank of the river, making useless any further resistance by the French Territorial troops who opposed their progress.

If they had succeeded in driving home the wedge in this section they might have undermined the French front as far as Reims, besides securing the other advantages already mentioned.

But the conditions had become far less favorable for the direct coöperation of the German Crown Prince than during the Battle of the Marne. At that time his army held a front of about twenty miles from Revigny northeastward to a point not far from Verdun, while detachments, at least, had advanced far enough to take part in the bombardment of Troyon. But the general retirement of the Germans carried the Fifth Army with it, as we have seen. The wedge-like obstruction of the Forest of Argonne separated the German armies as they retreated northward, the Crown Prince passing to the east and the Duke of Württemberg to the west.

This forest, rendered famous by the stubborn contests waged for its possession during many months, is a rocky, densely-wooded plateau, about thirty miles in length from north to south and eight miles in width, lying between the Aisne and its eastern tributary the Aire. About the middle of September the Germans were back at Varennes and

Vienne-la-ville about nine miles apart on opposite sides of the forest, while the French were installing themselves in the forest itself for the purpose of threatening communication between the German armies. Thus the Crown Prince's front had been pushed back to a distance of thirty miles from St. Mihiel and his right flank was threatened from the Forest of Argonne.

Nevertheless, the Germans advancing from St. Mihiel were on the point of debouching into the valley of the Aire, where they could have assailed the rear of the French army confronting the Crown Prince. The French chiefs realized the peril before it was too late. The Twentieth Corps, summoned in haste from Lorraine, marched westward throughout the night of the 25th and most of the 26th, crossing to the left bank of the Meuse at Lérouville. Its cavalry advance-guard came into contact with the enemy late in the afternoon of the 26th, and during the night the Germans were driven back to the Meuse. They retained at least a bridge-head on the left bank and intrenched themselves very carefully around St. Mihiel and the ruined Camp des Romains, a position of little value except for the constant threat which its possession implied.

The failure of the effort to isolate Verdun was evident by September 27th, and the first gun was fired at the defenses of Antwerp on the 28th. It is natural to assume that the continuity was the result of calculation and not of fortuitous circumstances. But this second undertaking, to which the Germans were in a measure driven by the obstacle on the Meuse in France, was rendered even more urgent by the northward movement of the French, whose lines had already been extended across the valley of the Somme. It was indispensable for the Germans to forestall the course of this new maneuver by taking Antwerp.

But before we consider the thrilling progress of the attack on Antwerp we must trace very briefly the fortunes of the Belgian army since we took rather unceremonious leave of it to follow the breathless course of the German armies in their race towards Paris and the heart of France.

At the beginning of the war fifteen year-classes had been called under arms in Belgium. The eight younger classes were enrolled in the field-army; the seven older formed the garrison or fortress troops. The field-army, divided, as we have seen, into six army divisions and one cavalry division, numbered 117,000 men at first, and was afterwards increased by 18,500 recruits. The Third Division was stationed at Liége, the Fourth at Namur. The others were concentrated on the line covering the capital on the east. Later, the Third Division, retiring from Liége, joined the principal mass of the Belgian army drawn up on the line Tirlemont-Jodoigne. The approach of half a million Germans forced the Belgians to abandon their position on August 18th. Thus far only one corps of French troops had arrived on Belgian territory, having reached the line of the Meuse and Sambre. To await the coöperation of the French would have been fatal and therefore the Belgian field-army took refuge within the fortified camp of Antwerp on the 20th. After the evacuation of Namur, 12,000 soldiers of the Fourth Division escaped into France and finally made their way to Antwerp, partly by sea, and joined their comrades in arms.

With the capital and a large part of Belgium in the hands of the Germans, General Field-marshal von der Goltz, famous for his writings and for his reorganization of the Turkish army, was installed as governor-general of the conquered territory and District-president (Regierungspräsident) von Sandt of Aachen was appointed chief of the civil administration, assisted by a council of five members.

After the bulk of the German armies had swept across Belgium and turned into France, an army of observation, composed of the Third and Ninth Reserve Corps and some Landwehr formations under General von Beseler, was posted before Antwerp to cover Brussels and the German communications on a line extending from Wolverthem to Diest.

In the absence of superior German forces the occasion seemed favorable for the Belgian army to strike at the German lines of communication in the general direction of Louvain. Four divisions took part in this operation on August 25th and 26th. But the Germans with their customary foresight and thoroughness had already intrenched their positions and taken the other necessary precautionary measures, so that the Belgians were unable to make any permanent impression on their lines. Finally, the Germans executed counter-attacks on the flank of the Belgians and forced them to withdraw within the line of their defenses.

An act was committed at this time in Belgium which has been the subject of more passionate discussion than any other event of the Great War. The Germans deliberately destroyed an important part of Louvain by fire, a city situated eighteen miles east of Brussels, containing 42,000 inhabitants, famous for its venerable monuments of beautiful architecture and for its other artistic treasures, the seat of a famous Catholic university and headquarters of the Jesuits. A treacherous outbreak of the population, instigated and engineered by the Belgian government with the complicity of the priests, and carefully timed with reference to the sortie from Antwerp on the 25th, was alleged as the motive for this stupendous act of retribution. The people of Louvain, who had presumably been disarmed several days before, are said to have opened fire on the soldiers from the houses in the evening when the garrison had been reduced to a single Landwehr (or Landsturm) battalion in consequence of the conflict with the Belgian forces from Antwerp. The claim was even made that it took the Germans twenty-four hours to quell the insurrection. Desultory conflagrations in Louvain on the 26th were followed by the systematic destruction beginning in the night of the 26th-27th. Most of the inhabitants were driven from their homes. Many persons were summarily executed. A large number of men of military age were transported as prisoners to detention camps in Germany. The Town Hall, 500 years old, an even more beautiful example of the Gothic style than the Town Hall of Brussels itself, was saved through the efforts of the Germans themselves. It served as their headquarters. But the ancient Cloth Market was consumed, and with it, the university library, which it housed, with priceless treasures, fell a prey to the flames, an irreparable loss to humanity.

The destruction of Louvain will doubtless remain notorious as long as the memory of the Great War endures. Rightly or wrongly, the imagination of mankind will associate it with the devastation of the Palatinate by Louis XIV as a conspicuous example of ruthless barbarity.

On September 4th the Germans occupied Termonde, at the confluence of the Dendre with the Scheldt, dispersing the Belgian detachment which guarded it, crossed the Scheldt and menaced the communications of Antwerp with the west. But the First and Sixth Belgian Divisions, which were therefore transferred to the left bank of the Scheldt, drove the Germans back and finally expelled them from Termonde. The Belgians executed an offensive movement from the intrenched camp of Antwerp on September 9-13. They occupied Aerschot on the left and reached Cortenberg on the right, but gained no permanent advantage, although they compelled the Germans to recall some forces which had been sent to France.

An important topographical feature of the neighborhood of Antwerp is the semicircular water-course to the southeast and south of the city formed by the succession of the Rivers Nèthe, Rupel, and Scheldt.

The construction of defenses for Antwerp after plans by the celebrated Brialmont was inaugurated in 1859. The ramparts of the city itself and the detached forts, 2,200 yards apart, forming a ring around the city, about two miles from the ramparts, were regarded at that time as an impregnable system of fortifications. In course of time, however, the gradual development of siege-artillery rendered these defenses inadequate, and the construction of outer forts, designed also by Brialmont, was begun in 1877. The southern ones covered the approaches to the bridgeheads on the Nèthe and Rupel, permitting the garrison of Antwerp to make sorties against an enemy coming from this direction. The recent expansion of the defenses, adopted in 1906, incorporated these outer forts in an exterior ring of modern fortresses and redoubts completely embracing the city, which was not finished until November, 1913. It formed the essential part of the system of defenses confronting the Germans in 1914. The course of this outer girdle lay considerably south of the line of the Nèthe and the Rupel at a distance of eight or ten miles from the city. Mechlin is only about two miles south of Forts Waelhem and Wavre St. Catherine in the outer circle. The general constructive principles of these forts were the same as those of the forts of Liége and Namur.

Upon the approach of the Germans the Belgians destroyed many villages and farms in the zone of fire of their

German soldiers in front of the Town Hall, Antwerp. The people on the left are waiting for permits to pass in and out of the city.

Ruins in the Rue de Peuple, Antwerp, after the bombardment.

forts, sacrificing without hesitation property of great value to the stern requirements of warfare, and flooded the low-lying fields bordering the Rupel.

The German army assembled under General von Beseler for the operations against Antwerp consisted of the Third Reserve Corps, two Ersatz divisions, a marine division, two Landwehr brigades, an artillery brigade, a pioneer brigade, and probably a Bavarian division, numbering in all probably 125,000 to 150,000 men, a somewhat stronger force than the Belgian army, but composed chiefly of troops of the second line. The Ninth Reserve Corps had been sent to France.

Numbers, however, were after all of secondary importance. The Germans are said to have concentrated about 200 guns, and a large number of their pieces far exceeded in range and destructive force the ordnance mounted in the forts. The Germans were evidently too weak to invest Antwerp so as to cut off the communications and eventual retreat of the Belgian army. Their method was to concentrate the fire of their powerful artillery upon a limited section of the fortified girdle, crowning the effect by the furious charges of their infantry. Thus, they counted on forcing their way to the heart of the city.

As a preliminary measure the Germans shelled Mechlin on the 27th compelling the inhabitants to seek safety in Antwerp. The operations against the actual defenses of the latter were inaugurated by the bombardment of Forts Waelhem and Wavre St. Catherine the next day. At first the 21-centimeter mortars were employed; later, heavier pieces were brought into action and the 28-centimeter howitzers and the Austrian 30.5-centimeter mortar-batteries rendered their effective service. At least two of the famous 42-centimeter pieces were probably used.

The Belgian field-army was posted along the threatened front covering the intervals between the outer forts in improvised trenches, faulty and inadequate in construction. The First and Second Divisions held the part of the front corresponding with the course of the Nèthe, the Third and Sixth covered the line of the Rupel, and the Fourth, with its headquarters at Termonde, guarded the line of the Scheldt southwest of Antwerp and thus protected the lines of communication through Belgian territory towards the sea. The Fifth Division acted as a general reserve. The Belgian field-guns were in masked positions between the forts and behind the course of the Nèthe and the Rupel.

Fort Wavre St. Catherine was silenced on the 29th. The concrete fabric and steel cupolas were smashed and the explosion of a magazine completed the demolition. Nevertheless, it seems not to have been occupied by the Germans until five in the afternoon of October 1st.

The Belgian authorities probably realized as early as the 29th that without such reinforcements as they could scarcely expect the defense of Antwerp could only serve the purpose of a delaying action, since the German artillery would eventually crush every obstacle in its path. It was decided to substitute Ostend as the base for the Belgian army, but the removal of the military stores already presented serious difficulties. Railway communication from Antwerp through Belgian territory was reduced to a single line starting from the left bank of the Scheldt and running westward through St. Nicolas. There was still a continuous connection by railway between St. Nicolas and Antwerp over lines in Belgian possession crossing the Scheldt at Tamise and the Rupel near Willebroeck. But the railway bridge at the latter point was within range of the German artillery. Nevertheless, the trains conveying military stores successfully traversed this part of the route by night with lights extinguished from September 29th until October 7th. While this operation was in progress the Belgian cavalry division patrolled the line of the Dendre to prevent an incursion of the Germans and the interruption of traffic further west.

The Belgians were driven back to the Nèthe on October 1st and counter-attacks on the 2d failed to recover their original outer positions. Fort Koningshoyckt, which had been partially destroyed on September 30th, had to be abandoned at 2.30 on October 2d and Fort Lierre, pounded by the heaviest artillery, was evacuated at six. Fort Waelhem was silenced the same day. The bombardment of Fort Kessel was begun at six A.M. on the 3d and by eight-thirty the same morning it was a heap of ruins.

The defenders of Antwerp had now only the 15-centimeter mortars and 12-centimeter cannon in two armored trains, besides the ordinary 7.5-centimeter field-pieces, with which to reply to the powerful siege-artillery of the Germans. There was manifestly only one possible outcome for a struggle under such conditions. The Belgians retired behind the Nèthe on October 2d.

It is said that the Belgian government had decided to leave Antwerp at ten on the morning of the 3d and had made the necessary arrangements, and that the foreign consuls had already embarked on a vessel at five on the afternoon of the 2d, when the plan was abandoned on receipt of the news that British reinforcements were approaching. This assistance, for which the Belgian government had made an urgent appeal, was hopelessly inadequate. A British marine brigade of 2,200 men arrived in Antwerp on the evening of the 2d and relieved a Belgian brigade in the neighborhood of Lierre. Two naval brigades arrived on the 5th. The expedition was commanded by Brigadier-general Paris and was accompanied by no less a personage than the Right Honorable Winston Churchill, First Lord of the Admiralty. The British naval forces were employed because they could be dispatched at shortest notice. They brought with them some naval guns.

The doom of Antwerp was unmistakably sealed. Forts Kessel and Brochem were silenced and the town of Lierre was occupied by the Germans on the 5th. The Germans first gained a footing north of the Nèthe at four on the morning of the 6th. A reflection on the general situation at the time will doubtless convince us that a prompt retirement of the Belgian army had become absolutely indispensable. Since the Belgians were manifestly unable to withstand the attacks of the German forces now concentrated against them, their only salvation was to unite with their allies. But the corridor of unconquered territory stretching around the north and northwest of Belgium formed a very precarious connection with the French and British armies. Not only was it menaced by the repeated attempts of the Germans before Antwerp to force the passage of the Scheldt in the general vicinity of their positions, but the northward progress of the German forces in France threatened to intercept it completely. This was the moment when the Germans seemed about to outflank the French to the north of Arras. They were extending their front to La Bassée and collecting large forces near Lille. Their cavalry was active in the neighborhood of Armentières, and they had occupied Ypres. By throwing a barrier across the intervening space to the North Sea they could sever the territorial connection and isolate the Belgian army.

Even assuming that the Germans were not in force beyond Lille, the distance from Lille to Nieuport on the Yser, the nearest seaport and natural goal for the advance of the German flank, is hardly forty miles. But the distance from Antwerp to Nieuport is more than twice as great, and it was now a matter of life and death for the

Barbed wire entanglements used for defense in the streets of Antwerp.

Method of barricading street in Diest, Belgium.

Belgian army to forestall the Germans in reaching Nieuport and the line of the Yser. Nieuport became the converging point for strategic movements of great significance.

The passage of the Nèthe by the Germans under cover of a heavy fire of artillery showed that the Belgian army had no time to lose. King Albert gave orders for the departure of the main part of the field-army on the night of October 6-7. The garrison troops, Second Belgian Division, and the three British brigades remained within the intrenched camp to continue the defense. The crossing of the Scheldt by means of a bridge of boats at Antwerp was accomplished in good order by the morning of the 7th. The king left at three in the afternoon and accompanied his army on a race to the sea and along the sea to the line of the Yser upon which its very existence depended. On the same day the Germans forced a passage of the Scheldt at Schoonaerde.

At the same time the Fourth Corps of the British army, so far as it was already mobilized, under Lieutenant-general Sir Henry Rawlinson, the Seventh Division and the Third Cavalry Division, disembarked at Ostend and Zeebrugge, October 6-8. Indian and Territorial troops were afterwards to be incorporated in this command. Parts of the Seventh Division and a force of French marines proceeded to Ghent to reinforce the garrison. The possession of Ghent, a very important center of communications flanking the line of retreat from Antwerp, was indispensable for the safety of the Belgian field-army. The total force of 25,000 to 30,000 Allies thus concentrated at Ghent prevented the Germans from penetrating northward to the Dutch border and intercepting the narrow strip of territory which formed the outlet from Antwerp westward. German forces advancing on Ghent were repulsed at Melle on the 9th.

Meanwhile, on the 7th, the Germans installed their heavy guns in positions north of the Nèthe, where they could train them effectively on the inner forts and the city itself. Until the 6th calmness and a hopeful spirit had prevailed among the population of Antwerp. But disquieting reports spread on the 6th and civilians began to leave the city in large numbers. The bombardment of Antwerp began at midnight. The water supply had failed because the reservoir situated just inside Fort Waelhem had been damaged by the enemy's shells. Gas and electricity were likewise cut off. Panic-stricken people rushed to the Central Station and found that no trains were running. Departing tugboats crowded to their utmost capacity made no perceptible impression upon the size of the despairing mass that thronged the river-front. The complicated system of habits, associations, conventions, and intercourse, which forms the basis of society and support for the normal life of the individual had suddenly collapsed. Here was the tragedy of Belgium in all its horror. Every available avenue of escape from Antwerp was crowded with dense columns of refugees, especially on the 8th when the greater part of the population departed. Such a sudden and complete interruption of the normal activity of so large a city and flight of the inhabitants had never been witnessed in modern times. The population which poured forth wherever an egress was open had already been swelled by thousands of homeless refugees from the ruined towns and villages within the range of the military operations. Thousands crossed the Scheldt by ferry, but a far larger number, possibly a quarter of a million, made their way by road to the Dutch frontier. Vehicles of every class and description had been brought into service and loaded with the most necessary or valuable household articles. But the greater number of the fugitives were forced to walk, carrying their burdens as best they could. To peasants and laborers, accustomed to the ruder tasks, the physical exertion presented no unusual hardship; but invalids and persons habituated to a life of ease and refinement suffered untold misery and discomfort. Fortunately the weather was fine, for most of the fugitives had to bivouack in the open. The Dutch had hastily improvised the necessary arrangements for the shelter and nourishment of the pathetic multitude, constrained to become their guests, with admirable efficiency and unlimited, but unpretentious, generosity.

The military authorities in Antwerp set fire to the petroleum tanks on the left bank of the Scheldt so that their contents should not be utilized by the Germans, and the dense black columns of smoke rose all day on the 7th and 8th. But the appearance of Antwerp under bombardment at night was a spectacle of terror unsurpassed in human record. Masses of seething flames rising from the burning oil-tanks illuminated the foreground, making the shadows blacker and more spectral by contrast, and were reflected with a strange, portentous glow in the undulating volume of smoke above. Conflagrations had broken out in different parts of the city and the buildings were silhouetted against the ruddy background of flame. The incessant roar of guns, meteoric shower of fiery projectiles, and bursting of shells completed the frightful impression of a stupendous outbreak of baneful, unearthly forces.

Antwerp, with all her historic buildings and precious possessions; her noble cathedral and its incomparable tower, a marvel of elegant proportions and exquisitely beautiful tracery, delicate as Mechlin lace; her Town Hall and stately Grande Place, and her wonderful collections of art; the city of Rubens and repository of his greatest masterpieces; the seaport rivalling Hamburg, London, and New York, with its miles of granite quays, warehouses, and exceptional harbor facilities; the embodiment of opulence and splendor;—lay as a prostrate victim in passive expectation of her fate.

Antwerp has become renowned for her sieges. After the splendid era of her prosperity under Charles V, she suffered the savage violence of the Spanish soldiery who mutinied from lack of pay in 1576. She was besieged for two years by the Duke of Parma in 1584-1585, and finally, in the nineteenth century, her Dutch governor held out for two years in the citadel after the Belgian revolution until expelled by the intervention of the French in 1832.

The most powerful artillery was probably not employed for the bombardment of the city in 1914, while the larger part of the projectiles used in this final stage of the operations was shrapnel, so that the actual destruction of property in Antwerp itself, while considerable, proved not to be so great as was feared.

At five o'clock on the afternoon of the 8th, when the fall of Antwerp was plainly but a matter of hours, the military governor gave orders for the departure of the Second Division of the Belgian field-army and most of the British troops, who began to cross the Scheldt by the bridge of boats in the evening. But the order failed to reach some of the British in time, so that they did not begin their retreat until the morning of the 9th.

The First Belgian Division had been transported from St. Nicolas to Ostend by rail on the 8th, while the other divisions which left Antwerp on the night of the 6th-7th proceeded in the direction of the Ghent-Terneuzen Canal on foot. The main part of the forces withdrawing from Antwerp passed this canal by the morning of the 9th, after completing what was the most exposed stage of their march, because a considerable Allied force, stationed at

Belgians in flight from Antwerp.

Belgian refugees taken to England on fishing boats.

Ghent, as we have seen, still covered their further retirement westward.

Even those who were traversing the first section of the journey on the 9th used the railway, in part at least. But the Germans were now in force to the west of the Scheldt and they intercepted at St. Nicolas many of those who departed last, forcing them to seek refuge across the Dutch frontier. In all, 1,560 out of the 8,000 British sent to Antwerp and about 20,000 Belgian soldiers were interned in Holland at this time.

Several of the inner forts at Antwerp were taken by the Germans on the morning of the 9th. The bombardment ceased about noon and the Germans entered the city towards evening, but the formal capitulation did not take place until the 10th.

The Germans captured between four and five thousand prisoners in the course of the operations and at the final surrender of the city, and took as booty 500 guns, considerable railway material, including an armored train, many motor-vehicles, about 4,000 tons of wheat, together with supplies of flour, coal, and wool valued at $2,400,000, copper and silver worth about $120,000, and many cattle. They found four British, two Belgian, one French, one Danish, thirty-four German, and two Austrian steamers in the port. The engines of the German vessels had been damaged. The harbor was intact except that the gate of the great sluice had been obstructed by large stones.

An irritating source of distraction to the Germans was removed by the expulsion from Antwerp of the Belgian field-army, which had been prodding them in the rear whenever a critical situation demanded their undivided energy and attention elsewhere.

Besides the great advantage of securing communications in the north, the fall of Antwerp released a large force of men for service in the field. It enabled the Germans to close in on the still unoccupied Belgian territory so as to reduce very greatly the necessary length of their own front and thus to make their position more solid. Historical and sentimental causes combined with the practical advantages to render the capture of Antwerp an exploit which created enormous enthusiasm in Germany, and the manifest futility of the assistance sent by Great Britain whetted the feeling of satisfaction.

But the loss to the Allies, aside from the detriment inherent in advantages won by the Germans, was more of a contingent than positive nature. For the actual benefit which they had derived from Antwerp was slight as compared with the service which the position might eventually have rendered them. Their chagrin must have been greatly assuaged by the successful escape of the Belgian field-army and its junction with the British and French. It was essential that the Belgians should join with their allies in presenting a common front to the enemy, since the most violent struggle in the whole campaign was soon to begin and the united forces of the Allies would be much stronger than equal numbers acting separately. And since, in the actual situation, Antwerp was too remote to be included in a common front, the withdrawal of the Belgian field-army had become a necessary operation.

Besides the German army which had been operating against Antwerp, four reserve corps had been concentrated in Belgium for the proposed offensive movement towards Calais and Boulogne. These corps were now advancing in the direction of the coast.

The Allies evacuated Ghent on the 12th, and the Seventh British Division took up a position covering Ypres in contact with the other divisions of the British army on the 14th. The Third British Cavalry Division referred to above, commanded by Major-general the Hon. Julian Byng, which had been posted at Bruges, was also transferred to Ypres, arriving there on the 14th. The Belgian field-army was proceeding to Nieuport and the Yser by way of Eecloo and Ostend, while the two opposing lines in France were being pushed northward towards the same point as rapidly as possible.

All eyes were suddenly turned upon an obscure corner of Belgium bounded by the sea as the goal towards which great armies were hurrying from all directions, and places heretofore almost unknown abroad soon became forever memorable as the scene of the most sanguinary struggles in the world's most terrible drama.

The Belgian government, which had been transferred from Antwerp to Ostend, was now compelled to accept hospitality on foreign soil. It left Ostend on the 13th and arrived at Havre the same evening, where it was installed as the guest of the French Republic.

There was a frantic rush of fugitives endeavoring to obtain passage from Ostend to England on the 13th. This became a veritable stampede in Ostend on the 14th, when the Germans occupied Bruges, only twelve miles away. The last steamer departed on the same day, leaving a great crowd of terrified refugees in a frenzy of despair. In their desperate efforts to embark at the last moment several persons were pushed into the water and drowned. A Taube dropped a bomb into Ostend when the panic was at its height. German patrols entered Ostend on the 15th and the Third Reserve Corps was quartered in the vicinity on the 16th. Meanwhile, the streams of fugitives were moving along the muddy roads towards the French frontier, drenched by the continual rain, sleeping in the fields, a woful spectacle.

The Belgians prepared to coöperate with their allies by defending the last corner of their national territory. On the 15th they took up a position along the River Yser and the Yser-Ypres Canal from the sea to Boesinghe, a distance of about twenty-three miles, with their right resting on the British. The latter continued the front past Ypres to the vicinity of La Bassée where the French line commenced. Thus for the first time, the Belgians were ranged along the side of the British and French on a common front, and the human rampart was complete from the North Sea to the boundary of Switzerland.

CHAPTER X

The Stemming of the Tide in Flanders
(*October 16-November 11, 1914*)

The situation on October 16, 1914. The revised plan of the Germans. The nature of the battlefield along the Yser. The German forces before the Yser front. Belgians attacked in their outposts and driven back. The German attack on Nieuport, and on Dixmude, which is defended by the "soldiers of Liége," October 19th. Renewed attack, terrible bombardment and burning of the town. Passage of the Yser by the Germans near Tervaete, night of October 21-22; the Germans west of the river. Arrival of the French to reinforce the Belgians, October 23d. Bombardment of the German positions by British warships. Violent renewal of the attack on Dixmude, October 24th. Belgians at the limit of their resources and endurance. The gradually rising inundation. The culminating moment and German retirement, November 2-3. The situation on the British front. Desperate combats in the region of Ypres with repeated attacks of the Germans in dense masses, October 20-23. The contest at Neuve Chapelle. The very critical moment before Ypres on the 31st. Renewal of the battle on November 1st. Storming of Dixmude. Supreme effort of the Prussian Guard to crush the British lines, November 10th and 11th.

The 16th of October, 1914, saw the beginning of a distinctly new stage of the struggle in the West. Then for the first time the Allies presented a solid front on an unbroken line from the North Sea to the frontier of Switzerland. Running northward from the Oise the Allied front passed west of Roye, east of Albert, west of Bapaume, east of Arras, west of Lens and La Bassée, and east of Armentières and Ypres.

The French held the line as far as the Béthune-La Bassée Canal. From there the British, supported by the forces of General d'Urbal, mainly Territorial troops, prolonged the front across the boundary into Belgium. The Second British Corps operated between the French left and the

Lys. The British Third Army Corps and the Cavalry Corps with two French Territorial divisions and a brigade of marines occupied Ypres and the adjacent section of the front. There they were joined by the Seventh British Army Division and the Third Cavalry Division. Thus the British with their French supports established the front as far north as Zonnebeke, while cavalry covered the interval between this point and the right flank of the Belgians. The First British Corps detrained at St. Omer on October 17th, reached the front in the section of Ypres on the 21st and extended the line of the British still further north.

The original position of the Belgians in this period of the contest corresponded, as we have seen, with the line of the Ypres-Yser Canal from Boesinghe northward to its confluence with the River Yser at Fort de Knocke, and then with the course of that river to the North Sea. Dixmude stood at the center of the Belgian position, and the most important part of the Belgian front lay between Dixmude and Nieuport, a seaport on the Yser, about two miles from its mouth.

Again, a very brief recapitulation of the course of events may serve to illuminate the purpose of the German attack and the significance of the sanguinary struggles which the present chapter will describe.

After the Allies had been driven for a time before the irrepressible fury of the initial German dash into France, they recovered their grasp of the situation before it was too late, collected their forces, and faced the enemy in the interior of the country. Whether the Germans overpowered and scattered their opponents or recoiled before the human rampart stretching from Paris to Verdun had been, as we have seen, a matter of deep concern for the whole human race. With dauntless determination the

Allies repelled the tremendous onslaught on the Marne, grasped the initiative, forced the Germans to recede, and tried by enveloping one extremity of their front to make their defeat irremediable. At this moment of greatest opportunity for the French, lack of material preparation and equipment rather than available men probably curtailed the measure of their success. The Germans collected their resources, recovered their assurance, and strove to regain the initiative. Their strength rose higher than ever, and now, like a vast returning tide, they threatened to burst every barrier that obstructed their progress. The campaign was approaching its second crest of highest tension. Flanders was now the storm-center towards which the destructive elements converged from all directions.

In a strictly technical sense the return of the Germans to a violent aggressive action may be regarded as a counter-offensive, since the Allies were still for a time unwilling to relinquish the initiative. The offensive movement started in the region of the Lys and of Ypres on the arrival of the British army was still being pushed after the Teutonic whirlwind had broken with terrible fury on the line of the Yser further north. But the dominating factor in this, as in the earliest part of the campaign, was the passionate resolve of the German chiefs to obtain a decision in the western theater at the earliest possible moment, and all aims and efforts of friend and foe alike were subordinate to the prodigious exertions put forth for the attainment of this single purpose.

The aim of the German offensive at this time has been the subject of much speculation and discussion. The view that Calais was the objective for the renewed attacks was extensively published and eagerly accepted by the Germans. But sound principles of strategy as well as the logic of events demanded that the main object of the German offensive should be the destruction of the Allied armies by the most direct and effective means.

Perhaps the German authorities intentionally obscured the main purpose and ultimate direction of the renewed offensive so as to avoid the appearance of insincerity in their earlier declarations, which announced, or at least implied, the accomplishment of the original plan in the West. They seem to have encouraged the impression that the initial campaign against the French, whose resistance had now been reduced to a practically negligible factor, had passed by a normal transition into a campaign against the British army, now become the principal adversary. The German press apparently responded to the suggestion and the impending march to Calais became the watchword of popular enthusiasm. In the imagination of the German people and in frequent rumors the capture of Calais was naturally associated with the prospect of a terrifying combined attack by sea and air against the British, of a great naval battle, and of a victorious landing on the English coast. Many speculative schemes for the invasion of England appearing at this time animated the spirit of the German people, but served, on the other hand, no doubt, to stimulate recruiting in Great Britain, where they excited rather a feeling of curiosity than consternation.

But it is simply inconceivable that the German leaders, believing themselves to be involved in a life and death struggle and in a war in which time was a factor of the greatest importance, should have employed the larger part of their mobile forces in the West for any enterprise which did not offer the prospect of decisive results. It is regarded, moreover, as an incontestable principle of strategy that the reduction of cities and strongholds has in itself no final effect in modern warfare. Only the destruction of the enemy's field armies is decisive. We must assume,

General Maunoury, commander of the French Sixth Army at the Battle of the Marne.

General Sarrail, commander of the French Third Army at the Battle of the Marne.

General d' Amade, commander of the first French expedition which invaded Upper Alsace.

therefore, that the Germans struck at the northernmost section of the Allied line because they believed that this would be most conducive to the destruction of their adversaries' armies. The German offensive, based upon the well-fortified triangle, Antwerp-Namur-Liége, was facilitated by the dense network of Belgian railways as means of communication. Success in the initial stages would open the way for the renewal of the turning movement on an imposing scale, the crumpling up of the Allies' left wing, and a new dash for Paris.

One important factor in the situation is often overlooked, and that is, that the German blow at this time was first launched against the only section of the Allied front where almost insurmountable obstacles were not to be expected. The initiative had deserted the front between the Oise and the Meuse because a situation had developed there making progress impossible. For experience had shown that suitable trenches were practically impregnable from the front, and the completion of such defenses from the Oise to the Meuse had produced a hopeless deadlock in that quarter. Activity turned to the valley of the Somme and then to the region of Arras, but in each of these the paralyzing tendency soon made its influence felt. It was a contest of defensive against offensive methods, a race of the spade against the gun, in which the latter's original lead was constantly diminishing. Thus in each successive extension of the front mobility was soon followed by stagnation. Hence the Germans delivered their present blow at the most recent section of the front, where the defenders had only just arrived, and where artificial defenses, if prepared at all, would be least effective.

It is not to be assumed, however, that the German leaders were actuated exclusively by a single, supremely significant motive in framing their revised plan of attack. Lesser aims and incentives were doubtless mingled in the considerations that produced the general design. For strategists usually have in view a minimum as well as a maximum objective. Calais, like Paris in the original offensive drive, would serve as a convenient point of convergence for defining the direction of the movements, a sort of topographical peg on which to hang the general scheme of operations. The capture of Calais and the other Channel ports would confer important specific advantages. It would presumably create uneasiness and apprehension in England and hinder thereby the sending of British reinforcements to the continent. It would embarrass, although not completely interrupt, communications between England and the British army in northern France. The harbor of Calais would furnish a convenient base for submarines operating along the English coast, and for mine-layers infesting the Straits of Dover. The most powerful German artillery planted on the French coast would create a zone of safety for the operations of German warships extending almost across the straits at the narrowest point. Even a partial success in a movement southwestward along the coast would reduce the length of the German front from the Meuse to the North Sea and so effect a corresponding saving in the number of troops required to maintain it. Furthermore, the presence of a German army on the English Channel would exercise a tremendous moral effect in neutral countries as well as in the belligerent nations.

The German offensive failed in the extreme northern section and was renewed in the region of Ypres. The struggle in Flanders may be conveniently divided, therefore, into two general phases, which might perhaps be regarded as distinct battles. The first of these, the Battle of the Yser, beginning about October 18th, passed through its culminating stage of violence from the 24th to about the 30th and died away about November 3d. It overlapped the second, the Battle of Ypres, which developed during the last days of October, rose to very great intensity, and subsided rapidly after November 11th. But to speak of battles as distinctive episodes at this stage of the struggle is apt to be misleading, for there were scarcely any intervals in the warlike operations that could set off individual battles. We give our attention almost exclusively to the course of the most violent and determined offensive efforts, neglecting the places where operations were desultory and comparatively aimless. But the fact should not be overlooked that hostilities were practically continuous along the entire front. Only in the sense that in certain sections and at certain times the action rose to a relatively very much higher degree of intensity were there separate battles.

The whole region where the Germans encountered the Belgians is extremely flat, except for the dunes along the coast, and in some parts it is lower than the level of the sea at high tide. The Yser has been canalized and is confined by dykes, and forms a serious obstacle for an attacking army, although it is only about twenty yards broad. The country on both sides is interspersed with streams, canals, and ditches, and frequent rows of willows furnish cover for the movements of troops. Aside from its petty meanderings the Yser forms in its general course from Dixmude to Nieuport the arc of a circle swelling out towards the northeast, with a railway line connecting the two towns as the chord. The embankment of this railway from one to two yards in height forms a second defensive barrier, reinforcing the line of the river.

The total German forces finally concentrated for the offensive in the north were divided into three distinct groups. The army of the Duke of Württemberg nearest the sea comprised the Fourth Ersatz Division, Third, Twenty-second, Twenty-third, Twenty-sixth, and Twenty-seventh Reserve Corps, a division of the Twenty-fourth Reserve Corps, and probably a naval division; the army detachment of General von Fabeck, next in order southwards, is said to have contained the Fifteenth Corps, two Bavarian corps, and three other divisions; and, finally, the army of the Crown Prince of Bavaria included the Fourth, Seventh, Fourteenth, and Nineteenth Corps, the Guard, parts of the Thirteenth Corps, the Eighteenth Reserve Corps, and the First Bavarian Reserve Corps. The formations mentioned were supported by numerous cavalry formations. The forces of the Duke of Württemberg and of General von Fabeck and part of those of the Crown Prince of Bavaria, in all about twelve corps, together with four cavalry corps, operated between the sea and the River Lys.

Forces amounting to seven divisions, namely, the Fourth Ersatz Division, and the Third, Twenty-second, and Twenty-third Reserve Divisions, disposed in the order mentioned on a front extending from the sea towards the southeast, were advancing against the Belgians, who could scarcely muster 60,000 for the defense, including the force of 6,000 French marines under Admiral Ronarc'h, who had retired from Ghent to Dixmude. The Germans, furthermore, concentrated 400 pieces of artillery, many of them of heavy caliber, against the Belgian position between Dixmude and the sea, while the Belgians could reply with only 300 field-pieces and twenty-four mortars saved from Antwerp. The Belgians were fatigued from their long marches and dispirited from their constant succession of losses and the apparent futility of all their exertions. The confident expectation that the Allies would speedily triumph and deliver Belgium from the hands of the enemy had been dissipated. Battered and war-worn, homeless

Cavalrymen asleep on heaps of straw in a French town.

Infantrymen in a trench near Ypres.

and disappointed, they gathered courage for the final effort to defend a pathetic remnant of their national territory. The French commander-in-chief asked them as a supreme performance to hold out for forty-eight hours until considerable reinforcements of the French could come to their relief, and even this, at the time, seemed an almost hopeless request. But in reality the Belgians held out for a week against the German onslaught with the sole assistance of the French marine brigade, and even then bore the brunt of an unequal contest with unsurpassable courage and determination for at least two weeks longer.

However, in spite of the absence of clearly-marked physical features in this part of the country, the ground did offer some distinct advantages for the defensive, as we shall see. The convex curve of the river from Dixmude to Nieuport was a possible element of danger for the Belgians, it is true, since it made the escape of the defenders along the bank more difficult, in case the extremities of the arc were captured by the Germans. Furthermore, a protruding loop-shaped bend in the river midway between the towns just mentioned was a very positive source of peril, because, if held, it exposed the Belgian lines within its fold to cross-fire from the opposite bank, and if evacuated or captured by the enemy, it opened a breach in the defensive line formed by the stream. But the dyke along the left bank of the canalized Yser formed a suitable rampart for the Belgians. Yet the fact that the river is much higher than the adjoining fields, the relative elevation of the dyke being consequently much greater on the land side, made it all the harder to dislodge the Germans when once they had gained a foothold upon it.

Possession of the crossing-points at the different towns and villages along the river was extremely important because the low country, interspersed in large part with canals and dykes, ditches and willow thickets, hedges and fences, was very difficult to cross except by the roads which led to the bridges. The marshy character of the soil near the river, moreover, interfered with the preparation of artificial cover, since trenches very quickly filled with water. But the most effective resource supplied to the defenders by the physical conditions of the region will be described at the point where it first became operative.

The immediate purpose of the Germans was to break the enemy's line at Nieuport and Dixmude, converge on Furnes, the Belgian headquarters, and thus enclose and annihilate the Belgian army. The Belgians occupied a line of advanced positions in the villages east of the Yser. The first shots were exchanged on the 16th, but the German attacks on the Belgian advanced posts did not become serious until the 18th.

The Belgians were driven from their outer line on the 19th, retaining only the positions on the right bank of the Yser near Nieuport, Schoorbakke, and Dixmude. The heavy artillery of the Germans had now arrived from Antwerp and the bombardment of the Belgian positions commenced in earnest.

But the appearance off the coast of a British flotilla under Rear-admiral Hood, which took part in the action, reminded the Germans quite forcibly that without command of the sea their lateral movement for outflanking the Allies had reached an impassable limit. This squadron included three monitors which had been built for the Brazilian government and were intended for river operations. While mounting 6-inch guns their shallow draught permitted them to approach so close to the shore that the seamen employed even their small-arms with effect. The larger vessels maneuvered at a much greater distance. Aircraft directed the fire of the British, which inflicted severe losses on the Germans who were attacking in the direction of Nieuport in the vicinity of the coast. Nevertheless, the latter pressed on, and after thirteen hours of uninterrupted exertion took Lombaertzyde, the defenders retiring to positions already prepared a little further back.

Dixmude was defended by the French marines under Admiral Ronarc'h, who was in local command, the 11th and 12th Belgian regiments of the line under Colonel Meiser as chief-of-brigade, forming part of the Third Division, known as the "soldiers of Liége," and a regiment of artillery containing twelve batteries of field-pieces.

The Germans charged in dense masses about three on the afternoon of the 19th, joining in a hand-to-hand struggle with their antagonists, but retired at nightfall. Dixmude, set on fire by incendiary shells in the evening, burned for several days, so that between the conflagration and the converging fire of powerful artillery, the situation of the garrison, harassed day and night by the attacks of greatly superior forces, became almost unendurable.

Rain produced a comparative lull on the 20th; but the 21st was a very critical day for the Belgians and their allies. The Germans, whose concentration in the north was now complete, assailed their opponents all along the front from La Bassée to the North Sea. They hurled themselves upon the trenches protecting Dixmude in eight distinct attacks, charging in dense masses sixteen or twenty rows in depth. One after another the formidable gray-green waves rushed forward, urged on by an unshakable determination to succeed at any cost, impelled by a veritable frenzy of self-immolating patriotic devotion, crossing the deadly zone swept by the fire of rifles and machine-guns right up to the wire entanglements or even to the foot of the Belgian intrenchments, only to waver and stagger back, with ranks thinned and torn by the awful streams of lead.

The townhall in Dixmude, which had been converted into a temporary hospital, was threatened with immediate destruction by fire and the enemy's projectiles. With self-sacrificing courage the members of the Red Cross removed the wounded while shells were crashing all about and transported them to the base hospital at Furnes.

A correspondent of the London *Telegraph*, who visited Dixmude while the contest was raging on the 21st, has given a vivid account of his impressions, and a few passages from his account will serve to illustrate the terrible violence of the struggle.

"No pen could do justice to the grandeur and horror of the scene. As far as the eye could reach nothing could be seen but burning villages and bursting shells. . . .

"Arrived at the firing line, a terrible scene presented itself. The shell fire from the German batteries was so terrific that Belgian soldiers and French marines were continually being blown out of their dugouts and sent scattering to cover. . . .

"Dixmude was the objective of the German attack, and shells were bursting all over it, crashing among the roofs and blowing whole streets to pieces. From a distance of three miles we could hear them crashing down, but the town itself was invisible, except for the flames and the smoke and clouds rising above it. . . .

"The battle redoubled in fury, and by seven o'clock in the evening Dixmude was a furnace, presenting a scene of terrible grandeur. The horizon was red with burning homes."

At sundown the Germans crossed the Yser south of Dixmude, but were confronted by machine-guns and driven back. On this day the length of the Belgian front was contracted to about twelve and a half miles, the southern extremity being withdrawn to St. Jacques-Cappelle

Field Marshal Horatio Herbert Kitchener, Earl of Khartoum, English secretary of state for war.

about two miles south of Dixmude, the interval evacuated by the Belgians being filled with French forces.

On the night of the 21st-22d the Germans captured a bridge near Tervaete in the darkness and poured into the loop formed by the river, bringing so many machine-guns to this position that the Belgians were unable to dislodge them in repeated attacks on the 22d. The presence of the Germans at this point on the left bank of the river compelled the Belgians to evacuate Schoorbakke on the right bank on the morning of the 23d, since the trenches there were exposed to enfilading fire.

The situation was becoming hourly more critical for the Belgians. A new and very much more serious difficulty was now added to their other preoccupations. The intensity of action required of the Belgian field-guns to compensate for the tremendous superiority of the German artillery had made many of the pieces unserviceable and reduced the supply of ammunition to less than 100 rounds for each gun. Everywhere assailed day and night by superior numbers, with scarcely any available reserves, the Belgians still held the essential points in their defensive line with the feverish grasp of shipwrecked sailors clinging desperately to a wave-swept raft.

The situation had been one of such uniform gloom for the Belgians, stubborn resistance ending always in retreat, unrelieved by any cheerful circumstances, their hopes had been so often deceived, that they probably heard the joyful rumor that reached the front on the evening of the 22d with an instinctive feeling of incredulity. The Forty-second French Division, commanded by General Grossetti, had been transferred from Reims to Belgium by rail and late in the afternoon of the 22d the advance-guard marched in review before King Albert and General Joffre in the market-place of Furnes. For the first time the Allies were coming in force to fight by the side of the Belgians, and the latter, who may have been embittered at times by the thought that their terrible losses and hardship had been a gratuitous sacrifice for others who ignored and deserted them, again took heart. The comradeship in arms of the western nations was a visible reality. And yet, although the effect of the French reinforcements in stiffening the resistance of the Belgian army was unmistakable, the situation became even more critical and the climax was reached several days after their arrival.

The French marched to Nieuport on the 23d to relieve the Second Belgian Division, which was to be brought back into reserve for partial recuperation. The French forces crossed the canal bridges under a shower of German projectiles, traversed Nieuport, drove the Germans from Lombaertzyde, and attacked Westende.

M. Émile Vandervelde, Chairman of the International Socialist Bureau, who was appointed Belgian Minister of State by royal decree on August 4th, so that all parties should be represented in the government, witnessed the operations in the vicinity of Nieuport on the 23d. He relates that while he was standing only thirty feet from Belgian field-guns in action he could hear nothing but the thunder of the guns of the British squadron, probably 12-inch pieces, then two miles away.

The Germans continued their bombardment of Nieuport on the 24th and French heavy artillery which had been brought to the Belgian front replied from behind the town. At the same time the Germans attacked with violence all along the front. The Belgians gave way at St. Georges where the Germans captured a crossing-point on the river. On the morning of the 24th one of the brigades of the Forty-second Division came to the relief of the hard-pressed Belgian troops who were struggling against serious odds to confine the Germans at the loop in the river near Tervaete. During the day, however, the Allied forces in this part of the line were forced back to the Beverdijk, a canal running midway between the Yser and the railway.

The Duke of Württemberg was now making a supreme attempt to capture Dixmude. Fourteen furious attacks were repulsed by the Belgians and the French marines during the night of the 23d-24th. The effort was renewed during the day. The trenches guarding the bridge-head were lost for a time, but afterwards recovered. In some places the foremost opposing trenches were no more than fifty feet apart at this time. It meant unremitting tension, practically uninterrupted physical exertion, for those who occupied them.

On the 25th the French resumed unavailingly their attack on Westende. The Germans had now mounted heavy artillery along the dunes, which kept the British fleet at a distance.

By the 26th the German attack against the center of the Nieuport-Dixmude line had advanced so far that the Belgian General Staff even withdrew temporarily from Furnes. The Allies were being forced back to the railway embankment, the last line of defense in this region.

As early as the 25th the Belgian chiefs decided to resort to an inundation as a final expedient for arresting the forces which threatened to overwhelm them. Most of the ground which was being so desperately contested lay below the level of the sea at high-tide. Usually the sluices at Nieuport were opened at low-tide for drainage and closed at high-tide to prevent an overflow into the canals which terminated there. By reversing this process and closing the culverts and other openings in the railway embankment, the area between the railway and the river could be gradually flooded. The process would be materially aided at this particular time by the unusual rainfall which swelled the streams supplying the canals.

It was natural that the Flemish in their present extremity should appeal to the sea for assistance. The Low Countries have been the scene of struggles in behalf of liberty which are scarcely less celebrated than Marathon, Thermopylae, and Salamis, and every schoolboy has heard of the heroism of the burghers of Alkmaar and Leyden who foiled the arrogant Spanish invaders and preserved their independence by the heroic measure of opening the dykes and flooding the country. But this final analogy was hardly necessary to make men's minds revert to the days of the Duke of Alva and the "Spanish Terror."

Charles-Louis Kogge, the aged sluice-master, afterwards created Knight of the Order of Leopold as reward for his service rendered at this time, Captains Thys and Ulmo of the engineers, and about twelve others toiled several nights in succession under cover of the darkness, often between the lines, raising and lowering the gates according as the tide rose or fell, closing the outlets from the area to be flooded, and executing the other necessary preparatory measures. Their efforts seem to have escaped the attention or suspicion of a vigilant enemy.

The Germans gave no heed to a slight, but ominous, rise of water in their trenches on the 27th. Later, the realization of what was going on intensified the fury of their attacks and fortified the determination of their commander to forestall the impending obstruction. They drove the French from Lombaertzyde on the 28th.

During the night of the 29th-30th in a violent storm of wind and rain, the Germans drove the Allies from a section of the railway embankment and captured Ramscappelle on the southwest side of it. This brought the conflict to its climax and final stage. The Germans had penetrated the last barrier. Nieuport and Dixmude might at any moment become untenable. The Belgians might be dispersed or swept into the sea. Their fate, now tottering on the brink of calamity, involved immediate peril for the whole left wing of the Allies, since the destruction of the Belgian army would open the way for the envelopment of the Anglo-French positions. Motor-vehicles of every sort lined the highways of France for miles speeding northward with thousands of reserves. The incessant reverberation of the British naval guns and the German heavy artillery replying from the shore shook the coast of Belgium and was even heard in Holland. But the real decision was at hand in the furious contest for the possession of Ramscappelle with the advance of the flood as the dominating factor. It was a race of time and tide against the progressive effect of superiority in forces. The waters were rising behind the Germans, converting an area from two to three miles wide and nearly ten miles long into a shallow lagoon.

By the morning of the 31st the cannonading of the Allies made Ramscappelle untenable and the Germans advanced to the west of it. Then the French and Belgians charged them with impetuous violence, convinced that the fateful moment had arrived. The Germans wavered before the furious onslaught and were forced back. By nine o'clock they lost Ramscappelle after a desperate engagement in the streets. An hour later the Allies were over the railway embankment. One more service was demanded of the overworked Belgian "seventy-fives." They were mounted on the embankment to riddle the Germans now floundering in the slimy pool. The water was only three or four feet deep, but this was enough to make operations practically impossible. Many of the Germans were drowned in the canals, the course of which had been hidden by the inundation.

On November 2d the Germans withdrew hastily beyond the Yser, abandoning wounded, cannon, and stores of munitions. Eventually they succeeded in capturing Dixmude, as will be related, but it was too late for them to derive any distinct advantage from its possession, since they were unable to debouch from it towards the west.

It must not be forgotten that operations in the plain of the Lys and around Ypres not only accompanied, but preceded, those on the Yser. Convenience and clearness require that the two series should be treated separately, although their intimate relationship makes the division an arbitrary one. Precedence has been given to the treatment of events in the northern section because the action on the Yser had almost ceased several days before the struggle around Ypres reached its final climax.

The operations between La Bassée and the Yser, to the beginnings of which allusion has already been made in Chapter IX, grew steadily to the intensity of a great battle, while the balance shifted from side to side following the alternate accessions of strength to the opposite forces. The thoughtful observer must be impressed by the nervous delicacy of the general equilibrium maintained throughout this period by the rigid application of all the elements of strength as rapidly as they became available, the utmost tension of every resource, and the feverish employment of every moment of time. The slightest acceleration or delay in the arrival of the successive masses of troops on either side might have altered profoundly the relationship of the essential factors and destroyed the general balance of the contending forces.

As late as October 11th definite lines of demarcation between the combatants had not been drawn beyond

La Bassée. But the Second British Corps already encountered serious opposition in its advance towards Lille in the region between La Bassée and the Lys after its detrainment at Béthune on that day.

The Third British Army Corps and British Cavalry Corps coöperating with the Eighty-seventh and Eighty-ninth French Territorial Divisions and French cavalry further north encountered only slight resistance. As we have observed, they dispersed the German detachments and occupied Ypres on the 13th, and were joined the next day by the Seventh British Army Division and Third British Cavalry Division, which had marched from Ghent and Bruges *via* Roulers. The turning of the tide in this section scarcely set in before the 19th. An Allied army of normal size might have turned the German flank without difficulty and thrust itself forward like a wedge between the German bases in Belgium and the armies in northern France. The inability of Great Britain to throw an additional army into a field of operations not eighty miles from the British coast at this exceedingly opportune occasion remains a most striking illustration of the country's lack of preparedness. On the other hand, an acceleration of four or five days in the arrival of the main German forces must have produced alarming results for the Allies.

After the capture of Lille, German forces pushed northwestward, from French on to Belgian soil, as rapidly as possible, while the German army released by the fall of Antwerp was pursuing the Belgian field-army southwestward along the coast. By the 19th the Twenty-sixth and Twenty-seventh German Reserve Corps were pushing westward from Courtrai. These, with the Twenty-second and Twenty-third Reserve Corps a little further north, filled the gap in the German front between the forces which had followed the Belgian army from Antwerp and those which were moving northward from France.

The formation of this combination, and the consequent German offensive, inaugurated the Battle of Ypres in the more restricted sense. It shifted the balance at once to the side of the Germans. The First British Corps, which reached the front on the 21st, was unable to make headway against the German attacks. The curving form of the front on the western sections in France complicated the movement of the Allied forces from the line east of the Oise to the new field of action. But French reinforcements were nevertheless hurrying northward. The arrival of the Forty-second French Division at Nieuport on the 23d has already been mentioned. The Ninth French Corps came into line east of Ypres on the 24th. The initial vehemence of the German offensive spent itself without accomplishing significant results and a period of diminished tension followed from the 24th to the 28th.

Then General von Fabeck's army detachment, containing many first line troops from other sections of the front, came as support for the wearied reserve corps. The Germans resumed their attacks with far greater vigor around Ypres, as along the Yser, on the 28th. The contest raged with increasing intensity and the German offensive made noteworthy progress. But a stream of reinforcements was also augmenting the strength of the Allies.

The first Indian contingent of the British forces, consisting of two divisions, commanded by Lieutenant-general Sir James Willcocks reached France in September. They were greeted with indescribable enthusiasm upon arrival at Marseilles. As the first occasion that an Indian expeditionary force had set foot in Europe, this was an event of incalculable potential importance, at the same time an impressive demonstration of the loyalty of Britain's imperial

Indian troops watering mules at canvas troughs.

Indian troops of the British forces in France.

possession and another token of the reality and unbounded significance of the Franco-British alliance.

The Indian troops made a long sojourn at Orléans, since a month was required for their acclimatization and the necessary adaptation of their equipment. Finally, the Lahore Division reached its concentration area behind the British Second Corps on October 19th and 20th and the Meerut Division arrived in the same vicinity shortly afterwards. The Sixteenth French Corps took its place in the battle-line south of Ypres on the 31st. The German attacks waned again about November 3d and there was another period of comparative calm while the Germans effected a new adjustment of forces for the final effort. The renewed offensive reached its climax on the 11th in the sensational charge of the Prussian Guards, followed quickly by its collapse and failure.

Thus the action as a whole was marked by three distinct points of culminating intensity occurring on October 23d and 31st and on November 11th, respectively.

It is now important to return for a more detailed examination of the operations. About October 17th the Second British Corps supported by Conneau's French cavalry corps extended from Givenchy northeastward to Radinghem. The front of the Third, passing east of Armentières and terminating at the Bois de Ploegsteert near Le Gheir, was broken by the course of the Lys at Frelinghien. The British Cavalry Corps, now dismounted, stretched from the Bois de Ploegsteert to Zandvoorde. The Seventh British Army Division, soon to bear the brunt of the German assaults, extended from Zandvoorde to Zonnebeke. From there, the British Third Cavalry Division, four French cavalry divisions under General de Mitry and the two French Territorial divisions carried the front northward as far as the right wing of the Belgian army.

From the region of Arras to the plain on the north side of the Lys the army of the Crown Prince of Bavaria confronted the Tenth French Army under General Maud'huy and the British. The Fourteenth German Corps, a division of the Seventh, a brigade of the Third, seven Jaeger battalions, and four cavalry divisions faced the Second British Corps with its cavalry supports, while the Nineteenth Corps, a division of the Seventh, and three or four cavalry divisions faced the Third British Corps.

The Second British Corps advanced in the teeth of stubborn resistance, fighting its way from house to house in the villages along the ridge northeast of La Bassée.

The Third British Corps under Lieutenant-general Pulteney captured Capinghem, lying between two of the isolated forts of Lille, less than three miles from the city.

But the Allies met with a determined counter-offensive on the 19th, which arrested their progress. The Second British Corps suffered a serious reverse on the 20th east of Neuve Chapelle. Fierce encounters continued during the 21st and 22d and the British were driven from most of their positions on the ridge between the La Bassée-Lille Canal and the plain of the Lys.

North of the Lys the Allies expected to continue without interruption the forward movement which had brought them into possession of Ypres. Their immediate purpose was to occupy the main line of communication between Lille and Ostend. With this in view, the Seventh British Division advanced in the direction of Menin on the Lys, but suffered a repulse. Meanwhile, on the 18th, De Mitry's cavalry took Roulers, situated on the railway and highway connecting Lille with Ostend and Bruges at the point where they cross an important route leading to Dixmude.

The Germans began their counter-offensive in this region also on the 19th, concurrently with attacks on the

Yser in the neighborhood of Nieuport, and drove the French from Roulers and from the vicinity of the Lille-Ostend railway. This opened the road to Dixmude, where the first serious attacks were made by the Germans on the 21st. The presence of the four new German Reserve corps, to which reference has already been made, seems to have been a surprise to the Allies. Far from recognizing the strength of the German forces, Sir John French intended that the First Corps upon its arrival should push northeastward by way of Roulers to Bruges and thus cut off the German forces which had been following the Belgian army along the coast.

The position of the Allies in front of Ypres took the form of a triangle, with the general line of the Yser-Ypres and Ypres-Comines Canals as base and the apex at Westroosebeke pointed towards Roulers. On the 20th the Germans struck this apex and shattered the triangle as far down as Passchendaele and Poelcappelle. The British First Corps, which was passing through Ypres on the same day, was ordered to recapture these two localities. A violent encounter with the Twenty-sixth German Reserve Corps ensued on the 21st. The British repulsed the German attacks, but halted on the line Zonnebeke-Bixschoote in consequence of the retirement of the French on the left, whose position made them more sensitive to the fierce attacks which were being delivered by the Germans at this time on the line of the Yser.

The French cavalry abandoned their attempt to occupy the Houthulst Forest east of the Yser-Ypres Canal, an especially advantageous position for the Germans, because it covered the highway from Roulers to Dixmude and afforded convenient concealment for the emplacement of heavy artillery and for the concentration of their forces.

On the same day the Twenty-seventh Reserve Corps attacked the Seventh British Division between Zonnebeke and Zandvoorde with great violence, but without effect, and further south the Nineteenth German Corps took Le Gheir. This was a crucial point because it lay in the line of advance to a region of wooded eminences south-west of Ypres terminating towards the west in the Mont-des-Cats, which commanded the plain north and south. The occupation of these hills by the Germans would have severed the British front, threatened the forces around Ypres in the rear, and probably compelled the Allies to abandon all their positions between La Bassée and the North Sea. Naturally, throughout the entire course of the operations in this quarter, the Germans never lost sight of this objective. But the British retook Le Gheir the same day by bringing troops from across the Lys.

On the 21st and 22d the Third British Cavalry Division was shifted to the section of the front between Hollebeke and Zandvoorde, permitting the Cavalry Corps to restrict to this extent the inordinate extent of its front.

The British Cavalry Corps, commanded by General Allenby, was the principal guardian of the approaches to these supremely important heights, because it held the front from Le Gheir to the Ypres-Comines Canal, and fortunately for the Allies the soldiers of this corps displayed no less energy and skill on foot than when mounted. Scarcely 4,000 in number, stretched out to a perilous degree of attenuation across a space of more than six miles, they offered day after day a determined resistance to the attacks of very much superior forces. On the 22d they received the sorely-needed reinforcement of nearly a brigade of the Lahore Division of the Indian contingent.

On the same day the German attacks against the now semicircular projection in the Allied lines before Ypres increased in violence. The Germans had brought their heavy howitzers within range and opened a destructive fire on the British positions. In the evening they forced a breach in the line of the First Corps near Langemarck, but a brigade held in reserve filled the gap before morning. In this period of their offensive the German attacks reached their greatest intensity on the 23d. In many places the British were blown out of their trenches or buried alive in them by the explosion of powerful shells. The Seventh Division, covering the unusual space of eight miles, was in a very critical situation. Depleted in number by the losses already sustained in the desperate contest, it had no longer a single man in reserve to support the firing line, which threatened to give way at any moment. For seven days the men had been almost incessantly engaged and the strain had become unendurable. Reinforcements from the First Corps reached them just in time to prevent a disaster.

The Germans charged repeatedly in dense masses on the 23d, singing *Die Wacht am Rhein.* They rushed forward each time as if to submerge the British lines completely, only to recoil, broken and torn, before the fire of rifles and machine-guns operated with deadly accuracy at close range. Unbelievable courage,—courage that advances in the face of machine-guns or endures unflinchingly the risk of sudden dismemberment in a shower of high-explosive shells,—has become a commonplace phenomenon in the Great War, while the most astonishing feats of daring have lost much of their effect by frequent repetition.

From the very first days of the campaign the imagination of the world has been made to shudder at the practice of the German commanders in urgent situations of hurling their forces at the enemy in repeated, headlong attacks, in compact masses, regardless of the dreadful effect of the fire from the opposing intrenchments, and this method has been characterized as reckless prodigality, as proof of ruthless indifference for the lives of the common soldiers. But to assume that actions of such fundamental importance were not guided by an earnest consideration of all the circumstances is to misinterpret the whole spirit of German conduct. This method of attack was undoubtedly the consequence of a thoughtful computation of probable advantage and loss.

Only the thorough discipline and organization of the German army made it possible to carry out attacks in very close formation in spite of the appallingly rapid depletion of the ranks under concentrated fire, and German tacticians were undoubtedly convinced that the swifter attainment of the object of the attack would compensate for the terrible exposure and make the total wastage less. Their theory had a reasonable basis. For such attacks can be launched much more quickly, since deployment from the marching column to a narrow, compact battle-formation requires far less time than the transition to a far-flung line in open order, and blows delivered in dense masses, having more weight, would presumably lead to a more rapid decision.

Nevertheless, these fearful operations were frequently unsuccessful, the heavy losses having been incurred in vain. The Germans, however, assert that their enemies likewise squandered the lives of their own men in precisely this fruitless fashion. Each side, it would seem, strives to impute to the other the reproach of inadaptability to changed conditions, a peculiarly ignominious defect in the eyes of the present age.

The British field-artillery was operated on the 23d with feverish energy. A single battery consumed 800 rounds of ammunition. The German howitzers exacted heavy toll from the British trenches. But the German losses were undoubtedly severe, and the exhaustion of the new reserve corps after these days of terrible strain must have been very great.

The reserve corps which attacked the British positions were composed of the Landwehr, Ersatz Reserve, and volunteers, with probably a small nucleus of troops of the first line. The prisoners taken were either below the regular military age or else belonged to the older classes, often men thirty-nine or forty years of age, fathers of families. This circumstance strengthened the groundless opinion that Germany was already approaching the limit of her available recruits. These troops were all newly and faultlessly equipped. But one is apt to suspect that instinctive contempt for the British had again led the Germans to commit a fatal error by sending comparatively raw formations against professional soldiers. Although it is clear in this terrible ordeal that the British excelled their opponents in endurance, the Germans waited until the last to bring up their most stalwart contingents. They counted, no doubt, on wearing down the fortitude of the British by repeated attacks in relays, and the recent formations would serve as well as the well-seasoned troops of the first line for the early stages of such a process.

The 23d, as we have seen, was a turning point in the struggle in Flanders. On that day the Forty-second French Division reached the Belgian front at Nieuport, and in the course of the following night a division of the Ninth French Corps took over a section of the trenches held by the First British Corps northeast of Ypres.

But the situation was constantly becoming more difficult for the British. The Second Corps had been almost pierced at the center, the Third had been dangerously pressed back, and the exhaustion of the Seventh Division was hourly increasing.

On the 25th the Indian contingent relieved the greater part of the Second Corps, the Lahore Division taking over the section of the front which included Neuve Chapelle. Two days later Sir John French made a readjustment of the dispositions further north. He suppressed the Fourth Corps, merging the Seventh Army Division and the Third Cavalry Division with the First Army Corps under the command of Lieutenant-general Sir Douglas Haig, and sending Lieutenant-general Sir Henry Rawlinson back to England to supervise the mobilization of the Eighth Division. In consequence of the arrival of the Ninth French Corps, and of the changes in the disposition of the British forces, the front of the Seventh Division was now reduced to the space between Zandvoorde and the road from Ypres to Menin, with the First and Second Divisions in successive alignment with it towards the north as far as Zonnebeke.

The British were now ranged along a front about thirty miles in length, with the Indians and a part of the Second Corps in the south, followed in order by the Third Corps, the Cavalry Corps, and the First Corps, including its recent accessions, as far as Zonnebeke. This arrangement did not undergo any fundamental change for many months.

The first encounter between Indian and European troops took place on the 28th. The result of this event was watched with the keenest interest. For, although the Indians were fearless in their native warfare, the absolutely strange conditions of the European struggle, the hours of anxious waiting in the trenches, the deadly showers of shrapnel and shells of terribly destructive force, might have bewildered and unnerved them. After a series of furious attacks the Germans had taken Neuve Chapelle on the evening of the 27th. To relieve a very threatening situation it was necessary for the British to retake this position at once, and the 7th brigade and about an equal number of

Indians were assigned to the undertaking. Together they stormed the village under heavy cannonading and expelled the Germans at the point of the bayonet; the Indian soldiers acquitted themselves with gallantry and distinction.

On the 29th, *Minenwerfer*, or trench mortars, were first employed by the Germans in the plain of the Lys. These inconspicuous, but formidable, engines, placed at the bottom of pits, hurled bombs with a bursting charge of 150 to 200 pounds at a high angle and close range into the enemy's trenches. The system of trench-fighting now commonly employed in the western theater had created an unprecedented situation and new problems. The regular types of artillery were unserviceable for contests at close range between the troops in the foremost opposing trenches, and with the lines drawn so close, sometimes approaching to within fifty feet, the shelling of the most advanced positions on either side by the artillery of the other, placed at a suitable distance, would have been destructive to friend and foe alike. Now for the first time a weapon had been devised with special reference to these revolutionized conditions.

The Seventh Division repelled a spirited attack by troops of the German Twenty-fourth Reserve Corps at the extremity of a salient in the British front near Kruiseik on the 27th. In the evening Prince Maurice of Battenberg, youngest son of Princess Henry of Battenberg and brother of the Queen of Spain, was mortally wounded during a surprise attack by the Germans. He was lieutenant in the King's Royal Rifle Corps and only twenty-three years of age.

At this stage of the battle General von Fabeck's army detachment came into action east of Ypres. The actual composition of this body remains somewhat doubtful. According to *The French Official Review of the First Six*

Months of the War it consisted of the Fifteenth Corps, two Bavarian Corps, and three (unspecified) divisions. There is other evidence for the presence of at least parts of the Thirteenth Army Corps, of an Eighteenth Corps (probably the Reserve), and of the Twenty-fourth Reserve Corps. The French troops transferred to the region of Ypres noticed that some of the Germans taken prisoners at this time belonged to the regiments from which they had already taken prisoners in Alsace during August and near Reims in September. These must have belonged to the Fifteenth Corps.

The supreme importance ascribed to the contest around Ypres in the last days of October is shown by the presence near the battlefield, not only of Generals Foch and Joffre, but of President Poincaré and the Kaiser.

It was announced on the 25th that the Prussian Minister of War, General von Falkenhayn, had assumed the duties of the Chief of Staff in consequence of the illness of General von Moltke. Later it appeared that the latter's illness was the same conventional kind of indisposition which had probably afflicted General von Hausen, when he was superseded in the command of the Third Army by General von Einem soon after the Battle of the Marne. Von Moltke quietly disappeared and General von Falkenhayn received definite appointment in December to the post of which he was already exercising the functions. It was rumored that the retirement of von Moltke was due to fundamental differences of opinion as to the proper aim for the German offensive, the former Chief of Staff insisting against the view of the Kaiser that the main strategic objective should be the crushing of the French lines at Verdun.

The Kaiser came to Thielt and Courtrai in Belgium and every moral incentive was employed to inflame the ardor of the German troops. Consistently with the supposed change in the object of the campaign, the instinctive Anglophobia, a baffling compound of envy, suspicion, and disappointment, flamed forth with increased violence at this time in Germany. Many observers noticed the comparative absence of aversion to France as contrasted with the intense bitterness displayed against England. France was regarded with a feeling of compassion, or of respect for her chivalrous, though misguided, devotion to an ideal. But Great Britain was looked upon as the invidious, malignant antagonist, who, unable to meet her principal commercial rival in honorable competition, had secretly organized the war for the purpose of ruining Germany, treacherously concealing her own nefarious conduct until the decisive moment had been reached. The British Empire stretched its unwieldy mass and impudent pretensions across the path of Germany's legitimate aspirations on every side. The irritating assurance with which the British continued the normal course of their affairs, "doing business as usual,"—"during alterations in the map of Europe," as Mr. Asquith added,—while Germany's trade was profoundly disturbed and her splendid merchant marine was rusting in the docks, exasperated the Germans. But they were filled with contempt at the spectacle of a government advertising for recruits, stooping to coax and conciliate the different classes of its own people, and dependent upon the favor of the labor unions for the maintenance of production in the most vital branches of industry; and at the prospect of an empire which claimed authority over more than a fifth of the earth's surface and nearly a fourth of its inhabitants pretending to oppose the Fatherland's organized might with a smaller force than was deemed suitable for Belgium, and trusting in an assumed superiority of individual resourcefulness to "muddle through" the supreme trial of strength and efficiency.

All influences conspired to fill the German soldiers with a fury of resentment, and to impress upon them the conviction that, for the present, one issue alone was paramount and that England was the "one and only foe."

The spirit in which the Germans threw themselves upon the British trenches may be conceived from the following order of the day by the Crown Prince of Bavaria to his soldiers, made public on October 28th.

"Soldiers of the Sixth Army! We have now the good luck to have also the English opposite us on our front, troops of that race whose envy was at work for years to surround us with a ring of foes and to throttle us. That race we have to thank especially for this war. Therefore, when now the order is given to attack this foe, practise retribution for their hostile treachery and for the many heavy sacrifices. Show them that the Germans are not so easily to be wiped out of history. Prove it to them with German blows of a special kind. Here is the opponent who chiefly blocks a restoration of peace. Up and at him!"

Another communication, emanating from the same source on November 11th, revealed a similar sentiment in the following expressions:

"Soldiers! The eyes of the whole world are upon you. It is now imperative that in the battle with our most hated foe we shall not grow numb, and that we shall at last break his arrogance. Already he is growing pliable. Numerous officers and men have surrendered voluntarily, but the great decisive blow is still to be struck. Therefore you must persevere to the end. The enemy must be downed; you must not let him loose from your teeth. We must, will, and shall conquer."

At the same time copies of Ernst Lissauer's famous *Hymn of Hate against England*, which first appeared on

FIELD MARSHAL

SIR JOHN DENTON PINKSTONE FRENCH

Commander of the British expeditionary forces.

From the painting by John St. Helier Lander.

September 1st, were distributed as an order of the day among the soldiers of the Sixth Army.

On the 30th, the day on which the Germans took Ramscappelle in the north, the British trenches were subjected to a bombardment fiercer than any thus far experienced. In places the lines were temporarily broken, and along the Ypres-Comines Canal the Germans advanced to within three miles of Ypres, which was being shelled. The fighting on the 30th and 31st was favorable to the French from Zonnebeke to the Ypres-Yser Canal, but the struggle in which the British were engaged reached a crisis on the 31st in which the safety of all the Allied forces was involved. The Germans directed a tremendous effort against the entire British front. Terrible cannonading preceded each attack of the infantry. General von Deimling directed the assault on Gheluvelt, which was quickly reduced to a mass of bloodstained ruins. Several British regiments were practically annihilated in the trenches in this section. The bombardment of Ypres had now commenced in earnest and the terrified inhabitants were fleeing westward. The resistance of the First Corps seemed on the point of collapsing under the formidable pressure of overwhelming numbers. The troops had been driven from the trenches and were being steadily forced backward.

Then, by an almost miraculous revulsion of spirit, the British rallied in the woods behind the lost trenches. Impelled by a spontaneous outburst the 2d Worcestershire Regiment and the 42d brigade of the Royal Field Artillery drove the Germans from Gheluvelt at the point of the bayonet in a sensational charge, and by ten P. M. practically all the British positions had been regained.

In the meantime, west of the canal, the British Cavalry Corps, supported by the Indian brigade, four battalions of the Second Corps, and the London Scottish Territorial

Battalion, together with the small part of the Third Corps north of the Lys, were struggling desperately to arrest the onslaught of two German army corps which had captured Messines and Wytschaete on the very edge of the hilly district which was so vitally important.

The London Scottish Territorial Battalion, the first complete unit of the British Territorial Army to take its place in the battle-line by the side of the regulars, intrenched itself in this quarter on the 31st and repelled several frontal attacks during the night. Finally, under cover of the darkness, the Germans got into position on its flanks. When daybreak showed that a prompt retirement alone could preserve the battalion from capture or annihilation, it retreated in good order across an area devoid of any cover, swept by the galling cross-fire of German machine-guns.

At this critical moment a part of the Sixteenth French Corps arrived and relieved the pressure on the British Cavalry Corps.

But, in general, the German offensive continued on November 1st and 2d with unabated fury. The British were again forced to evacuate Gheluvelt. Their lines were pressed back towards Ypres, which was riddled with shells. However, when the lull came on the 3d, they had lost no vital position. A portion of the First Corps was now relieved by eleven battalions of the Second, and the Seventh Division, which had exhibited such conspicuous proof of indefatigable courage, received a well-merited respite for a few days. From the 4,000 men in the 21st brigade of this division, who had landed at Zeebrugge about a month before, there remained only 750 effectives, while the officers had been reduced from 120 to 8. The 22d brigade could muster only four officers and 700 privates, and the second battalion of Royal Scots Fusiliers had been reduced in the same time from 1,000 to 70. These examples will typify the appalling losses suffered by the troops which bore the brunt of the terrific German attacks.

A momentary recrudescence of German activity on the Yser was probably timed with reference to the final blow which was being prepared against Ypres. A bombardment of Dixmude, commencing at two A. M. on November 10th, more terrific than any yet experienced at that point, served as prelude for an infantry attack of overwhelming force. No fewer than 40,000 Germans rushed upon the Belgian brigade and the French marines about eleven in the morning, drove them from three successive lines of trenches back into the ruined town, where a hand-to-hand encounter in the streets terminated with the expulsion of the Allies. By five in the afternoon Admiral Ronarc'h had withdrawn all his forces to the west side of the river.

The prestige gained by the capture of a place so stubbornly contested palliated the subsequent failure of the German offensive before Ypres. But the possession of Dixmude afforded very little actual advantage to the Germans, since they were not able to debouch from it towards the west.

The Germans prepared for their final blow in the direction of Ypres by transferring two divisions of the Guards from the front near Arras to the neighborhood of Gheluvelt. These picked troops were chosen for the most difficult operations. It will be recalled that they were engaged in the fierce combats in the streets of Charleroi at the commencement of the German dash for Paris, that they were roughly handled in the brilliant counter-offensive of the French between the Oise and the Somme on August 29th, and, finally, that they bore the brunt of the furious attempt to drive a wedge through the French center in the Battle of the Marne and were decimated while stubbornly resisting in the marshes of St. Gond when the tide turned on September 9th. It was reported that they were reviewed by the Kaiser and received his personal exhortation before marching to deal the finishing blow at the British.

At daybreak on the 11th the German batteries near Gheluvelt raked the British positions with a hurricane of shrapnel and high-explosive shells. Suddenly, fifteen battalions of the Prussian Guard emerged from a curtain of fog, rushing forward with imposing momentum. Sheets of rifle and machine-gun fire flashing from the British trenches mowed down their foremost rows. Shrapnel and shell tore bloody gaps in the surging mass. But the effect was lost in the rapidity and wild enthusiasm of the charge. They closed ranks, pressed on with irresistible vehemence, and swept over the British trenches in three places, penetrating for some distance into a belt of woods that lay between the British lines and Ypres. But, as is usually the case when one side breaks through the front of the other, they were assailed on three sides and enfiladed by machine-guns. The vehemence of the general attack was lost in a welter of small groups struggling desperately at close quarters in the forest.

Gradually the British rallied, stood their ground, and recovered most of their lost trenches. The last great effort of the Germans failed and the Battle of Ypres could be regarded as finished.

War's savage lust for destruction had nowhere raged with more remorseless fury than in this devoted tract of Flanders. The peaceful region of Ypres, smiling with comfort and contentment had become an inferno of slaughter and ruin. The seductive charm and picturesqueness of this old-world city had been rudely defiled. Its venerable relics of the life and art of an age long past had been destroyed, an irretrievable loss to humanity.

Prominent among them was the imposing Cloth Hall. Its stately façade, composed of splendid rows of arcades, wherein strength and elegance, dignity and beauty, were intimately linked, was surmounted by a majestic belfry of noble, commanding proportions. This sumptuous old building, the pride and glory of Ypres, was shattered by projectiles, devastated by fire, and reduced to a melancholy, empty shell.

Throughout the fighting zone the villages had been burned, the soil had been mauled and lacerated by repeated showers of explosive shells, and the forests had been shorn and scarred.

Dixmude had probably suffered a more intense bombardment than any other town. Every house had been perforated, every street had been torn and pitted with shells. The plain from Dixmude to the sea, once densely populated, cheerful, and prosperous had become a dreary spectacle of water-logged desolation. The toil of many centuries had been obliterated in a smaller number of days.

The movement of the campaign from the Marne to Flanders had brought the battle-lines almost 150 miles nearer London, and the greater proximity of the struggle had undoubtedly affected the attitude of the British people. In August and September they had regarded the war as an event of absorbing interest, but yet with a certain feeling of detachment. Now it had gained the position of a fundamental fact of the national life.

To those, however, who regarded the future with impatient solicitude, the British democracy was exasperatingly slow to realize the imminence of peril and to bestir itself to a degree commensurate with the emergency. Like the passengers on a steamship, whom the thin plate of steel alone separates from eternity, the people of England pursued the uninterrupted tenor of their callings and pastimes, relying with stolid confidence on the protection of the narrow Channel.

To the Germans such insensibility was unmistakable evidence of the fatal torpor into which the British nation was sinking. But the perception of both contestants was probably obscured by the hazy medium of prejudice and preconception created by the sharp contrast of their traditions and mental habits.

The inhuman character of warfare is partially obscured in the Battle of the Marne by the brilliant generalship, rapidity of movement, and palpable results; but in the sanguinary struggles in Flanders, it reveals itself to our imagination in all its undisguised reality of loathsome cruelty, wallowing in apparently useless carnage, unrelieved by the dash of spirited maneuvers or the glory of definite achievements. The contest raged incessantly for nearly a month, with alternate assaults and counter-charges carried out at appalling sacrifices. But in spite of the "satisfactory progress" persistently heralded on both sides, the lines remained at the close practically where they had been at the beginning. With such an outcome, each party disclaimed the offensive. For the offensive would imply a purpose which failed, and failure carries with it the discredit of defeat. But in its larger aspect the conflict may unquestionably be regarded as the stemming of the fierce German tide in Flanders. Dashing repeatedly, but in vain, against the dogged resistance of the Allies, the Germans were frustrated in their final great endeavor to retrieve the disappointment on the Marne and consummate the campaign as originally planned.

The fact that the British bore the chief brunt of the onslaught suggests a reflection of very general significance.

The two peoples who now stood forth as the most conspicuous antagonists had sprung from a common stock

But the course of their development had diverged very sharply from the time that the forefathers of the English left the ancestral home by the Elbe and the Weser.

Protected by their insular position from foreign interference, the development of the English people was a consistent evolution from the free customs of the ancient German wilderness. The greatest political experiment in history was here worked out and England conferred upon humanity the inestimable gift of representative institutions, her parliament becoming the mother of parliaments. The passionate instinct for personal liberty, here unsuppressed, devised the constitutional guarantees and safeguards which have become a model and an emblem wherever freedom is held in repute.

But the development of the Germans who remained in the homeland, denied the privilege of immunity from interference, was dominated by intercourse with other peoples, by the vicissitudes of subjugation and conquest, by political disunion, and by the introduction of doctrines of Roman law, creating an atmosphere in which the old-time freedom could not survive.

Elements of freedom and authority are inevitably present in every normal community, but to every independent people comes a time for determining which principle is to be paramount. England's choice was made when her king laid his consecrated head upon the block. Germany came to the same parting of the ways two centuries later. She rejected the liberal doctrines of the hour, fortified her organization, and exalted the power and authority of the state.

Without pressing the contrast too far we may say that in England the rights of the individual, in Germany the prerogatives of the collective authority, are the fundamental elements of political feeling and doctrines. And thus, without intending it, Great Britain and Germany appeared as the respective champions for two principles of supreme significance; the first for liberty, a noble, uplifting ideal, but unfortunately often powerless to deal with injustice, inefficiency, and wastefulness; the second for intelligent authority, guiding and correlating the various activities of men for the common good, but disfigured often by arrogance, intolerance, and inflexibility. An ultimate victory for either power would not create a universal empire, but a triumph for Germany would inevitably win the world for compulsory efficiency.

The invasion of America would follow; not the physical invasion which battleships and coast defenses might possibly repel; but the subtle, imponderable, furtive invasion of ideas and methods, which eludes the vigilance of the strictest guard, and which conquers by the very means which are devised to combat it.

CHAPTER XI

The Campaigns in Poland and Serbia

Poland and the war. General situation in the eastern theater about October 1st. Operations on the East Prussian frontier. Von Hindenburg and the Teutonic general offensive. The culmination of this great movement: the failure to take Warsaw, the conflict before Ivangorod, and retreat of the Austro-German armies. The fluctuating course of the struggle in Galicia; Peremysl relieved and reinvested. The German retreat in Poland, lightning change of front, and counter-offensive in November. The combats around Lodz. The increasing deadlock. Operations in Serbia. Failure of the first Austrian invasion; Battle of the Jadar, September 15-20. The renewed invasion in November and the occupation and loss of Belgrade and severe defeat of the Austrians in December. Atrocities in the Austro-Serbian hostilities. Reflections on the results of the Eastern operations.

We turn from the West at the culminating point of the Kaiser's persistent effort to blast a way to the Channel ports, postponing for the moment consideration of the slowly diminishing current of events that followed, and complete our picture of the crucial weeks just passed by tracing the progress of operations in the eastern theater, which developed to a climax of magnitude and intensity that will illuminate the frantic determination to force a decision in Flanders at any cost. The conflict in the West had been prolonged with undiminished energy until, instead of dealing with their antagonists in turn, as the General Staff had expected, the Germans were confronted by formidable forces on both their fronts, and the indispensable condition for the accomplishment of their original design had vanished. The second act in the great world-drama had begun before the actors of the first had evacuated the stage.

In spite of the exposed situation of Poland, only the margin of the country had been disturbed by the violence of war during the first two months of the struggle. But at length the inevitable march of events made this central region between the principal masses of the contestants the theater of the most furious encounters. Vast armies crossed and recrossed the plain of western Poland in every direction, despoiling and destroying, and this unhappy country, the passive victim of her own location, with little heart in the struggle, was devastated to an extent and a degree of thoroughness which were absolutely without parallel.

The fundamental strategic importance of Poland, the value of her resources, and the passionate longing of her people for the restoration of their unity and rights of nationality made the favor of the Poles the object of the most ardent and unwonted courtship by all parties at the commencement of the contest. Teutonic aviators flew over the country scattering broadcast an invitation for the Poles to unite with the German and Austrian armies, which were coming as their deliverers, for the restoration of the ancient unity and autonomy of Poland under a Catholic, German prince. The appointment of a Polish archbishop at Posen under the Kingdom of Prussia at just this time was probably not foreign to the German propaganda for the support of the Poles.

On August 15th, the Grand-duke Nicholas, as commander-in-chief of the Russian armies, issued the following celebrated proclamation:

"Poles! The hour has struck in which the sacred dream of your fathers and forefathers may find fulfilment.

"A century and a half ago the living flesh of Poland was torn asunder, but her soul did not die. She lived in hope that there would come an hour for the resurrection of the Polish nation and for a brotherly reconciliation with Russia.

"The Russian Army now brings you the joyful tidings of this reconciliation. May the boundaries be annihilated which cut the Polish nation into parts! May that nation reunite into one body under the scepter of the Russian Emperor. Under this scepter Poland shall be reborn, free in faith, in language, in self-government.

"One thing only Russia expects of you: equal consideration for the rights of those nationalities to which history has linked you.

"With open heart, with hand fraternally outstretched, Russia steps forward to meet you. She believes that the sword has not rusted which at Grünwald struck down the enemy.

"From the shores of the Pacific to the North Sea the Russian forces are on the march. The dawn of a new life is breaking for you.

"May there shine, resplendent above that dawn, the sign of the Cross, symbol of the Passion and resurrection of nations!"

The calculating spirit that prompted these effusions must have been plain to all but the most simple.

But the war had created an especially deplorable situation for Poland in the fact that at least a million Polish young men were subject to compulsory military service under the three belligerent powers among which the former national territory had been divided and were liable, therefore, to be dragged into fratricidal slaughter. Mobilization destroyed in a moment of time the chance and the will for national concerted action.

The Poles were joined with the Austrians in a common faith, and with the Russians by racial affinity; but with the Germans they had no moral bond of association. The Austrian government had treated its Polish subjects mildly and considerately, allowing them local self-government, and Polish universities and schools, and they were generally contented with the existing situation. The Russian government had repeatedly broken the promises made to its Polish subjects. It had persistently withheld local self-government and repressed their nationality. But Prussia's relentless policy of crushing the language and traditions of the Poles and of driving them from their land, although a conspicuous failure, had made her the object of the most intense execration.

The concentration of all the Poles under a single government was unquestionably preferable to the existing partition which paralyzed all combined action, and the Russian Poles were convinced that the surest way to obtain this partial success was through a Russian victory. The economic development in Poland has tended to strengthen the ties uniting the country with Russia. Allusion has already been made to the increasing industrial importance of Poland, which is largely due to her coal supplies and the natural intelligence of her laboring class. The annual product of the factories and workshops amounted to about $500,000,000 in value, and two-thirds of this was absorbed by the markets of the Russian Empire. The rise of Socialism, concurrently with the growth of an industrial proletariat, created purposes which ran athwart the aspirations of nationalism and tended to identify the interests of the laboring classes in Poland and Russia.

The idea of conciliating Poland had never become entirely extinct in Russia, and although the Poles had regarded the Panslavistic program with suspicion, the grand-duke's appeal to the sentiment of racial community and brotherhood made at this moment of expectancy and exaltation was received by a large part of the nation with sympathetic approbation. Noteworthy was the the support of prominent Polish authors, artists, and musicians, among them

Crowds in the streets of Lodz, a city of five hundred thousand inhabitants, awaiting the entry of the German troops.

Officers outside the headquarters of General von Mackensen at Lodz.

the famous Sienkiewicz, who even exhorted his fellow Austrian Poles to adopt the cause of Russia.

In view of all these facts it is not surprising that on August 16th the leaders of the several political parties assembled at Warsaw responded to the archduke's proclamation in the following resolution:

"The representatives of the undersigned political parties, assembled in Warsaw on the 16th of August, 1914, welcome the proclamation issued to the Poles by his Imperial Highness, the Commander-in-chief of the Russian forces, as an act of the foremost historical importance, and implicitly believe that upon the termination of the war the promises uttered in that proclamation will be formally fulfilled, that the dreams of their fathers and forefathers will be realized, that Poland's body, torn asunder a century and a half ago, will once again be made whole, that the frontiers severing the Polish nation will vanish.

"The blood of Poland's sons, shed in united combat against the Germans, will serve equally as a sacrifice, offered upon the altar of her resurrection."

(*Signed by*)

THE DEMOCRATIC NATIONAL PARTY.
THE POLISH PROGRESSIVE PARTY.
THE REALIST PARTY.
THE POLISH PROGRESSIVE UNION.

The loyalty of the people of Russian Poland was an essential factor in the general situation, at least as long as the Russian armies maintained themselves on Polish soil. The Russian government itself was doubtless surprised at the apparent unanimity of sentiment, and a corresponding wave of enthusiasm found practical expression at Moscow and other large Russian cities in contributions for the war-sufferers in Poland. It is doubtful to what extent the

Teutonic powers had counted on the active coöperation of the Russian Poles in a military sense. But they doubtless expected that the attitude of the population would be a substantial support for their own operations and a constant source of embarrassment for those of the Russians; hence the indifference to their own benevolent professions must have been a bitter disappointment.

The concentration of the Russian forces was carried out, as we have seen, on the borders of East Prussia and of Galicia and in the eastern half of Poland under cover of the Vistula and of the great fortresses. As none of the Russian army corps had been stationed west of the Vistula, the projection of the Polish salient was practically reduced by about one-half. For Western Poland was only partially covered by feeble detachments, largely cavalry, and was therefore exposed to invasion. This situation may have been regarded as an acknowledgment of weakness, but events will show that it served the purpose of a snare.

The Russians had launched their attacks against East Prussia and Galicia concurrently and with startling rapidity. It was evidently their purpose to clear the positions on the flanks of Poland before undertaking a decisive invasion of Germany and Austria. While the incursion into East Prussia ended in an inglorious failure, the sweeping success of the Russian operations in Galicia profoundly disturbed the plans and expectations of the central empires.

But the course of events towards the end of September tended to modify somewhat this striking disparity in the fortunes of the campaign in these two opposite regions, as will be explained.

Von Hindenburg's attempt to prolong his brilliant course of victories by penetrating eastward into Russia seems to have terminated in a disappointing failure, as has already been mentioned. For Rennenkampf's forces, shattered by the early disasters, retired behind the Niemen under cover of the great fortresses, Kovno and Grodno, for recuperation, and later took the field again reinforced and refreshed. On September 26th the Germans attacked on a broad front along the Niemen so as to engage the attention of the Russians while they forced the passage of the river at a particular point. But after a bitter conflict continuing four days they were forced to retire to a line running through Augustof. The effective coöperation of the extensive German forces had probably been hindered or paralyzed by the swamps and tracts of almost impenetrable forest. The course of the ensuing operations remains obscure. The German communications offer us the same paradoxical succession of victories on a constantly receding front which marked the early stages of the fighting in August: the total defeat of the left flank of the Russian army with the loss of 3,000 prisoners near Augustof on October 1-2, the repulse of the chief mass of the Russian army against East Prussia with the loss of 2,700 prisoners near Suwalki on the 5th, the checking of a Russian column near Lyck on the 8th, and the discomfiture of a Russian turning movement in the north by way of Schirvindt with the loss of 4,000 prisoners on the 11th. At the same time the Germans announced that the great victory in the vicinity of Augustof and Suwalki proclaimed by the Russians was purely fictitious.

It is clear, however, that von Hindenburg abandoned the siege of the fortress of Osovietz at the crossing of the Bobr and left the principal field of hostilities east of the Prussian border in the enemy's possession. The most that we can concede to the German pretensions is that the Russian achievements were only Pyrrhic victories. Convinced of the uselessness of pushing the campaign in this quarter, von Hindenburg moved his front back to the eastern line of the Mazurian Lakes and the Angerapp, where strong intrenchments had been prepared, and withdrew most of the troops of the first line for the new offensive movement in another quarter, replacing them with Landwehr and Landsturm contingents.

Meanwhile, the movement of the Russians westward through Galicia spent its initial force about September 22d. In fact, the Russian commander was probably removing some of the troops from the Galician front as the prelude for a renewal of the offensive in the north.

But at the end of September the eastern theater presented a singular appearance with the bulk of the Germans in the extreme northeast, the Austro-Hungarian armies concentrated in the extreme southwest, with the chief masses of the Russians grouped before these two, and the long intervening space but feebly defended on either side. This was in striking contrast with the situation in the West, where the contestants, filling all the spaces along the front, struggled with the utmost vehemence for every available position of advantage.

Here was a condition of affairs that favored the development of the warfare of maneuvers on an imposing scale, with rapid movement, dashing offensive and sudden counter-offensive. The general staffs of the central empires were convinced that a vast offensive movement in the East, before the termination of decisive operations in the West, and therefore contrary to the original intention, had become an indispensable measure at this time.

The Russian invasion of Galicia and menace to Cracow would not alone account for such a radical departure. The advance of the Russians in Galicia had, temporarily at least, been arrested. In seeking the more general motives we may safely assume that the most urgent features of the situation were the wide dispersion of the Teutonic forces and the excessive length of the eastern frontier of the central empires to be defended in view of the actual and ever increasing superiority of the Russian forces. There is reason to believe that with most of Galicia in their hands, the Russians were preparing to push their offensive in East Prussia with redoubled energy, and were already redistributing their forces with this purpose in view. In these conditions it was perilous for the Teutonic armies to await passively the Russian blow.

Once the great strategic line of the Vistula with all its crossing points were in their power, the Russians could utilize their superior forces for a sweeping offensive restrained no longer by concern for the safety of either flank. The Austro-German staffs decided to forestall such a dangerous emergency once and for all by a sudden offensive from Posen, Silesia and the region of Cracow eastward across Poland and Galicia to the central course of the Vistula and the line of the San. In this way a straight line would be substituted for the awkward concave section of their front, a formidable natural barrier would be interposed between themselves and their antagonist, and economy and compactness in the disposition of their forces for defensive purposes would be obtained. These were the minimum advantages which the new project was expected to secure. There was in addition the possibility of entrapping the enemy and inflicting upon him a disastrous defeat.

The intervening region of western Poland which it was necessary to traverse is an undulating plain, interspersed with forests, where there are few good roads but much marshy ground in rainy weather. Winter was expected to create solid tracks for an advance of the Russian hosts across this region, and the Germans were determined to deprive their opponents of such an opportunity for the offensive by securing for themselves the line of the Vistula before the ground had become frozen. It was an enterprise that involved a serious hazard and demanded the utmost rapidity of execution, because a period of rainy weather might at any time entangle the German armies in an impassable quagmire.

Von Hindenburg left the frontier of East Prussia and assumed chief command of all the German armies operating in the East on September 25th. His strategic conceptions and the views of the German General Staff doubtless controlled the operations of the Austrian armies also during the period of strenuous efforts which followed. Von Hindenburg proceeded to carry out the new offensive with characteristic energy and unhesitating determination.

The forces of the Teutonic empires, now for the first time ranged shoulder to shoulder for common action, were drawn up on a general line that diverged slightly towards the southeast from the Silesian frontier near Kalisz, passed near Czenstochowa, Cracow, and Neu Sandez, and finally coincided with the barrier of the Carpathians. The field of action of the German forces or group of armies in the north, possibly composed of as many as twelve army corps or about 500,000 men, was the region between the lower Vistula and the Pilica, and that of the mixed group of Germans and Austrians in the center, which was nearly as large, was the territory between the Pilica and the upper Vistula, while the principal Austro-Hungarian armies still operated within the natural confines of Galicia. The Russians had probably about 3,000,000 men under arms at this time, while the Austro-German forces on the eastern frontier seem to have amounted altogether to about 2,000,000.

The Teutonic offensive developed slightly more rapidly in the south, where the Russian forces which had penetrated the Carpathian passes were compelled to retire from Hungary. It was probably intended that the Austrians should force the passage of the San as soon as possible so as to assail the forces defending the line of the Vistula in Poland on the left flank or even in the rear while the German and the Austro-German forces attacked them on the front.

This maneuver involved the relief of Peremysl which was closely invested by the Russians. It is said that the besieging army consisted of five army corps, but this number may have included bodies of troops which protected the attacking operations against interference from the outside. This besieging force may be compared in size with the German army of 125,000 to 150,000 which captured Antwerp. The Russians completely surrounded and isolated Peremysl, but the Germans, with their superior artillery, confined their attacks to a restricted section of the fortified perimeter of Antwerp, where their sledge-hammer blows crushed every obstacle.

Threatened in their siege-operations against Peremysl by the advance of the Austrians, the Russians intensified their efforts to take the fortress by storm. Furious attacks in the night of October 8-9 failed to accomplish their purpose, and on the 9th the Russians were repelled on the south front with heavy losses. By this time the approach of the Austrian army compelled the Russians to relinquish their positions on the west of Peremysl and communication was established between the relieving force and the garrison. Gradually the Russians were forced to withdraw from the north and south, and finally, on the 11th and 12th, they were overpowered and driven from the east side as well. The departure revealed to the Austrians gruesome evidence of the reckless determination with which the Russians had pushed their assaults and the deadly efficiency of the defenders' weapons. Apparently whole battalions lay in contorted masses as death had suddenly overwhelmed them. Shallow excavations with shovels scattered among the bodies of the dead showed where the Russians had been enveloped in a curtain of fire while intrenching themselves. Hundreds had sought shelter under cover of a step-like scarp separating the two planes in which the glacis of the girdle fortresses of Peremysl had been erected, only to be mowed down by the lateral fire of hidden machine-guns disposed in such a way as to sweep this deceptive zone of safety.

The Russian forces withdrew behind the San and were concentrated along the right bank of that river and of the Vistula below their confluence. The German and Austro-German forces advancing eastward from the border of Silesia had to march nearly 200 miles across western Poland and then attack armies of uncertain strength. It was impossible for the Germans to determine accurately the rapidity at which the Russian forces were being redistributed to meet the altered situation. It was to be assumed that the lack of direct communication by railway between the Russian front in Galicia and that in Poland would fatally impede this necessary movement. But in this as in other situations the Russian commanders showed themselves capable of shifting immense forces, in spite of primitive conditions, with a skill which their opponents had not anticipated.

The German and Austro-German armies advanced with remarkable expedition. On October 8th the Germans occupied Lodz, the great industrial center of western Poland, a city which had grown with phenomenal rapidity from 32,600 people in 1860 to 450,604 in 1912, thanks to the fostering protection of the imperial Russian fiscal policy. Its textile industry, in which it stands foremost in Poland, with 650 plants producing goods to the value of about $75,000,000 annually, was mainly in the hands of Germans

General W. A. Sukhomlinoff,
Russian Minister of War.

General Vaivode Putnik,
Chief of Staff of the Serbian Army.

Field-marshal Alexander, Ritter
von Krobatkin, Austrian Minister of War.

who, by establishing themselves at Lodz, competed in the vast market of the Russian Empire without tariff restrictions.

By the 11th the German army on the left wing had reached the Bzura, a tributary of the Vistula, the eastern boundary of Prussia between the second and third Partitions of Poland, less than forty miles west of Warsaw. No serious opposition had thus far been encountered and about seven German army corps were now advancing against the Polish capital and the adjacent section of the Vistula, which was still inadequately guarded. Further south the mixed Austro-German forces advanced through Kielce and Radom and were approaching the section of the Vistula in which the fortress of Ivangorod is situated.

The struggle before Warsaw began on the 11th. The city itself passed through days of the most intense anxiety, of violently fluctuating impressions and emotions, of throbbing excitement. For a time its evacuation by the Russians seemed inevitable. Most of the foreign and Russian residents departed and the funds of the national bank were transferred to a place of greater security in the interior of Russia.

The tragic history of Warsaw still in the making, its majestic situation commanding the beautiful Vistula, the vivacious temperament of its inhabitants, and its contrasts of prodigality and poverty, splendor and squalor, gaiety and gloom, are well-suited to captivate the imagination. A central position where many important lines of communication converge and the nearness of the productive mineral region of southeastern Poland have made it the seat of an extensive and varied industry, with manufactures of iron and steel predominating, and have contributed to the remarkable growth of its population from 161,008 in 1860 to 872,478 in 1911,—Warsaw was a prize worthy of persistent effort.

Thousands of fugitives from the villages west of Warsaw fleeing in terror before the swiftly moving storm of invasion poured into the city, filling halls, warehouses, and all other available places of shelter. There were the aged and infirm and many who had abandoned everything in their hasty departure and were absolutely destitute and in the direst misery.

Day by day the cannonading became louder as the Germans forced their way towards the city, until the buildings vibrated with the tremendous roar. By the 16th they were only seven miles away and on the 17th German shells exploded within the municipal limits. Hostile aëroplanes made their daily visits dropping bombs upon the city, by which a number of civilians were killed or injured.

The first reinforcements for the hard-pressed, greatly-outnumbered forces defending Warsaw were a body of Siberian troops who detrained on the 18th in Praga, the suburb on the right bank, and marched across the Vistula by the imposing Alexander Bridge. The passage of these Russian troops through the streets of the Polish metropolis was greeted with an ecstasy of enthusiasm which might have befitted a triumphal procession, such a popular demonstration as would have seemed incredible only a few months before. It has been said that the fall of Warsaw had been scheduled for this very day, following the capture of Antwerp on the 9th, as double token of Germany's invincible expansive energy east and west. But from this time additional troops were continually arriving in this section until by the 21st the Germans were in full retreat and Warsaw was saved.

The decisive factor had been the advance of strong Russian masses from the vicinity of Gora on the Vistula south of Warsaw and especially from the great fortress of Novo Georgievsk at the confluence of the Bug and the Vistula below Warsaw, which threatened to envelop the wings of the northern German army. By the 20th the Russians attacking from the north had rolled back the German left as far as Sochaczef, subjecting the position of the whole army to a dangerous compression. The final episode in the fighting near Warsaw was a desperate struggle at Blonie, in which the Seventeenth and Twentieth German Corps, covering the retreat of the rest of the army, bore the brunt of furious attacks, but without arresting the Russian advance.

The Russian counter-offensive was being pushed simultaneously on a front of 200 miles. The German forces in front of Warsaw were the first to be compelled to retreat before it, but the other Teutonic army groups had to give way one by one in the order of their succession southwards.

Seven army corps, of which two were German, under General Dankl, had advanced against the Ivangorod section of the Vistula line. Several attempts to cross the river below Ivangorod on pontoon bridges were frustrated by the Russian artillery on the right bank.

Finally, the Russians themselves took the initiative, crossed the river below Ivangorod, traversed with unexpected resolution a treacherous, marshy zone along the left margin of the stream, and assailed the Austro-German positions on the higher ground beyond. There followed a week of terrific struggles in an extensive forest lying west of the Vistula, in which the larger organized masses on each side were dissolved into a confusion of minor groups often fighting at close-range and with the utmost ferocity and desperation. After they had finally been driven from the forest, the Austro-German troops had to retire across open country exposed to the Russian artillery, where they suffered heavy losses. Their retreat was continued by way of Radom and Kielce.

In Galicia the Austrians under command of General von Auffenberg and the Archduke Joseph Ferdinand had forced the Russians to abandon the entire left bank of the San. They had recovered Czernowitz in the extreme southeast, occupied Sandomierz in the north, retaken Jaroslaw, and raised the siege of Peremysl. The relieving column marched into Peremysl amid great rejoicing. The lifting of the blockade, which turned out to be only temporary, was an opportunity for replenishing the supply of provisions and for removing a large part of the non-combatant population, whose presence was only a burden.

The Austrians made many fruitless attempts to cross the San, and their offensive in Galicia waned about October 22d in response to the unfortunate turn of events in Poland. Sandomierz was retaken by the Russians, November 3d, on the 5th the Austrians were compelled to retreat from the San after a series of fierce conflicts, and the Russians resumed the siege of Peremysl on the 14th.

The Germans in Poland made futile attempts to check the victorious advance of the Russians at Skierniewice and Lovicz. The Austro-Germans engaged in a desperate rear-guard action near Kielce, in which they were defeated with heavy losses after a struggle of twenty-four hours.

In spite of the genius of von Hindenburg and the superior organization of the Germans the armies of the central powers were compelled to relinquish all the ground which they had seized. Their offensive had undoubtedly prevented or postponed the renewal of the Russian attacks against Cracow and East Prussia, but these were merely negative advantages. The great struggles during October along the Vistula resemble in some respects the Battle of the Marne. In both cases the Germans must have arrived on the field of battle with their energy impaired by forced

The Wawel, or Citadel, of Cracow. The mound in the center is a memorial to the Polish hero Thadeusz Kósciuszko.

The museum of Belgrade after the city had been bombarded by the Austrians.

marches. The difficulties of transportation must have increased enormously with the advance of the German armies in Poland, where the means of communication were very inadequate.

The offensive had been ventured in defiance of Napoleon's favorite maxim, never to do what the enemy wants. The Germans and Austrians advanced with feverish haste nearly 200 miles to throw themselves upon the front of an enemy of probably 50% superior strength, installed on the line of a strong natural barrier reinforced by powerful fortresses, and in the presence of his bases of supply.

The northern wing of the German army before Warsaw was "left in the air," since the westward course of the Vistula below the Polish capital, instead of protecting this flank, exposed it to attack, as long as the Russians commanded the crossing-point at Novo Georgievsk.

The great offensive was a failure. The popular expectation that the German armies would pass the winter in Warsaw, as well as Calais, was deceived. But with astonishing agility the Germans avoided a calamity and recovered the equilibrium.

The successful withdrawal of the German and Austro-German armies was chiefly due to their unusually extensive destruction of the railways and roads in their rear, which greatly impeded the progress of their pursuers. Von Hindenburg and the Germans retreated in a southwesterly direction towards the line Kalisz-Czenstochowa, the Austro-German forces to the line Czenstochowa-Cracow.

By the beginning of November the Russians had concentrated enormous forces, possibly as many as thirty-five army corps, in Poland and Galicia ready to pour into the upper valley of the Oder, which dominates the rich industrial region of Silesia, and to strike towards Vienna or Berlin. In spite of the escape of the German and Austrian armies, the development of the operations on the eastern front had reached an alarming stage for the central empires at precisely the period when the attempts to penetrate the lines of the Allies in Flanders, repeated with ever increasing fury and feverish determination, were about to end in failure. On November 9th Russian cavalry actually raided German territory near Pleschen.

But von Hindenburg made the apparently hopeless situation an opportunity for delivering a brilliant counter-stroke with such promptness and dexterity as even to give rise to the conjecture that the German retreat itself was a stratagem deliberately designed as a favorable preliminary for the renewed attack. The fortresses of Ivangorod, Warsaw, and Novo Georgievsk, with Brest Litovsk in the rear, constituted, as we have seen, a sort of vast citadel in the heart of Poland, the possession of which was a vital factor. And so von Hindenburg, after leading the chief masses of the enemy far afield in the course of his retreat, deftly slipped aside, and by a loop-shaped evolution launched a new blow from the northwest at the Russian line of communications, thrusting himself between a considerable part of the Russian forces and the portals of their great central fortified position.

Never did the Prussian railways offer more striking proof of the military utility of their unsurpassed efficiency. Enormous bodies of troops were spirited from place to place with an ease and celerity that seemed to violate the most elementary conceptions of mass and inertia. News that the tide of battle was approaching Silesia had scarcely been received, when it was announced that von Hindenburg's army had suddenly disappeared from this part of the border and then, almost immediately, that it had reappeared in the region of Thorn, and that the counter-offensive from that quarter was already in progress.

The success of this bewildering maneuver had been prepared by the direction given to von Hindenburg's retreat, and consequently to the Russian pursuit, which had swerved far to the left of a straight line towards the heart of Germany. This tendency gave the Germans ample room for their revolving movement and impeded the readjustment of the Russian forces through the lack of lateral communications on the Polish side of the border.

The Russian pursuit terminated about November 5th, and the German counter-offensive was fully under way by the 12th or 13th.

The initial disposition of the principal German forces for the new attack was the reverse of the arrangement for the offensive in October. Then the Teutonic front commenced north of Kalisz, followed the general direction of the border of Silesia, and trailed off towards the southeast, until it fell in with the Carpathian barrier. Now the German line beginning near the same point followed the Prussian boundary towards the northeast, stretching beyond the Vistula. But it was undoubtedly part of the general plan on this occasion that the Austro-Hungarian forces should rally on the old line, forming the complement, as it were, of the new German front of maneuver in the north, so as to envelop the Russian forces west of the Vistula on all sections of their semicircular position at the same time.

While the retreat of the Austro-German forces after the former great offensive was still in progress, six divisions of cavalry had been transported to the prospective theater of the renewed attack from various sections of the Teutonic fronts, notably from Flanders, where mounted troops could no longer be employed in consequence of the transition to trench-warfare. These divisions were thrown forward on both sides of the Vistula to mask the principal movement of the Germans. The most formidable masses of German troops were concentrated between the Warta and the Vistula, where the rivers covered their flanks, their left rested on the fortress of Thorn, and the main railway line from Berlin and Posen to Insterburg passed conveniently in their rear. From this position the advance was pushed with vigor and rapidity along the left bank of the Vistula and the railway line in the direction of Kutno and Lovicz. Since there were only scanty Russian forces in the space between the Vistula and the Bzura and the Warta, it was conceivable that the Germans could overwhelm the great Polish fortresses, which were probably garrisoned by troops of the second or third line, before the principal mass of the Russian armies could be withdrawn from the Kalisz-Czenstochowa-Cracow front in the extreme southwest.

The German forces advancing into Poland from the northwest were grouped in two armies. The left under General von Morgen, the right under General von Mackensen, with General von Hindenburg in chief command. Von Morgen's army won a victory at Kutno on the 18th, which opened the way to the Bzura, and on the next day von Mackensen drove a wedge through the Russian front between Zgierz and Strykof northeast of Lodz. The Twenty-fifth German Reserve Corps under General von Scheffer-Boyadel and the Third Guard Division commanded by General von Litzmann poured through the opening thus formed, wheeled to the right and advanced to attack Lodz from the east, but soon found themselves enveloped by Russian forces of superior strength. The troops defending Lodz stretched across the German front and around the right flank, where reinforcements approaching from the north formed a second hostile line. Russian forces returning from southwestern Poland threatened the

German left flank, and Rennenkampf with still another column was approaching in the rear from the southeast. Altogether the Germans were actually attacked, or at least threatened, by five Russian army corps.

After sanguinary combats on the 21st and 22d this German detachment suddenly broke camp during the night of the 22d-23d, marched rapidly eastward across the frozen ground until it crossed the Miazga River at Karpin, and then turned sharply to the north, and assailed the enemy's left wing at Gatkof, before the latter was entirely prepared to receive the attack. Breaking through at this point and pushing forward with the same impetuous tenacity the Germans stormed Brzeziny on the 23d.

Here they encountered the second enclosing line of their opponents and sustained repeated violent charges, but finally fought their way out after enduring very heavy losses. Their escape was facilitated by the approach of a German relieving column from the northeast on the 24th. This series of events remains one of the most thrilling episodes of the war.

Apparently the general plan of operations pursued by the Germans embraced a revolving maneuver, with their left flank near the mouth of the Bzura as the pivot, for the purpose of sweeping back or rolling together the Russian forces which were within reach. By November 22d the German left wing had advanced to the Bzura, less than forty miles from Warsaw, and Lovicz and Skierniewiece were again in the hands of the Germans.

But the most serious fighting took place nearer the extremity of the maneuvering wing, where the Germans pushed their offensive in the direction of Lodz with the utmost vigor. From their positions north of the city they subjected the defenders to a very severe bombardment, accompanied by many desperate assaults of the infantry.

When the conflict was at its height the night was illuminated by the flashes of exploding shells and the weirdly shifting gleam of the searchlights, and the thunder of artillery was said to have been faintly audible at Warsaw sixty miles away.

A striking example of heroism was exhibited during this bombardment by a Russian artillery colonel, who with some assistants made his way under cover of darkness to the vicinity of a German battery of heavy pieces which was keeping up a very damaging fire on the defenses of Lodz from a distance of seven miles at Zgierz. Creeping stealthily forward he laid a field telephone wire to within a mile and a half of the battery, and then, lying prostrate on the ground as the rays of a searchlight passed back and forth above him, he directed the action of the Russian guns, which finally silenced the German battery.

A new German army advanced from Kalisz eastward to coöperate in the siege of Lodz, and after a series of bitter struggles terminating in a battle lasting three days in which the Russians suffered very heavy losses from the German heavy artillery, Lodz was evacuated on December 6th. Its position at the extremity of a salient in the Russian lines made any further sacrifices for its defense strategically unsound.

The necessity of counteracting an advance of the Russians into East Prussia, where they occupied Soldau on November 10th, and the desire to prevent another fatal turning movement against the left flank of the German army in the heart of Poland were probably the chief motives for a German offensive undertaken from the north, on the right bank of the Vistula, about December 7th. Experience had shown that the Russians possessed one distinct advantage over the Germans in the Polish operations, the opportunity of shifting troops rapidly from one bank of the Vistula to the other, secured to them by their possession of the great fortresses on the river. The Germans might have been able to neutralize this advantage in part if they could have overrun the right bank of the Vistula as far as Novo Georgievsk and masked that fortress. But repeated German attacks for about ten days on the front Ilovo-Glovno produced no permanent results.

Just as the conflicts west of the Vistula in October were similar in their outcome to the Battle of the Marne, the subsequent struggles in the central Polish theater in November and December may be likened in their results to the Battle of the Aisne, since they inaugurated a period of stationary warfare, defining the general position of the fronts in this region for several months to come. The Germans, thrusting and struggling forward with titanic force and indefatigable energy, gradually drove the Russian front, like a massive door swinging on a hinge near the lower Bzura, from the line Sochaczef-Lodz back through the line Sochaczef-Tomasof-Novo Radomsk, until it rested finally in the position Sochaczef-Rava-Opoczno, along the Ravka.

The forward progress of the Russian armies in the Austrian dominions was not immediately checked by the renewed German invasion of Poland in November. The Russians were in possession of the Lupkow Pass through the Carpathians by November 25th and they cleared Bukovina of the Austrians before the end of the month. They reached a point only three and a half miles from the outer defenses of Cracow on December 2d. Then the Austrians, their formations stiffened by German contingents, took the offensive along the line of the Carpathians. They drove raiding parties from Hungary and marched northward over the Dukla Pass in great force, and although they were unable to make much headway behind the Russian lines in Galicia or relieve Peremysl, the threat conveyed by their presence helped to paralyze the Russian offensive westward. The Russians abandoned their operations before Cracow on December 12th, after suffering serious losses, and fell back on both sides of the Vistula, in conformity with the retreat further north, as far as the Nida and the Dunajec, where they were installed at the close of the year. Consequently, the Russian front, stretching across the Polish plain from the lower Vistula along the Bzura and the Ravka, reached the upper Vistula at the mouth of the Nida, and was prolonged across Galicia on the line of the Dunajec.

The initial state of hostilities between Austria-Hungary and Serbia, after creating a paroxysm of alarm and riveting the attention of the whole world, soon passed into obscurity beside the gigantic struggle of the Great Powers, so that the alleged cause for the whole war was apparently almost forgotten. The Austrians, compelled to face the Russians with most of their forces, restricted themselves during the greater part of the campaign to a temporizing course of operations on the southern frontier. But on three occasions they aroused themselves to convulsive efforts of considerable magnitude to rid themselves once for all from the goading activity of their diminutive neighbor.

The Austrians scarcely condescended to regard the operations against the Serbians as regular warfare. Their inroads into Serbia were "punitive expeditions," and they were calculated to reduce the country with impressive despatch to the position of another Belgium. This disdainful attitude invested with special ignominy the absolute failure of all the designs of the Austrians against Serbia in the campaign of 1914.

The military unpreparedness of Serbia and the serious inroads in the national resources resulting from the strain

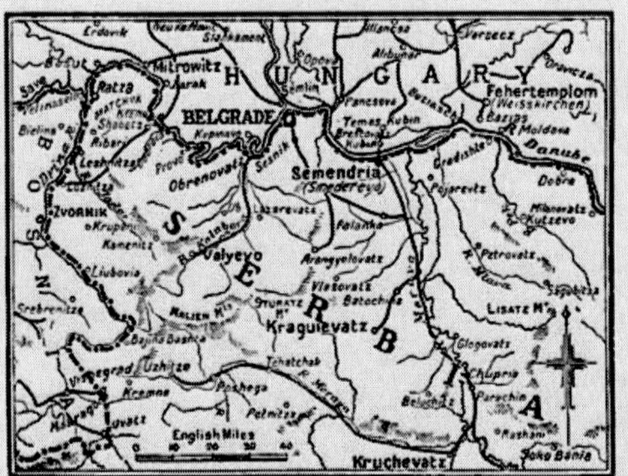

Map showing the Austro-Serbian frontier.

and wastage of the two Balkan Wars are striking evidence against the alleged complicity of the Serbian government in any scheme to precipitate a war with Austria-Hungary in 1914.

While the official declaration of war against Serbia was dispatched from Vienna on July 28th at 11.10 A. M., uncertainty as to the course that Russia would take restrained the Austrians from striking with determination and vigor before the Serbians had had time to concentrate the chief part of their available forces near their northern border. The boundary between Serbia and the hostile territory was almost entirely formed by rivers. The Drina, separating Serbia from Bosnia on the west, flows northward to the Save. The latter, with the Danube, into which it empties in front of Belgrade, separates Serbia from Hungary on the north. The Austro-Hungarian strategic railways give access to many available crossing points on the Drina and the Save.

An invasion of Serbia from the northwest, as contemplated by the leaders responsible for Austrian military policy, offered the opportunity of a concentric advance from several localities towards a single strategically important converging point. But the Austrians bestowed their chief attention on the line of the Drina, and planned their principal attack up the valley of the Jadar, a tributary of the Drina, southeastwards in the direction of Valyevo.

The Serbians, uncertain where the impending blow would fall, concentrated their main forces in the region of Palanka, Arangyelovatz, and Lazarevatz in the central part of the northern zone of the country, south of Belgrade, sending out strong detachments to points nearer the frontier.

The Austrian bombardment of Belgrade and the many attempts to cross the rivers on the north in the early days of the war were probably intended to distract attention from the region where the serious invasion of Serbia had actually been planned. The Fourth, Eighth, Thirteenth, and Fifteenth Austro-Hungarian Army Corps, and parts of the Seventh and Ninth, had been concentrated on the Serbian frontier. Three brigades of the Sixteenth Corps had been sent to restrain the Montenegrins, while the other three were held in reserve at Sarajevo.

The protruding northwestern extremity of Serbia, which is partially encircled by the Drina and the Save, is very flat. But the country south of this is rugged in character, with mountain chains and deep valleys and only a few serviceable roads.

The Austrians first penetrated Serbian territory at Loznitza on the morning of the 12th, crossing the Drina by means of boats and pontoons. The invasion was commenced about the same time at Shabatz and several other points, so that six different Austrian columns headed in the general direction of Valyevo. From the Drina in the general vicinity of Loznitza successive mountain ranges with intervening valleys extend laterally towards the southeast. From the north southwards the Tzer, Iverak, and Guchevo ridges occur in succession, the first and second separated by the River Leshnitza, the second and third by the Jadar. The Austrian commander, General Potiorek, planned his principal advance up the valley of the River Jadar towards Valyevo. The importance of the possession of the elevations confining this and the other routes can be readily appreciated.

The Serbian Chief of the General Staff, Field-marshal Putnik, who was responsible for the strategy of the Serbian armies, first served in 1876 in the war against Turkey and was captain of infantry during the Russo-Turkish War which followed. Lieutenant-colonel at the time of the ill-starred contest with Bulgaria in 1885, he was subsequently promoted to the colonelcy and named Chief of the General Staff, but fell into disfavor with King Milan for his Radical tendencies, retired and devoted his attention for several years to military studies and writings. He was promoted to the rank of general by King Peter and as minister of war he directed the reorganization of the Serbian army, which he commanded in the campaigns of 1912 and 1913. He is a man of plain, unpretending appearance and of few words, but a keen judge of men and of human nature.

At the beginning of the Great War he disposed of about 125,000 troops of the first line, or possibly 200,000 combatants altogether, including the second Ban, volunteers and recruits. But the unfriendly attitude of Bulgaria made it necessary to detach considerable forces to guard the eastern frontier and Macedonia, so that it is doubtful whether much more than 125,000 were available for the field armies which were to oppose invasion by the Austrians in greatly superior strength.

As soon as Field-marshal Putnik perceived that the greater part of the Austrian forces had been concentrated for the invasion of Serbia from the northwest, he dispatched the principal field armies toward the threatened quarter. All depended upon the rapidity with which these forces could traverse the intervening country and the skill employed in adapting their operations to the special physical features of the theater of hostilities.

It was of paramount importance for the Serbians to prevent the junction of the Austrian forces advancing southwards from Shabatz with the main bodies coming from the direction of the Drina, and accordingly the right wing of the Second Serbian Army (a group of three divisions) and the Independent Cavalry Division drove the Austrian advance-guards from the northern foot-hills of the Tzer range, the prospective field of contact, on the 16th. The tenacity with which the Serbians retained possession of this position on the northern flank of the battlefield, in spite of the threatened collapse of the resistance of their comrades further south, was the indispensable factor in the final victory.

In the meantime, the center of the Second Army was pressed back, but the left wing, coming into contact with the enemy after performing a march of fifty-two miles in twenty-four hours, repulsed the attacks of the Austrians along the Iverak range on the evening of the 16th. The Third Army in the valley of the Jadar was outflanked on the south and compelled to retreat from Jarebitzé on the road to Valyevo. This movement involved the left wing of the Second Army, so that on the 17th, while the Serbians prosecuted vigorously their offensive along the ridge of Tzer, their lines were everywhere on the defensive or in retreat in other parts of the battlefield. On the same day the Serbians made an unsuccessful attack on Shabatz, where the Austrians turned to the offensive on the 18th and repelled them step by step.

The 19th was the critical day. The Austrians, striving only to hold their own on the crest of Tzer were everywhere else pushing forward with alacrity and the Serbian Third Army was apparently at the limit of its endurance. It seemed inevitable that the Austrians, with their superior numbers and equipment, and with an adequate, if not equal, acquaintance with the territory, would sweep everything before them. But suddenly came a turning point. An Austrian flanking attack against the right wing of the Third Army encountered a fresh reserve division and was repulsed. The Third Army took the offensive at precisely the favorable moment, threw their opponents into confusion, and chased them down the valley of the Jadar. Meanwhile the Serbians, victorious on Tzer, attacked the Austrians on Iverak and dislodged them from the positions commanding the line of retreat down the valley of the Jadar.

The Austrians everywhere took to flight and poured from the lateral valleys and mountain routes towards the crossing points on the Drina. The total collapse of the invasion from the west permitted the renewal with greater strength of the Serbian attacks in the direction of Shabatz, which the Austrians evacuated on the night of August 23-24.

The losses in the Battle of the Jadar were heavy on both sides, but the possession of the field and the capture of 4,000 prisoners were palpable evidence of the Serbian victory, and the failure of the Austrians to overrun Serbia in this first attempt created a deep impression throughout Europe.

Shortly after this battle the Serbian First Army, two divisions of infantry and one of cavalry, crossed the Save at a place called the Kupinski Kut, where a tongue of land projecting from the northern side and almost enclosed by the river in a sharp detour, is commanded by the guns on the Serbian bank. The invasion of Hungary thus inaugurated progressed until September 11th, when the expedition was recalled on account of the second Austrian invasion of Serbia from the west.

Desultory operations had been proceeding on the Montenegrin front. One Montenegrin division under Prince Peter occupied Mt. Lovcen, engaged in artillery duels with the Austro-Hungarian ships in the Bocche di Cattaro, and held themselves in readiness to coöperate with the fleet of the Allies in attacking the Austrian positions on this valuable inlet. Another Montenegrin division commanded by General Martinovitch engaged the attention of the

Austro-Hungarian fortresses in Herzegovina. Montenegrin divisions are much smaller than those of the conventional type. The Montenegrins are born warriors who never lay aside their weapons. They are unsurpassable in guerilla fighting, but were unsuited by inclination and experience for military maneuvers on a comprehensive scale.

After the Battle of the Jadar the Serbians decided to operate concurrently with the Montenegrins in the invasion of Bosnia. A Serbian army crossed the upper Drina, occupied Vishegrad on September 15th, and effected a junction with General Vucovich at the head of at least two divisions which had advanced from the Montenegrin side across the frontier of the former sanjak, and together these forces made some further progress in the direction of Sarajevo. But in the meantime the Austrians resumed the invasion of Serbia.

While three or four battalions of the Sixteenth Austrian Army Corps with some contingents of Landsturm and recruits faced the Montenegrins, the remainder of this corps and the Eighth, Thirteenth, and Fifteenth passed the Drina somewhat above the earlier crossing places and forced their way into the mountainous region of Krupani, where fierce contests took place. Reinforced by the troops which had returned from beyond the Save, the Serbians drove the Austrians from many of their positions, but did not dislodge them entirely from Serbian territory. For many weeks the operations were almost stationary, with lines of trenches drawn through the rugged tracts and every point zealously contested in minor engagements.

About October 25th strong Austrian detachments overpowered the Serbo-Montenegrin forces near Kalinovik in Bosnia and compelled the Serbians to abandon Vishegrad in their retreat.

In consequence of the inadequacy of their forces, the Serbians found themselves reduced in November to the necessity of shortening their defensive lines by withdrawing their contingents from the positions in the extreme northwest and near the Drina, which fact furnished the occasion for a third invasion of their country involving a far more critical situation. The Austro-Hungarian forces swarmed over the boundary in the west and northwest in such formidable numbers that Valyevo had to be evacuated on November 11th and the headquarters withdrawn to Kraguievatz.

This conspicuous initial success may have engendered an excessive feeling of confidence on the part of the Austrians that impaired their vigilance. They probably regarded the Serbians as already demoralized and their own final victory as at hand. But the Serbians prepared to make a stand on the line of the Kolubara and of its tributary the Lug and along the crests of the mountains which cover the upper Morava valley.

Five Austro-Hungarian army corps were taking part in the new offensive movement. A formidable action against the Serbian positions, particularly those of the Second Army near Lazarevatz and of the so-called Uzhitze Army, which had been operating in the direction of Sarajevo, near Kosjeritchi, was commenced on the 15th; but for five days the repeated attacks failed to dislodge the Serbians. Finally, on the 20th, the Austrians captured Milanovatz and drove back the Serbian First Army, inflicting serious losses upon it.

By the 24th the fighting had become general, the Serbian Second Army had been dislodged from its positions, and an Austrian turning movement was in progress towards the Morava.

The problem of transportation became more difficult for the Austrians as they penetrated deeper into Serbian territory. But the Serbians were almost out of ammunition. Marauding bands threatened railway communication with Salonica and it was feared that Bulgaria might at any time intervene. The Austrians were advancing in seemingly overwhelming force, and the pressure was becoming too great for the Serbians, whose resistance seemed to be on the point of collapsing.

But suddenly, at the darkest moment, the ardor and determination of the Serbians were revived by an unexpected reaction of spirit. Fresh supplies of ammunition arrived and a vigorous counter-offensive was planned as a final effort. But before this was carried into effect, Belgrade had to be abandoned to the Austro-Hungarian forces on November 29-30 in the process of consolidating the Serbian positions.

Never was the essentially popular character of the great movements in Serbia more evident than in the impulsive effort which resulted in the expulsion of the invaders from the national soil. It was produced by a spontaneous revulsion of popular feeling and was guided by leaders who had sprung from the people. Colonel Givko Pavlovitch, for instance, General Putnik's principal collaborator as director of military operations, was the son of a farm laborer, and General Mislutch, who was placed in command of the First Serbian Army, was another self-made man, the son of a peasant. King Peter, despite his age and infirmities, came to the front to exhort his troops to a supreme effort by his presence and words.

The king's proclamation to his soldiers reflected the lofty spirit of Thermopylae and of the legendary age of early Rome: "Heroes, you have taken two oaths, one to me, your king, and the other to your country. I am an old, broken man on the edge of the grave, and I release you from your oath to me. From your other oath no one

Austrian siege-gun hauled by motor-tractor.

Serbian field hospital. *From a photograph made by a Serbian officer.*

can release you. If you feel that you cannot go on, go to your homes, and I pledge my word that after the war, if we come out of it, nothing shall happen to you. But I and my sons stay here."

It is reported that not one single man left the army.

The Serbian counter-attack took the Austrians by surprise just as they were trying to execute a double enveloping movement on December 2d. The First Army under General Mislutch stormed the positions in the region of Suvobor at the Austrian center and threw the Fifteenth and most of the Sixteenth Corps into headlong flight in the course of an encounter lasting three days. The Serbians advanced with increasing enthusiasm and momentum, driving the Austrians before them, reoccupying Valyevo, and taking thousands of prisoners. This series of actions resulting in the disorderly flight of the Austrians from Serbian territory in the northwest is known as the Battle of Suvobor.

As soon as success was assured in this part of the field, a portion of the Third Army was directed towards Obrenovatz on the Save, while an army group composed of the remainder of the Third Army, the Second Army, and the cavalry division, under Field-marshal Stepanovitch, advanced northwards for the recovery of the capital. The Serbians closed in gradually on Belgrade advancing in concentric formation, their wings extending to the neighborhood of the Save and the Danube respectively.

On the 14th they carried the defenses on Tarlak Hills outside the city on the south, which the invaders had greatly strengthened with earthworks and barbed wire. The evacuation of Belgrade began on the same day and continued all night. On the 15th the Serbians succeeded in destroying some of the Austrian pontoons, causing panic and much loss to the retreating army, while King Peter made his entrance into the capital.

In view of the animosity created on both sides by the Austro-Serbian controversies, and of the traditional practices of warfare in the Near East, it is not surprising that the campaign just described was conducted with great ferocity and that it was prolific in alleged excesses and violations of international law which recall the savage fury of the Croatian revolt against the domination of the Hungarians in 1849. The experiences of the recent Balkan wars loomed large in the imagination of the Serbian people. Their impulsive temperament was thrilled with elation and ardent devotion to their independence. Warfare with them was the liberation of a wild, elemental passion. It was hardly conceivable that they would confine themselves to the conventional restrictions of the kind of warfare which only was regarded as legitimate by their opponents.

The Austro-Hungarian authorities charged the Serbians with gross violations of the established usages of war and with many abominable atrocities, the treacherous use of the white flag, firing on Red Cross ambulances and hospitals, the poisoning of wells, robbing and killing the wounded and their prisoners and mutilating them in the most revolting manner, in some instances, as it would appear, before they were dead.

These charges were mainly preferred against civilians and the *comitadjis* or irregular troops, whose hostile activity and alleged barbarity were the constant subject of Austrian recrimination. The most shocking enormities, in so far as they were really committed, seem to have been due to spontaneous outbursts of individual savagery. But the general infraction, which was met by the systematic application of relentless measures of retribution, was the participation in acts of hostilities of those whom the Austrians refused to recognize as legitimate belligerents. This was the essential factor in the question of atrocities, all else was incidental.

The Austrians claimed that all classes of the Serbian population, including women and children, engaged in hostile action by firing on soldiers and convoys of wounded and provisions from concealment in towns and villages which were ostensibly peaceable, or lured their victims to destruction by treacherous artifices. The atrocities committed, or alleged to have been committed, in Serbia are associated with a fundamental distinction in the conceptions of the nature of warfare as entertained by the two parties. The Austrians regarded it as a strictly organized, professional activity, but this limitation was incompatible with the traditional habits and practices of fighting among the Balkan peoples. Irregular combatants, a survival from the struggles for independence, are a normal feature of every war in the Balkan peninsula. But the Austrians admittedly put to death all *comitadjis* as well as civilians caught carrying arms, and burned houses from which shots had been fired. These severities, which were practised in retaliation for the alleged infringements of the rules of war, were regarded by the Serbians as outrageous acts of brutality, and probably served very often to inflame the people to furious deeds of vengeance. The activity of the *comitadjis* amid the habitations of the civilian population must have involved many innocent persons as victims of the harsh reprisals of the Austrians.

According to Serbian reports, the operations of the Austro-Hungarian forces, particularly the Hungarians, were stained by the most shocking enormities and crimes, wholesale robbery and pillage, the useless destruction of property, the slaughter and mutilation of the wounded and prisoners, and the massacre of civilians. These accusations have been examined on the spot by Professor R. A. Reiss of the University of Lausanne, whose integrity there is no reason to doubt, and while the limitations of a single disinterested investigator were undoubtedly very great, the principal facts seem to be sufficiently substantiated.

In some important instances the indications contained in the Austro-Hungarian reports tend to corroborate the conclusions derived from the information presented by Professor Reiss. For instance, the Austrians declared that for twenty-four hours after the occupation of Shabatz, civilians persistently fired on soldiers from the rear and that the mutilated bodies of many Austro-Hungarians were found in the vicinity of the town, and, furthermore, that the village of Prnjavor, situated in the rich Matchva district in the extreme northwestern part of Serbia, excelled in atrocities against Austro-Hungarian soldiers. As these are precisely the localities where the Austro-Hungarian forces are said to have committed their most sanguinary atrocities, we naturally assume that such relentless behavior was intended as retribution for the refractory conduct of the local population.

According to a corporal of the 28th Austrian Landwehr regiment, who had been taken prisoner, more than sixty civilians were bayoneted by eight Hungarian soldiers by order of the general in command near the church at Shabatz. Local reports placed the number of the victims on this occasion as high as one hundred and twenty. Professor Reiss caused a pit to be opened behind the church where the bodies of at least eighty persons were found lying just as they had fallen. A report that more than a hundred women and children had been butchered and thrown into the burning house of a certain Milan Milanovitch in Prnjavor and that similar outrages had been perpetrated at the schoolhouse and in other parts of the village was confirmed by an inspection of the ruins, particularly by the bloodstains on the still extant walls of the buildings

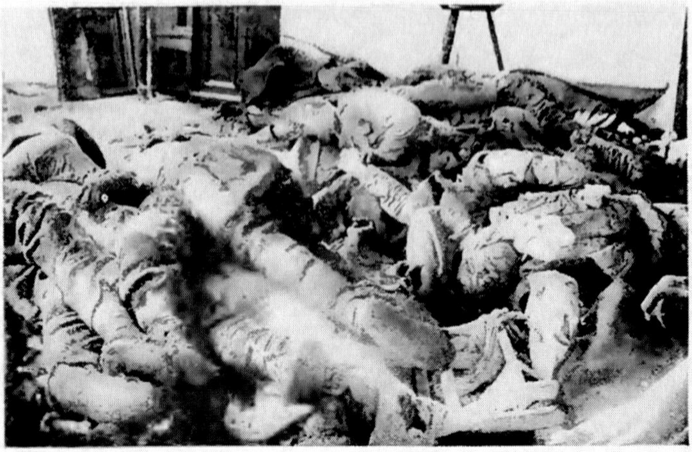

Dead in a room in a villa near Shabatz, where the Austrians are said to have bayoneted the wounded.

Barbarity of war in Serbia. Peasants of seventeen or eighteen alleged to have been massacred in the environs of Loznitza by order of the Hungarian commander Bazarek.

mentioned. Furthermore, a pit was opened at Leshnitza, in which one hundred and nine peasants, who had been collected as hostages from the neighboring villages, and later tied together and despatched by a volley, were lying in a confused mass as they had fallen.

According to the calculations of Professor Reiss, between three and four thousand civilians were slain by the Austro-Hungarian forces in Serbia. In pursuing their policy for the suppression of alleged transgressions of the rules of war, the invaders devastated much of northwestern Serbia, burning villages and farmhouses and making thousands of the people homeless. Moreover, they subjected open towns such as Shabatz, Loznitza, and Belgrade to prolonged bombardment, destroying factories, hospitals, public buildings, the university, and the national museum in the capital.

By an examination of the wounds, the testimony of prisoners and the ammunition found in their possession, Professor Reiss confirmed the alleged use of explosive bullets by the Austrians. These bullets explode upon contact, frightfully lacerating the flesh and thus producing far more serious wounds. Limbs struck by these bullets usually cannot be saved. The Austrians admitted that cartridges fitted with these bullets had been given to the soldiers, but claimed that they were intended only for determining the range by means of the flash or smoke produced by the explosion.

The conclusions to which we are led by the examination of the incriminations made by both sides in connection with the first campaign in Serbia will be useful as a clue for the interpretation of similar occurrences elsewhere. The conduct of the Austro-Hungarian forces in Serbia was undoubtedly marked by excessive cruelty, but they were actuated in this by the Serbian violations of the strict rules of legitimate belligerency. The brutality of the

Austrians, which intensified the animosity of the Serbians, the natural lawlessness of the irregular soldiery and the other special causes of irritation gave the campaign in Serbia a character of fierceness which was scarcely equalled in any other quarter.

The outcome of the campaign in the East was regarded with various sentiments by the three great powers which had been the principal combatants.

The Germans had won brilliant victories and exhibited sensational dexterity in their maneuvers. Nowhere had the wonderful efficiency of their organization appeared to better advantage. Compared with their achievements in the West, the results obtained by them in the eastern theater were highly satisfactory. The later weeks of the campaign had established even more firmly the immense popularity of von Hindenburg, and brought into prominence another leader whose fame was destined likewise to rise to a pinnacle of glory, General von Mackensen.

As reward for the illustrious exploits of the German offensive in November, in which 60,000 Russians had already been taken prisoners, Colonel-general von Hindenburg was elevated to the rank of field-marshal and his chief-of-staff and son-in-law, von Ludendorff, to that of lieutenant-general on November 27th, while the Order *pour le mérite* was conferred on General of the Cavalry von Mackensen for his brilliant leadership of the Ninth German Army. Lieutenant-general von Litzmann was made general of the infantry and commander of a reserve corps in recognition of his evasion of the Russian trap near Lodz, and on December 22d General of the Cavalry von Mackensen was raised to the rank of colonel-general.

But the futility of all predictions in the course of the great struggle, even of those made by the most distinguished authorities, is evidenced by some observations of von Hindenburg, who declared towards the close of the campaign that though the Russians were good soldiers and had learnt much since their war with Japan, they were already becoming listless, their food and munitions were giving out, and all the indications pointed to a speedy collapse of their efforts.

As for the Russian leaders themselves, it is difficult to penetrate their genuine expectations at the beginning of the campaign, so as to compare them with the results which were actually accomplished. The Russians suffered some serious reverses and the much-heralded offensive with overwhelming momentum failed to materialize. But in view of the incompleteness of Russian preparation and of the partly faulty generalship revealed by the course of the operations, the results of the campaign were undoubtedly as favorable as could have been reasonably expected. The Russian lines still held firm at the center in Poland, though opposed by the most powerful forces and the ablest generals, while on the wings, where the pressure against them was less formidable, they held a considerable slice of East Prussia and about two-thirds of Galicia and Bukovina.

We have already considered the inspiring personality of the Grand-duke Nicholas, the Russian generalissimo. But the service rendered by General Sukhomlinoff, the Russian Minister of War, while less conspicuous, was probably quite as essential, for the reorganization and development of the Russian military system in recent years was in large part his work. An impressive figure and personality, a contagious good-humor, a clear perception for reality, method, and industry made him appear as a Russian embodiment of optimistic efficiency. His reputation for administrative ability and his wholesome personal influence over his fellow officers date from the period of his headship of the Officers' Cavalry School in St. Petersburg. Retained in Europe during the Russo-Japanese War, he held various commands on the western frontier, devoted close attention to the annual maneuvers in those parts, and obtained an ample acquaintance with the future theater of hostilities. His activity in the war office, which began in 1909 and coincided with the rapid expansion of Russian military power before the war, was distinguished by two particular aims, the elevation of the standard of efficiency of the Russian officers and the development in Russia of the essential industrial basis for the nourishment of modern warfare.

He submitted the merits and failures of the officers to careful scrutiny; amplified the establishment for their higher training and urged them to frequent it; rewarded vigor, energy, and genius; and discreetly but systematically facilitated the retirement of those whose increasing years had not been matched by expanding talent. To him is due the credit for the foundation of schools of military aviation and of railroading for officers and the establishment of an effective auxiliary corps of automobiles. He strove persistently to ingraft into the financial administration of the army the straightforward, effective methods of successful business. The rumor that a German intrigue was launched in St. Petersburg for the removal of Sukhomlinoff just before the war, whether true or false, is proof of the popular esteem in which his ability was held.

Von Hindenburg declared, on the basis of the experience of the first campaign, that the Austrians and Hungarians were excellent soldiers and that their officers were spirited and courageous, and also that the relations between the chief commanders of the Teutonic empires were ideal,—a most fortunate situation, since it was evident from the first weeks that Austria-Hungary had to lean heavily upon the support of her more powerful ally. Infirmities inherent in the heterogeneous character of the realm were chiefly responsible for the disappointing situation of Austria-Hungary at the close of the campaign. Partly for the same reason, and partly perhaps as the effect of rival intrigues in Vienna, the Dual Monarchy had no generals who were popular idols like von Hindenburg and the Grand-duke Nicholas.

General of the Infantry Conrad von Hoetzendorf, Chief of the Austro-Hungarian General Staff, is an international authority of recognized reputation on military subjects. His treatment of the fundamental principles of tactics has been accepted as a text-book by all the war academies throughout the world. His distrust of Italy and advocacy of powerful fortifications on the Italian frontier almost produced an international crisis a few years before the war and led to his resignation as head of the army. He was reinstated in 1912 and enjoyed the cordial support of the Archduke Francis Ferdinand before the war and the confidence of the army generally at the time which we are considering.

But it was evident that a weeding out of some of the principal commanders was essential to the indispensable renovation of the Austro-Hungarian armies. General von Auffenberg was one of those whose retirement was deemed expedient. He had been superseded as minister of war in 1912 by Field-marshal Alexander Krobatkin, who held this office in 1914. General von Auffenberg will be recalled as the commander who was swept back before the first advance of the Russian forces into Galicia, evacuating Lemberg and suffering defeat on the Rawaruska-Grodek line.

Field-marshal von Potiorek's career underwent a similar eclipse. A Bohemian by birth, for a long time assistant to von Hoetzendorf as Chief of the General Staff, his post of head of the government of Bosnia and Herzegovina, a military position, in 1914, made him the natural commander of all the forces operating in the south. The Austro-Hungarian authorities insisted that the final repulse of their armies in Serbia was merely a temporary consequence of the failure to apprehend the true measure of the difficulties to be surmounted, the arduous character of the country, the inadequate roads, the fearful state of the weather, and the unexpected strengthening of the enemy's forces, and that it would not have any permanent influence on the outcome of the struggle. Yielding, however, to the malady which so often attacks unsuccessful commanders, Field-marshal von Potiorek petitioned to be released from his command, and General of the Cavalry Archduke Eugene was appointed to succeed him, with the general approbation of military circles.

CHAPTER XII

The Close of the Campaign in the West and the Land Operations Outside of Europe

The waning of the campaign in the West. Attempted Allied offensive in December. Some characteristics of trench-warfare. Christmas at the front. The war outside of Europe. The rally of the British dominions and dependencies: colonial and Indian troops sent to Europe; operations in colonial territory, the campaign in Togo, Kamerun, German East Africa, and German Southwest Africa; the insurrection in South Africa. Germany swept from the Pacific. The siege and fall of Tsingtau. Turkey's advent into the war; Turco-German designs, Egypt and the British Empire; Cyprus; events on the Persian Gulf.

There was no sudden break in the course of operations in the West after November 11th, the date which has been adopted for convenience to mark the termination of the Battle of Ypres. As late as the 13th the Germans penetrated the British trenches in several places. But the fighting subsided gradually into the colorless, mechanical routine of stationary warfare, a wearisome process of disillusionment for all the early hopes and eager elation. A period of four months from the close of the Battle of Ypres to the Battle of Neuve Chapelle was almost barren of eventful occurrences in the West. Juvenal, at one time an officer in the Roman army, asserted in praise of the military profession that a moment of time brought sudden death or glorious victory. But he had evidently gone through no experience resembling the agonizing monotony of pain and exhaustion in modern trench-operations.

There was a recrudescence of activity in the West about the middle of December. General Joffre issued an order of the day on the 17th declaring that, after all the attacks of the enemy had been repulsed for three months, with the strengthening of the Allies in men and material and the weakening of the Germans by the transfer of troops to the East, the time had come for striking a blow and clearing the French territory. British infantry and French marines coöperating with the Belgians had already taken the offensive in the extreme north. Barges mounting British naval guns took part in the operations on the lower Yser, where the British captured Lombaertzyde and the French and Belgians stormed St. Georges.

Encouraged by these small successes the Allies undertook to dislodge the Germans from positions west of Wytschaete which they had held since the Battle of Ypres. But two Scottish regiments and the Thirty-second French Division, attacking on the 14th, failed to secure any appreciable advantage.

On the 19th the Meerut and Lahore Indian Divisions attacked the Germans near Givenchy and gained possession of some trenches, but were afterwards driven back, suffering severe losses. On the next day the Germans took the offensive in this section, drove the British and Indians from some of their trenches, and engaged in a fierce struggle for the possession of Givenchy. It was captured by the Germans and afterwards retaken by the British and Indians, but the situation remained so serious for the Allies that one of the divisions of the British First Army Corps then stationed in reserve was brought up to reinforce the Indian troops on the 21st and the danger was not entirely averted until the 22d. Concurrently with these events the French assailed the Germans at various points along the line. Northeast of Châlons they captured

The façades and tower of the Cloth Hall and Town Hall at Ypres: on the right is the tower of the Cathedral of Saint Martin.

The ruins of the Town Hall, Cloth Hall, and Cathedral at Ypres on November 22, 1914.

a section of the German outer trench but were unable to maintain themselves in this position.

By the end of the year the war had become practically stationary in the East and West alike. To Napoleon and von Moltke, the great strategists of the nineteenth century, unassailable positions were unknown. If the enemy's front could not be broken, an attack could be directed against his flank. But now the extension of the fortified lines from the North Sea to Switzerland, a distance of about 350 miles, across the entire front in the West, and from the Baltic Sea to the northern extremity of Roumania, nearly 900 miles, in the East, had excluded all turning movements. In the East, where the disposition of the troops was generally less compact, there was still the prospect of advantages to be obtained by aggressive action without a wholly disproportionate expenditure of blood and munitions. But the course of events and the condition of affairs in the western theater, and to a lesser degree in the eastern as well, suggested the speculations of Frederick the Great in his military testament as to his future conduct in case Prussia were again attacked, as in the Seven Years' War, by a coalition of states.

Frederick declared that in such circumstances he would straightway advance far enough into his opponents' territory to live at their expense and to hold the hostile armies, when they should confront him, on lines chosen by himself and already fortified. He would reconnoiter the country as far as his patrols could be dispatched, so as to make himself perfectly familiar with all the lines by which his adversaries might advance to attack him. In places thus occupied and strengthened, he would calmly hold himself on the defensive,—not squandering his forces in assailing strong positions where the advantages would be all on the side of the hostile defenders,—until his antagonists wearied of the contest.

A brief description of some essential characteristics of the system of intrenchments which had now become the chief feature of all parts of the principal war-zones in the Great War may not be inappropriate at this point.

Two or more lines of earthworks extended along the entire front of the armies on both sides. The outer lines served for the protection of the troops who were regularly under fire, while the others sheltered the relief troops. The first and second trenches were usually from 800 to 1,000 yards apart, according to the local conditions and contour.

The troops on duty in the foremost trenches were commonly relieved after nightfall every twenty-four hours, when the substitution of fresh contingents from the inner trenches and the removal of the dead and wounded from the front were effected under cover of the darkness. The troops brought with them all the necessary provisions for their period on the outer line, where the preparation of food was usually impossible. They were kept ever on their guard against the enemy's surprises. There was scarcely any protection against the rigors of the winter climate, the rains and sleet, winds and cold. It was impossible in many places to drain the outer trenches and the soldiers stood for days or weeks in water, mud, or slush. The possibility was always present that a well-directed shell or bomb falling into these open trenches would blow the soldiers to pieces, while the exigencies of the situation required that the men should often expose their heads and shoulders to the searching fire of hostile rifles and machine-guns that raked the top of the earthen parapet. The service there was one of exhausting tension, incessant danger, and fearful hardship and privation.

But the inner trenches, often roofed over and protected in large measure from the enemy's projectiles, as well as from the severity of the weather, presented a spectacle of relative comfort. There were mats frequently for the floors, simple furniture, and arrangements for cooking and lighting. In some places dugouts or subterranean apartments, equipped with sleeping bunks, afforded an opportunity of repose with absolute freedom from danger, except perhaps from the explosion of mines and the shells of the heaviest siege-artillery. Officers' quarters of this kind, especially on the German side, were remarkable for the comfort, or even luxury, displayed in their furnishings. But generally, the absolute cheerlessness of the forward trenches made even the simple amenities of those in the rear seem unusually attractive and cozy by contrast.

At times, when the development of the conflict required the presence of the troops in the second line to support those in the first, it was impossible to relieve the latter at the regular intervals, so that they were often compelled to remain days at a time in their exposed position with scarcely any opportunity for sleep or relaxation, drenched or besmeared with mud or benumbed by the cold, and with only precarious nourishment.

Communication between the front lines and the sheltering trenches further back was provided, as far as possible, by excavated passageways, zigzagged and divided into sections by transverse earthen partitions, so as to localize the enfilading shell-fire of the enemy.

The Germans usually conducted their frontal attacks against trenches in the following manner. In order to shorten the space which the infantry had to cross in the open, exposed to direct fire, they advanced by sapping in narrow zigzagged trenches or in subterranean galleries to the proper distance for the final rush, where the various channels were connected by a lateral trench approximately parallel with the enemy's front. In this the forces were drawn up for the assault. But sometimes a sap was carried by night right up to the hostile parapet, which was blown in by the explosion of a mine, leaving a yawning breach. Except in the supreme moments of the charge the most noteworthy feature of a battlefield was the apparent absence of human beings.

The spirit of the Christmastide found expression in a very unexpected and remarkable manner. An unofficial truce was observed throughout a large part of the front in the West, and in many places the men on both sides issued from the trenches, mingled, exchanged gifts, and sang songs together. At one point there was a football match between British and Saxon soldiers, in which the former were defeated.

Very impressive was the Kaiser's celebration of Christmas among the soldiers who belonged to the General Headquarters. The walls and ceiling of a great hall were completely covered with evergreen. An altar was erected at one end of this hall, flanked by tall Christmas-trees, with a manger before it. Places for about 960 persons altogether, of all ranks from Kaiser to simple private of the Landwehr, were laid at long tables arranged lengthwise down the hall, to which smaller Christmas-trees with their many little lights added the customary, festive appearance. The gifts from home were found on these tables; and in addition, each guest received *Pfefferkuchen*, apples, and nuts, and a picture of the Kaiser. The privates received cigars and tobacco pouches also. After a short religious address and the singing of the appropriate hymns, the Kaiser addressed the assemblage as follows:

"Comrades! In a state of armed defense, we are here assembled to commemorate this holy festival commonly celebrated in the peaceful interior of our homes. Our thoughts go back to the dear ones, whom we thank for

French infantryman in act of throwing hand grenade, of which he has just lighted the fuse.

Trench after capture by the French forces and before the removal of the German dead.

the gifts which we behold in such profusion upon the tables before us. God let the enemy compel us to celebrate the festival here. Attacked, we are forced to defend ourselves. God grant that from this festival of peace, with God's favor upon ourselves and upon our land, ample victory may come from this bitter contest. We are on hostile soil with our sword's point turned to the enemy, our heart to God. We voice the words of the Great Elector: 'To the dust with all the enemies of Germany. Amen.'"

Many of the most extensive conflicts of the past have sprung from surprisingly petty or unworthy causes. As Macaulay once remarked of the rapacity of Frederick the Great, which precipitated the War of the Austrian Succession:

"The evils produced by his wickedness were felt in lands where the name of Prussia was unknown; and in order that he might rob a neighbor whom he had promised to defend, black men fought on the coast of Coromandel, and red men scalped each other by the Great Lakes of North America."

But the issues at stake in the present upheaval are certainly commensurate with its world-wide character, which will directly be partially illustrated.

The unanimous feeling of loyalty which pervaded nearly every corner of the British Empire, and the spontaneous rally of the colonies and dependencies in response to the danger that threatened the Mother Country, were one of the most impressive spectacles which the Great War afforded. And quite apart from every partisan sentiment, one may regard this phenomenon from the broadly human point of view with a generous feeling of gratification as a palpable demonstration that unison of heart and action of the most scattered communities may be created and preserved without the application of galling restrictions or of compulsion; in fact, without even the possibility of coercion. Even neutrals have been tempted to compare with a trace of malicious satisfaction this impetuous flood-tide of passionate loyalty with the pretentious but often superficial arguments by which the inevitable dissolution of the British Empire at the first serious shock was dogmatically predicted in German academic circles.

The Australian Laborite Ministry, voicing the feelings of all parties in the island-continent, proclaimed its unhesitating support of the Mother Country in the hour of trial. In the words of Mr. Millen, the Australian Minister of Defense, "Australia wishes the rest of the Empire to know that in this momentous struggle for liberty and national honor, the vigor of her manhood, the bounty of her soil, her resources, her economic organization, all she possesses to the last ear of corn and drop of blood is freely offered to help maintain the glory and greatness of the Empire, and to battle in the righteous cause wherein she is engaged."

Both Australia and New Zealand at once placed their own naval forces at the disposal of the British Admiralty and immediately offered contingents of 20,000 and 8,000 men respectively as a first instalment of troops, which were followed by several other contingents from each of these dominions.

The popularity of the Duke of Connaught, Governor-general of Canada, of the Duchess his wife, and of the Princess Patricia their daughter, made their presence an appropriate symbol and rallying point for the loyal enthusiasm of the great Dominion. The duke's zeal and experience in military matters stimulated and sustained the martial ardor of the country. The first Canadian contingent which set sail for England about the end of September, consisted of about 30,000 men, including a regiment of French Canadians, one of Irish Canadians, Princess Patricia's Light Infantry, and Strathcona's Horse.

Generous contributions of supplies for the Mother Country were raised by all the provinces, and the Dominion government itself made a gift of 100,000,000 bags of flour. Sir Robert Borden, Canadian Prime Minister, announced on October 7th that the government intended to raise and send forward a second contingent of the same strength, and later he promised that 30,000 men would be kept continuously in training while the war lasted.

At the very outbreak of hostilities the government of the Union of South Africa offered to undertake the responsibility for all the necessary operations in that part of the continent so that the imperial garrison would be available for service in the European campaign. Later, at the suggestion of the Imperial government, the Union accepted the task of carrying the war into German Southwest Africa. The Right Honorable General Louis Botha, who had been fighting against the British as commander-in-chief of the Boer forces scarcely more than twelve years before, now Prime Minister of the Union, announced his intention of commanding personally the forces in the field against German Southwest Africa.

The smaller colonies responded to the situation according to their resources, Newfoundland, for instance, equipping 500 men for foreign service and 500 for home defense, and Jamaica taking the necessary steps to provide for her own protection.

The unswerving allegiance of India, where the supineness of the British administration had been regarded by German observers with undisguised contempt, turned out in reality to be a most disconcerting element in the calculations of those who insisted that the British Empire was an unnatural association of incongruous elements which would fly apart on the first serious test.

The rulers of the native states of India, nearly 700 in number, offered their services and their resources at the outbreak of the war. A number of the native princes and nobles joined the expeditionary force at once, the corps maintained by the larger states as Imperial Service troops were immediately placed at the disposal of the Imperial government, and many other contingents and contributions in money and supplies were furnished by the native rulers.

The arrival and early exploits of the Indian Expeditionary Force in Europe have already been mentioned. It was an imposing armament of about 70,000 excellent fighting men, powerful Sikhs, the backbone of the Indian army, Punjabi Mussulmans, gallant Gurkhas, Pathans, Brahmans, and others, mostly battle-tried veterans and all thoroughly-trained warriors.

A glance at the map will show why Germany was very vulnerable, although for the most part rather insensible, to attack in her colonial possessions. The Allies' naval supremacy destroyed at once every chance of reinforcing the mostly feeble German contingents scattered in the dependencies and even shrouded their fate for a time in obscurity. But the Germans contemplated the almost inevitable loss of most of their colonial empire with comparative equanimity, convinced as they were that the deprivation would be only temporary and that the real issue would be decided exclusively on the European battlefields.

The operations in Africa during the present war, like the early colonial conflicts in North America, have been invested with a distinction out of all proportion to the number of the forces engaged, by reason of the incalculable importance of the eventual results for mankind and of the exceptional and varied conditions, which offered unusual

Intrenchments made by British native Indian troops in defence of the Suez Canal.

Artillery of British East African forces.

scope for individual initiative. In the majority of cases the forces were very small and consisted exclusively of native police troops with a few local European volunteers for special service and European officers. In some instances marines were landed from the naval squadrons, but these were available only near the coast.

Upon receipt of intelligence of the outbreak of hostilities in Europe, Captain F. C. Bryant, who held the temporary rank of lieutenant-colonel, as senior officer of the British Gold Coast station, without waiting for orders, led a small detachment eastward into Togoland, seized Lome, the coast town and capital, and coöperated with a small French force from Dahomey on the east in the pursuit of the small body of German police troops with their white officers and a few white volunteers, who retreated in the direction of Kamina about 100 miles inland. The powerful wireless station at this point, by which direct radiotelegraphic communication between Germany and the German dependencies in Africa had been maintained, was dismantled in the night of August 24-25 by order of the governor, Major von Doering, to prevent its use by the enemy. After some minor encounters and an unsuccessful attempt to obtain special terms, the Germans surrendered unconditionally and the Allies marched into Kamina on August 27th.

Soon after the commencement of hostilities the British and French blockaded the coast of Kamerun and invaded this German dependency at several points. One British column, crossing the frontier on August 25th captured a German fort at Garua, but was afterwards so heavily counter-attacked that it was compelled to retreat into British territory, losing its commander, Major (acting lieutenant) Maclear. Another British column from Nigeria occupied Nsanakang on August 25th. But the Germans attacked the British garrison posted there on September 6th and defeated them in a hotly-contested engagement, in which both parties suffered heavy losses in proportion to the numbers engaged.

In the meantime, a force of 300 Senegalese in the service of France took by surprise the German post of Singa on the Ubangi, a tributary of the Congo, in the eastern part of Kamerun, while French forces advanced from Libreville in the south. Late in September an Anglo-French expedition under Brigadier-general C. M. Dobell, operating in the coast districts of Kamerun with the support of an Allied squadron, captured Duala and Bonaberi, the former considered a very strong post, and took several hundred prisoners.

The native infantry and police and, in addition, a larger white population capable of bearing arms gave the Germans much stronger available forces in East Africa than in Togoland or Kamerun, stronger forces at first than any which their opponents could muster against them on the borders of the territory. In the conterminous British dependency of East Africa there were the native police and the East African Rifles, which were supplemented by local volunteers and reinforced, as soon as possible, by strong contingents from India.

The British promptly blockaded the coast of German East Africa and the Germans evacuated the port Dar-es-Salaam, destroying the wireless station. After several minor attempts to raid the British territory, a German column numbering about 400 crossed the border of Nyasaland on September 8th, but were defeated the next day in an attack on Karonga, near the northern extremity of Lake Nyasa. About the same time a Belgian force operating between Lake Tanganyika and Victoria Nyanza was defeated and expelled from the German territory.

The British campaign in the southwest undertaken by the Union of South Africa involved eventually by far the most considerable operations in Africa. For the German forces in Southwest Africa consisted of about 10,000 mounted infantry and artillery, all well trained and equipped, with a camel corps numbering about 500. This dependency, where the German government had found it necessary to deal with serious native uprisings, was very effectively organized for military purposes. Blockhouses, like nerve-centers of the administrative organism, studded the country and were connected with one another by telephone and with the capital, Windhoek, by wireless and underground telegraph, while an extensive system of roads and railways had been developed systematically in accordance with strategical requirements.

Early in the campaign a German force numbering about 2,000 entrapped two squadrons of the First South African Mountain Rifles and a section of the Transvaal Horse Artillery in a narrow defile and forced them to surrender after a gallant fight in which they suffered heavy losses. The first important advantage for the British was the capture of Lüderitz Bay on September 18th, the point where German authority was first established in that part of the continent, the only important harbor in the colony. But on September 24th the Germans occupied Walfish Bay, a port which had been retained by the British as an enclave in the midst of German territory.

A British troop advancing eastward from Lüderitz Bay was defeated by German forces near Garuab on December 16th.

The partial success of a dangerous insurrectionary movement among the Boers, which had been fostered by German intrigues, necessarily interfered with the further development of the British campaign in German territory.

The progress of this interesting minor outburst, produced as it were by sparks from the great conflagration, must be reserved for comprehensive treatment later.

The inauguration of the world-war was quickly followed by the effacement of all the German dependencies and stations scattered in the Pacific.

The Germans had just completed a wireless station at Tafaigata in Samoa, when an Australian war-vessel escorting a transport with an expeditionary force from New Zealand entered the harbor of Apia, the capital of the German colony, which was occupied on August 29th. The German governor and several officials were taken to New Zealand for internment.

A German wireless station was being completed at Bitapaka on the island of New Pomerania in the Bismarck Archipelago, when news of the outbreak of war was received on August 5th. The seat of government was immediately transferred from Rabaul to Toma in the interior, and the colored police troops, 300 in number, strengthened by a few German recruits or volunteers, prepared for defense. But an Australian expeditionary force occupied Herbertshöhe on September 11th and Rabaul a few days later, and, after some spirited bush-fighting in the neighborhood of Bitapaka, captured the wireless station and took Toma.

The Australian squadron seized Naura in the Marshall Islands about September 1st without any opposition, and destroyed the wireless station; and with the capture of the wireless station in the neighboring Caroline Islands, communication between Germany and her dependencies in the Pacific was completely abolished.

The crowning success of the Australian squadron and the expeditionary force was the capture of Friedrich Wilhelm, the seat of the government of Kaiser Wilhelm Land, the

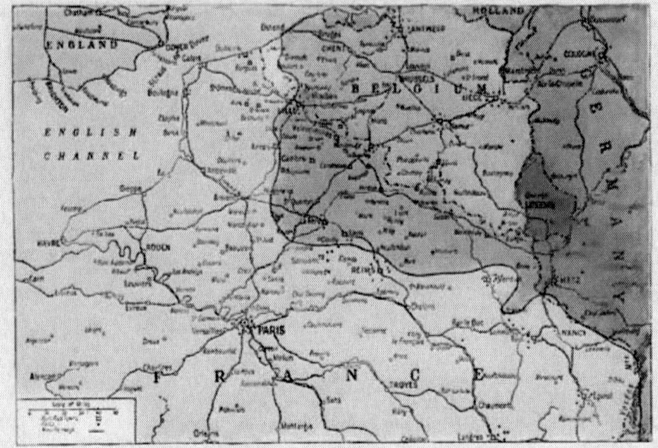

Map showing by the shaded portion the territory held by the opposing armies, December 31, 1914.

German part of New Guinea, which had long been regarded as a possible menace by the Australian States.

A Japanese squadron occupied Jaluit, the seat of government of the Marshall Islands, on October 3d, taking prisoner the chief official.

The feeling of annoyance with which Japan regarded the presence of the Germans in the leased territory of Kiau-Chau and the alacrity with which the Japanese grasped the opportunity afforded by the Great War to eject them are due to circumstances which have been explained in the first volume of this work. The Germans regarded this possession with peculiar pride as the most successful achievement of their modern expansionist policy. During seventeen years neither resources nor energy had been spared in making this vantage point a model colony, a pattern of German efficiency and thoroughness, and an impregnable stronghold. Tsingtau, the urban center, had developed rapidly along systematic German lines, with costly waterworks, fine streets and public buildings, and excellent harbor facilities. The leased territory was a prosperous offshoot of the Fatherland transplanted in the Far East.

Soon after the expiration of the time limit expressed in their ultimatum to Germany, August 23, 1914, the Japanese blockaded Tsingtau. The German defenses of Tsingtau were equipped with about 600 guns. The garrison, commanded by the governor, Naval Captain Meyer-Waldeck, numbered between three and four thousand, mostly marines. Five or six hundred German civilians or reservists hastened to Tsingtau from different places in China to offer their services. There were eight German war-vessels and the Austro-Hungarian cruiser *Kaiserin Elisabeth* in the harbor at Tsingtau at the time.

It is said that the authorization of Japan was asked for the removal of the *Kaiserin Elisabeth* to Shanghai, where she could be disarmed and interned, but that suddenly, despite the favorable attitude of the Japanese, instructions from Vienna directed the Austro-Hungarian ambassador to take his leave of Tokio and the commander of the *Kaiserin Elisabeth* to coöperate with the Germans in the defense of Tsingtau.

The expeditionary force from Japan landed at Tsimo, which was made its base, ten miles outside the limits of the leased territory of Kiau-Chau, on September 12th, and two days later the Japanese advanced against the Germans, forcing them to retire within their defensive lines, and the real siege of Tsingtau began. Twelve days later a British expeditionary force arrived at Laoshan Bay. The Japanese and British were repulsed in their first attack on October 6th, when their right wing was exposed to the fire of the cruiser *Kaiserin Elisabeth* and of a German gunboat, besides that of the forts, and the Japanese suffered considerable losses.

But after a fierce bombardment of the German positions by land and sea with heavy artillery, including several 28-centimeter mortars, continuing without interruption for nearly nine days, the Japanese and British advanced for the final assault on the night of November 6-7. The conflict raged with the greatest fury around the fort on Iltis Hill, the most important position in the German defenses. The capture of this final bulwark at the point of the bayonet necessitated the capitulation of Tsingtau.

Early on the morning of the 7th a white flag was raised on the observatory and later on the forts fronting the sea, and at nine German officers appeared within the Japanese lines to arrange the terms of surrender.

About 3,000 prisoners fell into the hands of the Japanese and were transported to Japan for internment. There were 436 German wounded in the hospitals of Tsingtau. Captain Meyer-Waldeck, the German commander, had himself been wounded in the defensive operations. The losses of the Japanese in storming the forts were reported to be fourteen officers and 426 men. The news of the fall of Tsingtau was received in Germany with a universal feeling of bitterness and chagrin.

The wide extent of the field reviewed in the present chapter is answerable for unusual demands upon the imagination of the reader, who must gird himself for another fanciful flight of several thousand miles, this time from the eastern to the western extremity of the continent of Asia.

As evidence of Turkish chivalry or of the timeliness of German diplomacy the circumstance may be mentioned that Turkey entered the war at a moment of comparative depression in the fortunes of her prospective allies, when a formidable Teutonic offensive in Poland had just been abandoned, the fruitlessness of the great effort in Flanders was becoming daily more apparent, and Austria-Hungary had accomplished nothing in her campaign against Serbia; a moment, in short, when the accession of a warlike ally was a peculiarly gratifying encouragement.

The reorganization of the Turkish army in fourteen army corps, as before the Balkan Wars, had undoubtedly progressed very rapidly since the outbreak of the European war in August, under the able direction of General Liman von Sanders. The plan of operations, prepared of course in agreement with the views of the German General Staff, contemplated a campaign against Egypt and another on the Caucasian frontier, while British initiative added a third field of action in lower Mesopotamia.

The most interesting feature of the situation created by Turkey's belligerency was the anomalous position of Egypt, which was virtually under British protection while acknowledging Turkish suzerainty. With the question of Egypt's allegiance was intimately associated the security of the Suez Canal, which was commonly regarded as an absolutely vital artery of traffic for the British Empire. The ominous but unobtrusive steps have already been described by which the Turks, with German encouragement and support, had been approaching this tempting and loosely-guarded prize. The apparent uniqueness of the opportunity stimulated the insidious activity of German secret agents in Egypt. Presumptive evidence that the activity of a certain Dr. Prüffer, who had been intriguing in Cairo, had the indorsement of the German government seemed to be afforded by his open, official connection with the German Embassy in Constantinople after the outbreak of the war. Lieutenant Mors, a German officer in the Alexandria police, who was arrested in October, confessed that he had just returned from a conference with Enver Pasha, to whom he had been conducted by a German official formerly in the German diplomatic agency in Cairo, and disclosed some of the intrigues fostered by the Germans and Turks against British authority in Egypt and India.

The Khedive Abbas Hilmi Pasha, who was visiting his nominal suzerain, the Sultan, at the time of the outbreak of hostilities, was won over to the Turco-German designs in respect to Egypt, and it was announced that he would soon return at the head of an army for the purpose of liberating his country from British domination.

On December 19, 1914, the establishment of a formal British protectorate over Egypt and the succession of Prince Hussein Kemal Pasha, with the title Sultan, to the throne of his dispossessed uncle, Khedive Abbas Hilmi Pasha, were officially announced by the British government.

Turkish positions of considerable strength on the Red Sea fell into the hands of the British. One of these was Akaba, situated at the northern extremity of the eastern arm of that sea, the intended base for one of the Turkish

Fort Iltis, at Tsing-Tau.

German twenty-eight centimetre howitzer and turret at Tsing-Tau destroyed by the Japanese gun fire.

army corps destined for the invasion of Egypt. Another was Sheich Seyd on the rocky peninsula threatening the southern approach to the Suez Canal. Three battalions of Indian troops landed under cover of a war-vessel on the low isthmus connecting this stronghold with the mainland, and captured all the Turkish positions.

The British government formally annexed the island of Cyprus on November 6th, in consequence of the state of war existing between itself and the Turkish government. The British had occupied and administered this island since 1878, paying annually £92,800 to the Sultan in acknowledgment of his ultimate rights of sovereignty. In return for this tenancy the British government had promised to defend the Asiatic possessions of the Sublime Porte against the encroachment of Russia. But now, by the strange revolution in policies, Great Britain, in league with the very power which was then the chief source of her apprehension, was striving to subvert the integrity of the Ottoman Empire which she had once been most determined to uphold.

It was natural that the British should hasten to forestall their enemies in the possession of the region of the lower Euphrates and Tigris, the possibilities of which had been emphasized by the construction of the Bagdad Railway. An expedition from India under Lieutenant-general Sir A. Barrett and Brigadier-general W. S. Delamain disembarked at the head of the Persian Gulf, defeated the Turkish forces in two engagements and occupied Basra situated at the confluence of the two great rivers, the contemplated terminus of the Bagdad Railway, on November 21st. The operations in this quarter, though commenced on a comparatively insignificant scale, held out the vague but seductive promise of an imposing development, the possible conquest of the most ancient seats of dominion and opulence, and the appropriation of the Garden of Eden.

CHAPTER XIII

War's New Aspects

Principles of strategy universal, their application variable. Changed international considerations as to Belgium. Plan of passing through Belgium and the German offensive. General advantages of the offensive. The tactics of the campaign: German mass attacks; the part of the infantry; the use of the cavalry; artillery support. Transportation: the railways and their various uses; motor-propelled vehicles. Air service: types of aircraft; anti-aircraft guns; improvements in air-machines; the modern Zeppelin: the captive balloon. Means of communication. The lesson of the campaign.

Wars are no longer confined to the operations of professional armies but are fought by whole nations. The "nation in arms" has become a reality and if in war the entire population does not actually stand under arms, it is at least mobilized for war. The art of war makes use of every product of industry, art, and science. The conduct of modern war absorbs every branch of human activity, coördinates, intensifies, and directs it for the sole purpose of subduing the enemy. The principles of strategy are universal and eternal and have for their object the destruction of the armed forces of the enemy, but the application of those principles varies according to the character of the peoples engaged in the conflict, with every war, in every theater of operations, and in every campaign. A young, vigorous, growing, progressive, and aggressive nation, conscious of its own strength, will as certainly carry the war into enemy territory as a poorly organized, or a merely fully developed and comfortably rich people will fail to take the initiative. The surprises of the Great War have been in the relative strength of the belligerents; in the employment of the machinery and materials of war; in the resources of the countries at war, which should have been better known; in the character of the man behind the gun; not in the numbers available, which were well known, nor in the broad lines of operations which had been foreseen by the students and writers of the several Great Powers engaged. That the armies of the central powers, under the leadership of Germany, would take the offensive was as well known to Great Britain, France, and Russia as to the Great General Staff which planned the campaigns; that they could take the offensive on only one front at one time was a condition imposed by the numerical superiority of the Entente armies; that the initial campaign of the war would be the German offensive in France was perfectly clear to all students of military affairs, and even to the casual reader. Just what form this campaign would take was known only to the German and Austrian leaders responsible for the conduct of the war.

An invasion of France through Belgium was a probability which received the consideration of all the powers concerned. Von Moltke, in outlining Prussia's primary military measures in case of war with France, in a memorial prepared in 1858, says: "Belgium sees in France the only actual enemy to her national independence; she considers England, Prussia, and even Holland as her best allies." He argued that the Netherlands lay outside of the probable theater of war, and that to occupy Holland at the very start would be an unjustifiable splitting up of the Prussian forces. Taking into consideration that the English army was in India and would be required there for years, he pointed out that Belgium could expect help only from Prussia in case she was attacked by France. He said: "If we respect Belgium's neutrality we will protect thereby the largest part of our western frontier." The conditions in 1914 were largely reversed, Belgium no longer considered France an enemy to her national independence. Great Britain, from being the traditional enemy of France had become her ally—an alliance directed against Germany. Belgium, from being a protection to Germany's western frontier in a war with France became, if not a probable enemy, at least a constant preoccupation. It cannot now be known when the campaign against France through Belgium was decided upon.

The plan being adopted, the campaign in France became at once the most gigantic as well as one of the simplest movements known to the Art of War. It may be described as an enveloping movement in which the holding attack extended from Switzerland to Verdun and the enveloping attack from Verdun to the frontier of Holland and in which the units employed were field armies. The object of the holding forces in the southern half of the frontier was to threaten the entire line so as to keep the French in ignorance of the direction of the principal attack, thus preventing them from shifting troops to other threatened points, and to offer effective resistance to any serious attack launched by the French against that part of the frontier. The main attack was designed to envelop the French left, including the Belgians and the British, and roll it back on the center, producing a congestion and confusion which would result in disaster, or to break through the lines, detaching the French left wing from the central group of armies. By very skilful maneuvering the French were able to withdraw the armies of the north without permitting them to be rolled up in confusion or having them cut off by the great German drive until the left rested on the intrenched camp of Paris.

French guns mounted on special railway trucks.

French one hundred and fifty-five millimetre guns.

It is difficult to overestimate the advantages of the offensive in the initial campaign of a war in which the numbers employed are so great. The aggressor virtually launches his campaign when he orders his mobilization, and his troops begin to move by prearranged schedule, each by the shortest line, to the points on the frontier from which they are to begin actual hostilities. The defender must order his mobilization at the same time. The mobilizations of the armies are based on the maximum capacity of the railroads of the country, in accordance with schedules prepared in time of peace, in which the day and hour of the departure of every unit from its home station and its arrival at its point of detraining are fixed. In the very nature of the movement any departure from the prepared schedules results, if not in confusion, at least in delays which may prove fatal. It is impracticable to change the zone of mobilization, once mobilization is ordered. Since the plans of the offensive cannot ordinarily be known, the attack must come in the nature of a surprise even though the offensive plans become known before the mobilization is complete. The army which assumes the defensive rôle in the initial campaign will naturally mobilize its great reserve in some central zone in the rear of the frontier, from which it may be thrown to the threatened points by a system of railways parallel to the frontier. Even this cannot be done until the original mobilization is complete. The defensive follows the lead of the offensive and much valuable time is lost. The defender will be very fortunate if his lateral rail communication is not cut before he can make use of it.

The Teutonic allies having initiated the war by a supreme effort in the West their defensive attitude in the East was a necessity. The duties of all the belligerents were at once clearly defined. In order to coöperate with her allies, Russia, as soon as her mobilization justified it and sooner than had been anticipated, took the offensive in East Prussia and in Galicia. Her prompt and vigorous action probably saved the Allied armies in France from disaster.

In the tactics of the several arms it can be said that there has been no violent upheaval as a result of the Great War. Much space in the daily press was utilized in describing the wasteful, mass tactics of the German infantry attack in the great drive through Belgium into France. This may be likened to the outcry from all sides about the use of dum-dum bullets. The reports were in both cases largely the product of the excited imaginations of observers unused to war. The success or failure of the attack depends on the ability or inability of the attacker to gain fire superiority over the defender. The defense will put in action the greatest number of rifles that his defensive lines will accommodate, which is one man per yard. The attack cannot be expected to succeed with fewer rifles on the line and more cannot be used. Reserves must be strong; the gaps in the firing line must be filled and the line maintained at its maximum strength until the moment of assault. A commander who undertakes an attack is not to be excused for failure so long as he has a formed reserve in hand. Reserves are provided to be used and if a position is carried without the employment of all the reserves they find their most important use in the pursuit. The attacker expects heavier losses than he can inflict on the enemy until the position is carried and the retreat begins. Then the successful force begins to reap the fruits of its victory. The vanquished loses according to the vigor of the pursuit and his own skill in withdrawal, but always heavily.

That the German mass attacks received the exclusive attention of the press is due to the fact that the Germans were on the offensive and attacking constantly up to and including the Battle of the Marne. It is likely that the

German commanders weighed carefully each situation, knew their own needs, estimated the cost in lives, and assaulted the desired position, or brought up heavy artillery to reduce it, or left an investing force, according to well understood principles of strategy and tactics as applied to the offensive. It is only the offensive that produces decisive results; a defensive attitude can only be justified as a temporary expedient.

Infantry remains the arm that decides the final issue of combat. The fact that in the trench warfare in France it can neither advance nor maintain its line without artillery support does not diminish the relative value of infantry; it simply makes greater demands on the artillery, as is always the case in siege warfare. The lines are covered by impassable obstacles which must be removed before they can be assaulted. Only artillery is able to clear away those obstacles and prepare the way for the assault. Only infantry is able to seize and hold intrenchments in the zone of siege operations. The strongest points cannot be held without artillery support, for the strongest fort of concrete and steel and earth may be destroyed by the modern siege-gun, once it is located. The number and power of the heavy field-guns now in use with the armies in the field is unprecedented. The Allies are said to be using 15,000 guns in France and Belgium. Since they have been unable to establish any superiority in artillery, their opponents must be using an equal number.

The first few weeks of the war in France saw the cavalry used to the limit of endurance. The German cavalry covered the right flank of the armies in advance, in retreat, and in the race for the coast which followed the Battle of the Marne. It was constantly in contact with the British and French cavalry, and cavalry combats were of daily, almost hourly occurrence. It is reported that a large force of French cavalry did not unsaddle their horses for five days and nights. Corresponding demands were made on the cavalry employed on both sides. The result was a partial destruction and complete exhaustion of the mounts of the cavalry. It was a heavy toll on the opposing armies, but horses can no more be spared than men when the safety of armies, the existence of nations even, may depend on them. As soon as the operations resulted in a deadlock from Switzerland to the coast, the German cavalry took its place in the trenches by the side of the infantry until it was required in another theater of war. The cavalry of the Allies had corresponding tasks consigned to it. Cavalry is a resourceful arm; it operates mounted habitually, but must be in every way the equal of dismounted troops when separated from its horses.

Not only the trench warfare in Europe but the increase in power and efficiency of guns and gunners has added greatly to the value of artillery. Prepared positions and intrenchments cannot be successfully attacked without powerful artillery support. Field-guns, before the war, depended largely upon shrapnel. Shrapnel have much greater efficiency against troops in the open than shell, but are powerless to destroy earthworks. Only by the use of high-explosive shells can the trenches and the obstacles which cover them be destroyed and the way cleared for the attack. Even with such ammunition the light field-piece is not very effective. Heavy field-artillery has assumed an unprecedented importance in warfare. One of the surprises of the war was that guns had been developed capable of transportation in the field and of operation from mobile or rapidly constructed platforms which were able to destroy every class of permanent fortification in a few shots. The French have perhaps the most efficient light field-gun employed in the war, but the superiority of

Photograph of enemy positions, showing one of the uses of flying machines in warfare. This view of the French trenches was made by a German aviator and was found by the French on the body of a dead German after an attack.

the German heavy guns easily offsets the advantage thus gained.

The Great War differs from previous wars first in the unprecedented numbers of the armies in the field. The mobilization, concentration, and supply of these armies is made possible only by the development of modern means of transport. First in importance is the railway, which takes a place in war second only to arms and munitions. The first great demand on the railroads was for the mobilization, in which France and Germany each used probably 5,000 trains, in addition to those used for concentration. Single-track lines accommodated twenty and even more trains daily each way, while the double-track lines were able to move an army corps per day, handling in some cases more than 200 trains. Troops should move by rail from 300 to 400 miles in twenty-four hours, while on foot they make from 12 to 25. The great mobilizations and concentrations could, however, with certainty be made by marching if it were not for the question of supply. The supply of rations and munitions to the enormous concentrations in limited areas is not practicable without mechanical transport, and only the railroad is adapted to the transport of the heavy material.

In addition to being a mere means of transport the railway coach and car are adapted to a variety of military uses. Water-cars and refrigerator-cars are extensively used; complete sanitary trains with almost every convenience of a stationary hospital transport the wounded from the front; armored trains provide protection against small-arms fire; and flat cars so arranged as to distribute the load over a number of axles and to take the weight off the wheels during action are used as gun platforms for siege-guns. Railroads are not usually so vulnerable that the ordinary means and time available for their destruction can render them incapable of rapid repair or reconstruction. The destruction of tunnels and large bridges are the only serious obstacles to the engineer troops that accompany armies. Unimportant bridges are hastily replaced by temporary structures and tunnels may often be avoided by laying track around them. Railroad construction corps are able to lay new track in fair country and keep up with the advance of the victorious armies against a stubborn foe.

An auxiliary to the railway, but one capable of a great variety of uses, is the motor-propelled vehicle. It transports all but the heaviest materials on the good roads of central and western Europe at a speed far greater than that of the horse-drawn vehicle and approximating for short distances that of the railway train. Every class of motor-car finds its use with the army; private cars are requisitioned for the use of staff officers and dispatch carriers, while large touring cars, auto-omnibuses, trucks, and even taxicabs have been extensively employed in the transportation of troops as well as of supplies. It has been estimated that 1,000 or 1,200 omnibuses or trucks capable of carrying thirty men can transport an army corps at the rate of seven to ten miles per hour, and that they would occupy a road space of only about two-thirds of the same force marching, thus adding greatly not only to the rapidity of movement but to the time required for concentration or deployment. Such a movement is practicable for infantry only. The cavalry and siege-artillery of an army corps cannot be transported by motor-trucks, and the heavy artillery can be conveyed only by heavy tractors at a much reduced speed. The movement of an army corps then without rail or water transport is limited to the rate of march of its artillery.

The most important as well as the most spectacular development in machines of war has been in the air service. The captive balloon has long been used for purposes of observation; free balloons have occasionally been used by a besieged force for communication with the outside world; dirigibles of various types have been experimented with in peace; and the heavier-than-air machine had demonstrated that it would find use in any future war. But the armored, fighting air-machine equipped with machine-guns and bomb-throwers is a creation of the war.

Before the war the Great Powers of Europe, notably France and Germany, had developed the American invention of the Wright brothers until the aëroplane was a familiar figure about the capitals and other large cities of the continent. The centers of aviation offered diversion and amusement daily to thousands of spectators who thronged the aviation fields. There were monoplanes and biplanes of almost as many types as there were inventors and builders who sought their fortunes in the production of craft for navigating the air. Flying became a popular sport, more perilous, more exciting than racing, and requiring greater expenditures without the corresponding sources of revenue. The air-machine had no earning capacity except as an exhibition feature and depended for its support largely on the governments and on popular subscription. Particularly in France, led by the press, popular subscription for the control of the air amounted to millions. France began by leading in the air. Conceding the control at sea to Great Britain and the greatest strength on land to Germany, supremacy in the air, as a national aspiration, became very popular in France. Germany followed in the use of heavier-than-air machines, but in a characteristic German manner, under the leadership of the government, which was quick to recognize its military value, the German flyers were not long in establishing world records, notably the endurance flight of twenty-four hours accomplished by a German aviator a few months before the outbreak of war. The same thoroughness which directed her industry and created her army made Germany superior to any of her enemies in the use of aircraft for war purposes at the beginning of the war.

The demands made on the aviation service by the armies in the field have resulted in the production of three types of craft based on tactical requirements. The defensive machine guards the front of the army of operations or important points in the interior against hostile aircraft. For this purpose a light machine offering a small target is used, in which armor protection is less important than extreme speed and great climbing power. The reconnaisance machine flies over the enemy's line, directs the fire of the artillery and photographs enemy positions. This requires a steady, safe motor, to secure which it is necessary to carry greater weight at a sacrifice of some of the speed possessed by the light anti-aircraft guard. The machine used to bombard hostile troops and positions must carry great weight in explosives, in guns, and in fuel; this requires a heavy, safe motor of great power, and results in a machine which develops less speed than the other two types. If at the beginning of the war it was recognized that aëroplanes could serve any other purpose in war than that of reconnaissance it was only the Germans who held such belief. The air attacks on Paris during the great drive in France, unimportant as they were, show that Germany had not overlooked the possibility of using her aëroplanes for attacking important points. The war soon demonstrated, however, that air scouts can no more obtain information without fighting for it than can cavalry. To prevent the enemy's reconnaissance it is necessary to attack him in the air. Defense by guns operated from the ground below are ineffective against the flyer, whether they are stationary or portable. Only important points can be guarded with guns and the guns are effective at only very limited altitudes. The pilot and the observer as well as the vulnerable parts of the machine are protected from fire from below, and have little to fear from small arms when flying at a height of 1,000 or 1,200 yards. On the other hand, to be protected from artillery fire the air-machine must maintain an altitude of two miles or more. However, neither observation nor bomb-throwing is very effective at such altitudes and the airman must take his chances. The aëroplane makes a very poor target. The vulnerable parts present a surface hardly greater than a square yard, while the largest machine looks about the size of a postage stamp at 8,000 feet elevation. Considering, then, that it moves in three dimensions at the rate of thirty yards per second the probability, or improbability, of hits by surface guns is apparent. For the gunner, reducing the range increases rapidly the difficulty of pointing.

At the beginning of the war Germany had 300 aëroplanes, Austria 100, France 300, and Great Britain 100. The British machines were poorly organized for war service; France had a number of squadrons of four machines and some experiments in reconnaissance had been made; the Germans had gone more thoroughly into the application of flying to military purposes and were employing many machines in long flights under conditions simulating actual war. The result was the remarkable efficiency shown by the German air service in the first months of the war, although fighting on two fronts. The number of heavier-than-air machines employed by the belligerents cannot at this time be ascertained, but it is estimated that the Allies are using more than 1,500 machines in France.

The French are reported to be making great efforts to create a type of mighty triplane carrying twelve men and four 1½-inch guns at a speed of eighty miles an hour.

They are designed to be used to bombard enemy positions in the actual zone of operations as an auxiliary to the land batteries. The air scouts for the heavy machines are small biplanes with great speed and climbing ability and armed each with a machine-gun. These scout machines can be manipulated without the use of the hands and carry only one man, who is gunner as well as pilot and observer. Great Britain has made great efforts to improve her air service regardless of the cost, which it is estimated will exceed one billion dollars for the year 1916.

The Germans have made great improvements in their machines. A British report credits them with having produced a type of tractor biplane driven by two motors, each of 100 to 150 horse-power; the machine is said to carry a pilot and two gunners armed with machine-guns, to have tremendous speed, and to be able to remain in the air for six hours. The Fokker, a Dutch invention used by the Germans, is a small, high-power machine of great speed, with a climbing power unequalled by any other machine produced. Its superior maneuvering ability makes it a powerful defensive weapon, but it is a small machine unsuited for reconnaissance or bomb-throwing.

Full information on the subject of the air service of the armies of the belligerents is naturally not available. A few points of importance are, however, well known. There are very few motors of less than 100 horse-power and, except for the very heavy machines, the minimum speed requirement is about 100 miles per hour. Before the war, aviation schools counted on six months to develop a pilot; the training camps of the armies now produce pilots in six or seven weeks, and the student aviator who does not become proficient in twelve weeks is declared inapt.

At the beginning of the war the efficiency of the German air service, like that of some of her other war services, had been underestimated by observers who should have been better able to foresee the result in war of German thoroughness, which was well known. Germany had an air fleet consisting of about ten Zeppelins. She had an equal number of smaller airships of the Parseval type with a speed of from twenty-five to forty miles per hour, but they have not sufficient power to fly in adverse winds and are so slow that they are easily attacked by aëroplanes. Only the Zeppelins have been efficient.

A modern Zeppelin is more than 600 feet long, 60 or 70 feet in diameter, and develops a speed of 60 to 75 miles per hour. It carries armor and armament which render it reasonably safe from attack by aëroplanes, except from above. It is said to be able to climb more rapidly than the aëroplane and possesses the distinct advantages of being able to rise vertically and to hover over the object of attack. Before the war Russia had a number of small airships, but they have not proved of any particular value, and neither France nor Great Britain had accomplished more than mediocre results with dirigibles. France has employed ten or a dozen small ships of the non-rigid and semi-rigid types, principally in coast guard duty, but none of these ships has the speed or maneuvering ability of a Zeppelin. Great Britain is said never to have produced an airship that has proved a success. The only air-machine that is capable of long distance raids is the Zeppelin. It makes use of the aids to navigation employed by ships at sea and may direct its course to distant points concealed by fog or clouds without reference to directing points below. In one of the bombardments of Paris the city is believed to have been invisible to the observers on the airship. One of the most important missions of the Zeppelin is believed to be scouting for the navy; its ability to remain stationary, receive as well as send wireless messages, and its possible destructive effect against submarines which are visible from above even when submerged, seem to clothe it with all the essential properties of an efficient naval scout. It is difficult to justify by results the expenditures of money and the vital energy which have gone into Zeppelins, but Germany continues to build them. Their raids have produced a depressing effect on the British public. The Zeppelin, which stands in a class by itself, is an offensive weapon, to oppose which the Allies have nothing.

For purposes of observation the captive balloon stands next in importance to the flying machines. The Germans use a sausage-shaped balloon that is much steadier than the spherical balloon, which is so unstable as to make observations very difficult and often produces such violent nausea that observation is impossible even for the most seasoned aëronaut.

Carrier pigeons still find their place in war. They have been used probably as extensively, though not so exclusively, as in former wars. Wire and wireless, aëroplane, heliograph, flag, automobile, motorcycle, mounted messenger, and dog are means of communication at once rapid and reliable; but they are not always practicable. An observer landed from an aëroplane behind the hostile lines may make use of carrier pigeons when no other means of communicating is available. If operating in friendly territory occupied by the enemy he may avoid detection and secure and transmit information of the greatest value. Dogs have been employed in the Great War to an extent not heretofore known. The "Dog of Flanders" drawing a machine-gun has been made a familiar figure by the photographer. To a less degree dogs have been used by sentinels on detached posts to give quick warning of approaching danger, and with patrols reconnoitering roads and by-ways in advance. Sled dogs from Alaska have been carried to

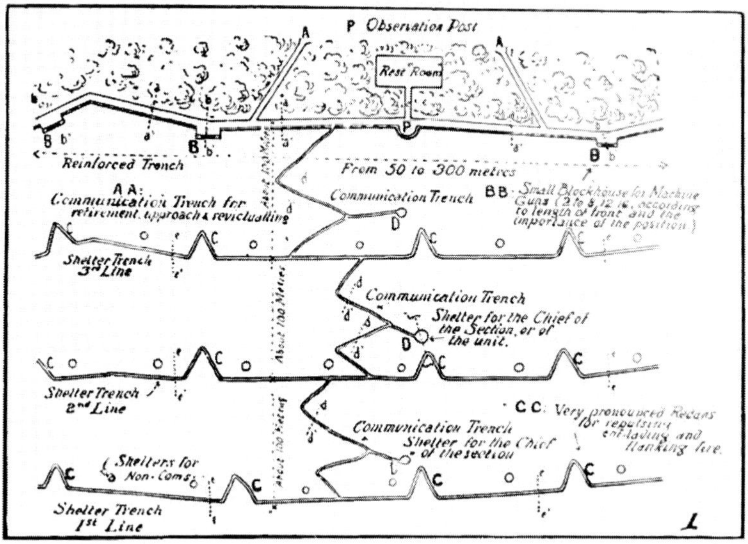

Arrangement of the German defensive and protective trenches.

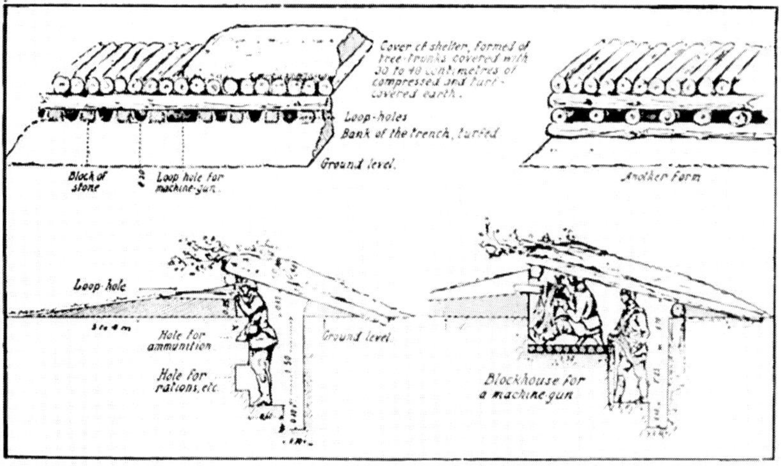

Reinforced trenches: Details of roofs, loopholes, and the form of the excavations.

France, where they are used in the mountains for transporting supplies, munitions, and the wounded.

The application of recent inventions in modern war is startling because of the enormous resources of the belligerents, which permits their employment on such a gigantic scale, but no machine of war has appeared which will cause an upheaval in the principles of strategy and tactics. Nevertheless, the Great War teaches many valuable lessons. Leaders of troops will await impatiently the detailed histories of operations which are to be their guides in future wars; but the responsibilities and duties of the nation are already clear. Modern war mobilizes the entire population and organizes every industry of the State in the service of national defense. "In time of peace prepare for War" states the problem; universal military training is the solution.

CHAPTER XIV

The Alleged Atrocities in Belgium and France

Accounts of outrages imputed to the German armies in Belgium and France received with amazement and horror, followed by an involuntary reaction of doubt. The committees for investigation and their indictment. The attitude and counter-charges of the Germans. Various forms of evidence. Destruction incidental to military operations; doubtful cases: Reims, Arras, Ypres, Mechlin. The essential distinction between isolated and irresponsible, and deliberate and systematic acts of brutality, and the paramount importance of the latter. Needful restrictions in the material admitted to discussion in the present chapter; the confinement of the argument to undisputed facts. Intentional destruction of property and acts of severity against the civilian population. The alleged organized people's war in Belgium; Aerschot, Dinant, Louvain. The international conventions relating to the people's warfare and the contrasted attitude of the combatants respecting the conditions for the possession of the rights of belligerency. Conclusion.

Scarcely had the nations been launched upon the seething flood of the world-war, when the most appalling rumors of cruelties and atrocities began to emanate from the seat of hostilities in the West, vague and desultory at first, but quickly increasing in vividness and persistence, tales of pillage and of the wanton destruction of property, of the violation of women and children and of the indiscriminate slaughter of innocent civilians, and finally of every kind of abominable crime and brutality which had ever defiled the blood-stained annals of warfare.

The world stood aghast, bewildered with amazement and horror. The imagination was staggered by the enormity of the offenses reported. Many persons whose judgment was not swayed by prejudice or passion refused to admit without indisputable evidence that a people whom they had always regarded as preëminently kindly, cultivated, and orderly could have lost so suddenly all restraining sense of compassion and humanity. The nations directly interested in the victims of the German invasion took systematic steps to collect and publish the evidence for the varions acts of cruelty and lawlessness which were said to have been committed.

Space will hardly permit the rehearsal of even a representative selection of the occurrences recorded in these reports. But a summary account of a few of them, chosen partly at random, partly for their connection with events already narrated, and partly for their special importance in connection with the general conclusions, which will be developed later, will afford a general idea of the nature of the entire series. The fundamental contradiction which complicates the whole question of atrocities should be stated clearly at the outset. In practically all cases where property was deliberately destroyed and civilians were killed, the Germans declared that they had been fired upon or otherwise attacked in violation of the accepted usages of war, while the Allies just as universally denied this allegation. Indorsement of the accounts presented by the different belligerents is not implied in the direct form of discourse which is used for the sake of brevity in the summaries of them that follow.

Within a few days after the outbreak of the war, the Belgian Minister of Justice, M. Henry Carton de Wiart, appointed a Commission of Inquiry composed of prominent statesmen and jurists, charging them with the task of examining into the violation of the rules of international law and of the established usages of war. This commission sat in Brussels. But after August 18th, when the seat of government was transferred to Antwerp and communication with Brussels was impeded, the minister of justice appointed a sub-committee of the Commission of Inquiry with headquarters at the new capital. In a series of reports the commission thus constituted has published evidence, derived from the sworn statements of eye-witnesses, calculated to show that the invaders of Belgium resorted to proceedings which violated the most elementary conceptions of humanity and are prohibited by the rules of warfare sanctioned at The Hague; that they maltreated and massacred the peaceful population; sacked and burned open and undefended towns and villages; reduced to dust historical and religious monuments, and gave to the flames the famous library of Louvain; and that they practised a deliberate policy of terrorism. On account of their exceptionally crucial character, some of the most conspicuous examples of the alleged German outrages in Belgium are reserved for a later part of the chapter, where they will be treated on the basis of a comparison of the testimony from all the sources.

We commence our survey of some of the other characteristic incidents with the occurrences at Andenne, which is situated on the right bank of the Meuse between Namur and Huy and was connected by a bridge with Seilles on the left bank, a circumstance which invested it with an ill-omened significance in the early days of the German irruption. An advance-guard of Uhlans arrived here on August 19th but found that the bridge had been blown up. A pontoon bridge was substituted and a column of German troops began to defile through the town on the 20th.

According to one of the Belgian reports a shot heard at six P.M., followed by an explosion, threw the German soldiers into a fury of excitement. They began firing wildly in the streets, sacked the town, set it on fire in several places, and shot many persons whom they encountered.

Next morning the inhabitants were driven from their houses and gathered together, and forty or fifty were singled out and executed in expiation of a pretended attack on the German troops.

Tamines on the Sambre between Namur and Charleroi was one of those populous, prosperous villages which had been the distinction and strength of thrifty Belgium. It was occupied for several days in August, 1914, by detachments of French soldiers, and these, supported by a party of the Garde Civique from Charleroi, resisted a German patrol on the 20th, killing several Uhlans. On the next day the Germans occupied the place, sacked and burned 264 houses, and arrested many of the inhabitants. On the 22d, between 400 and 450 of the latter were collected near the bank of the river and summarily executed by the fire of rifles and of a machine-gun, some of the wounded being finished off by bayonet thrusts, while others, it would appear, put an end to their sufferings by rolling into the Sambre.

The German troops who entered Namur at four P. M. on August 23d, conducted themselves in an orderly manner until the evening of the next day at nine, when suddenly, without the slightest warning, they set fire to the city in several places, shot many of the defenseless inhabitants as they attempted to escape from their burning houses, and engaged in extensive plundering.

The Germans entered and sacked the village of Hastière-par-dela on August 23d, killed and wounded a large number of persons, burned the greater part of the houses, and executed the parish priest, a professor of the University of Louvain, the local schoolmaster and others upon condemnation of a court-martial composed of officers, some of whom were intoxicated. About the same time eighteen men, including several priests, were executed by a volley at Surice, because, as a German officer alleged, a girl of fifteen fired on one of the German commanders.

By decree of the French government on September 23, 1914, M. Georges Payelle, First President of the Court of Accounts (Cour des Comptes); M. Armand Mollard, Minister Plenipotentiary to Luxemburg; Georges Maringer, Counselor of State; and Edmond Paillot, Counselor of the Court of Appeal (Cour de Cassation) were appointed a committee for the investigation of atrocities said to have been committed by the German armies in the portions of France which they had occupied. This committee professedly submitted the testimony, which had been received under oath, to a severely critical examination and made a report on December 17, 1914, presenting the facts which it regarded as established beyond dispute.

In addition to a large number of infamous crimes of an isolated character committed in all parts of the territory occupied at any time by the Germans, the report contained the account of many systematic outrages.

At Triaucourt, for example, in the Department of the Meuse, the Germans were said to have given themselves up to the worst excesses, burning thirty-five houses and killing indiscriminately many of the inhabitants.

During the first day or two of the occupation of Lunéville, they were content with robbing the inhabitants. But about 3.30 P.M. on August 25th their attitude suddenly changed. Claiming that the population had fired upon the hospital and had made an attack by ambuscade on the German columns and transports, they began shooting in the streets and setting fire to houses. Although M. Keller, the Mayor, made a tour of the streets with German officers and soldiers to prove the absurdity of this allegation, the synagogue, Hôtel de Ville, and about seventy houses were burned, several persons were shot, and a large contribution was levied on the town as an indemnity.

At Gerbéviller, in the Department of Meurthe-et-Moselle, the German troops, chiefly Bavarians, infuriated by the stubborn resistance of sixty French infantrymen, took vengeance on the civilian population. They rushed into the houses with savage yells, pillaging and destroying, and killing men, women, and children. About 450 houses were partially or totally destroyed.

After the population of Baccarat, in the same department, had been assembled at the railway station on August 25th, the town was first systematically pillaged under the supervision of German officers and then set on fire by means of torches and pastilles, 112 houses being destroyed.

The Germans entered Senlis in the neighborhood of Chantilly, as already related, on September 2d, where they were greeted by rifle-fire from African troops. Claiming that they had been fired upon by civilians, they sacked the town and set fire to it in two different quarters with grenades and rockets, destroying 105 houses, killing many of the inhabitants, and executing the mayor, M. Odent, on the ground that he had participated, or instigated others to participate, in acts of hostility.

The report claimed that arson was a favorite means for inspiring terror and cowing the inhabitants, that the German armies were provided with a complete outfit for producing conflagrations, comprising rockets, torches, grenades, petroleum pumps, fuse-sticks, and little bags of pastilles made of compressed powder which is very inflammable, and that thousands of houses were burned in France. German officers, whose supersensitive conception of honor was unable to endure the remotest implication of an affront from a comrade, were presumptuously charged by their adversaries with committing the basest and most ignominious forms of iniquity. They were represented as commonly indifferent to the licentious fury of their troops, as sharing in the exercise and profits of organized pillage, and as utilizing their irresponsible authority for perpetrating among a defenseless population the crimes of extortion, rape, and murder.

The German soldiers had been systematically encouraged to keep individual diaries of the campaign by the military authorities, who little suspected the purpose for which many of these documents were destined to be employed by their opponents. For a large number of them were taken by the French from the bodies of the German dead or from German prisoners and preserved for the sake of the incriminating information which they contained.

Professor Joseph Bédier, the eminent authority on medieval French literature at the *College de France* in Paris, carefully examined a number of these diaries, and published extracts from about thirty, nineteen of them accompanied by facsimile reproductions of the original text, in a pamphlet entitled, *Les crimes allemands d'après des témoignages allemands*—German Atrocities from German Evidence. Professor Bédier proposed to establish the guilt of the Germans by the testimony of the Germans themselves, and he declared that in the diaries employed for this purpose "the German soldiers depict themselves or their companions in arms as incendiaries, robbers, and murderers, who, however, in ordinary circumstances, only burn, rob, and murder to order and in the course of their military duty."

This pamphlet, which appeared about the first of January, 1915, was translated into most of the leading languages and spread throughout the world in thousands of copies. It made a very great impression at once, and with very good reason; for the evidence of this class is the most significant of all, because its ingenuousness and authenticity are beyond

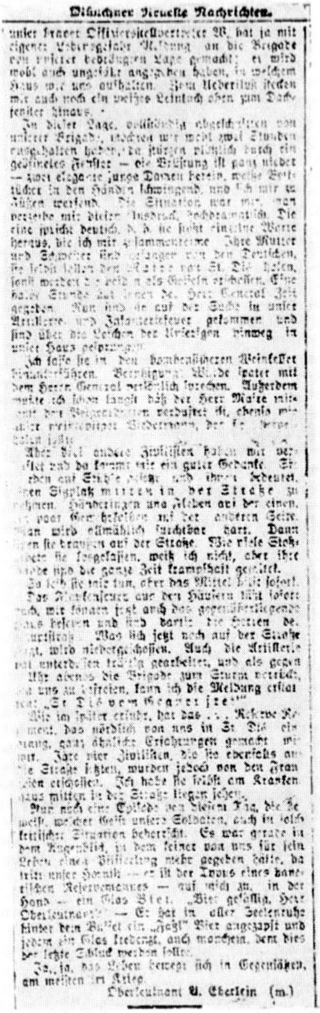

Münchner Neueste Nachrichten.

unser braver Offizierstellvertreter W. hat ja mit eigener Lebensgefahr Meldung an die Brigade von unserer bedrängten Lage gemacht; er wird wohl auch ungefähr angegeben haben, in welchem Haus wir uns aufhalten. Zum Ueberfluß stecken wir auch noch ein weißes Leintuch oben zum Dachfenster hinaus.

In dieser Lage, vollständig abgeschnitten von unserer Brigade, mochten wir wohl zwei Stunden ausgehalten haben, da stürzen plötzlich durch ein geöffnetes Fenster — die Brüstung ist ganz nieder — zwei elegante junge Damen herein, weiße Tücher in den Händen schwingend, und sich mir zu Füßen werfend. Die Situation war mir, man verzeihe mir diesen Ausdruck, hochdramatisch. Die eine spricht deutsch, d. h. sie sucht einzelne Worte heraus, die ich mir zusammenreime. Ihre Mutter und Schwester sind gefangen von den Deutschen, sie beide sollen den Maire von St. Dié holen, sonst werden [illegible] als Geiseln erschossen. Eine halbe Stunde hat ihnen der Herr General Zeit gegeben. Nun sind sie auf der Suche in unser Artillerie- und Infanteriefeuer gekommen und sind über die Leichen der Unseligen hinweg in unser Haus gesprungen.

Ich lasse sie in den bombensicheren Weinkeller hinunterführen. Beruhigung: [illegible] später mit dem Herrn General persönlich sprechen. Außerdem wußte ich schon längst, daß der Herr Maire mit [illegible] verduftet ist, ebenso wie [illegible].

Aber drei andere Zivilisten haben wir verhaftet und da kommt mir ein guter Gedanke. Sie werden auf Stühle gesetzt und ihnen bedeutet, ihren Sitzplatz mitten in der Straße zu nehmen. Händeringen und Flehen auf der einen, ein paar Gewehrkolben auf der anderen Seite. Man wird allmählich furchtbar hart. Dann sitzen sie draußen auf der Straße. Wie viele Stoßgebete sie losgelassen, weiß ich nicht, aber ihre Hände sind die ganze Zeit krampfhaft gefaltet. Sie tun mir leid, aber das Mittel wirkt sofort. Das Flankenfeuer aus den Häusern läßt sofort nach, wir können jetzt auch das gegenüberliegende Haus besetzen und sind damit die Herren der Hauptstraße. Was sich jetzt noch auf der Straße zeigt, wird niedergeschossen. Auch die Artillerie hat unterdessen kräftig gearbeitet, und als gegen 7 Uhr abends die Brigade zum Sturm vorrückt, um uns zu befreien, kann ich die Meldung erstatten: „St. Dié vom Feinde frei."

Wie ich später erfuhr, hat das ... Reserve-Regiment, das nördlich von uns in St. Dié eindrang, ganz ähnliche Erfahrungen gemacht wie wir. Die vier Zivilisten, die sie ebenfalls auf die Straße setzten, wurden jedoch von den Franzosen erschossen. Ich habe sie selbst am Krankenhaus mitten in der Straße liegen sehen.

Nur noch eine Episode von kleinem Krieg, die zeigt, welcher Geist unsere Soldaten, auch in solch kritischer Situation beherrscht. Es war gerade in dem Augenblick, in dem keiner von uns für sein Leben einen Pfifferling mehr gegeben hätte, da tritt unser [illegible] — er ist der Typus eines bayerischen Reservemannes — auf mich zu, in der Hand — ein Glas Bier. „Bier gefällig, Herr Oberleutnant?" — Er hat in aller Seelenruhe hinter dem Buffet ein „Faßl" Bier angezapft und jedem ein Glas kredenzt, auch manchem, dem dies der letzte Schluck werden sollte.

Ja, ja, das Leben bewegt sich in Gegensätzen, am meisten im Kriege.

Oberleutnant A. Eberlein (m.)

Translation.

"We had arrested three civilians, and suddenly a good idea struck me. We placed them on chairs and made them understand that they must go and sit on them in the middle of the street. On one side entreaties, on the other blows from the butt-end of a gun. One gets terribly hardened after a while. At last they were seated outside in the street. I do not know how many prayers of anguish they said; but they kept their hands tightly clasped all the time. I pitied them; but the device worked immediately. The shooting at us from the house at the side stopped at once; we were able to occupy the house in front, and became masters of the principal street. Every one after that who showed himself in the street was shot. The artillery, too, did good work during this, and when towards seven in the evening, the brigade advanced to free us, I was able to report that 'St. Dié is free of the enemy.'

"As I learnt later on, the . . . regiment of reserve which had entered St. Dié more from the north had had similar experiences to ours. The four civilians that had been made to sit in the street had been killed by French bullets. I saw them myself, stretched out in the middle of the street, near the Hospital."

Facsimile of an account of a company of Germans sheltering themselves behind non-combatants, written by Lieutenant Eberlein and published in the *Münchner Neueste Nachrichten*, October 7, 1914.

question, and it depicts the critical occurrences in the artless, confidential words of the actors themselves. In the hearing of the momentous action brought in the name of humanity against the German army, this is the defendant's own unpremeditated deposition.

A few of the most striking of these narratives will serve to illustrate the whole series.

The first quotation presented by Professor Bédier, a passage from the diary of a soldier of the first infantry brigade of the Guard, describes very graphically the frightful retribution inflicted at night upon a village near Blamont on September 1st:

"The inhabitants fled through the village. It was horrible. Blood was plastered on all the houses, and as for the faces of the dead, they were hideous. They were all buried at once, to the number of sixty. Among them many old men and women, and one woman about to be delivered. It was a ghastly sight. There were three children who had huddled close to one another and had died together. The altar and the ceiling of the church had fallen in. They also had been telephoning to the enemy. And this morning, September 2d, all the survivors were driven out, and I saw four little boys carrying on two poles a child five or six months old. All this was horrible to see. A blow for a blow. Thunder for thunder. Everything was pillaged. The poultry and everything else was killed. (There was a) mother with her two little ones; one had a large wound in the head and had lost an eye."

The diary of an officer of the 178th regiment, Twelfth (Saxon) Army Corps, relates the destruction of a village in the Ardennes in Belgium:

"The beautiful village of Gué d'Hossus has been destroyed by fire, although entirely innocent, as it seems to me. Apparently a cyclist fell from his machine, and in the fall his gun went off of itself. Straightway there was firing in his direction. The male inhabitants were simply consigned to the flames. It is to be hoped that such atrocities will not be repeated."

The diary of a soldier of the first battalion of the same regiment contains the following entry under August 23d:

"In the evening at ten o'clock the first battalion of the 178th regiment went down to the village that had been burnt to the north of Dinant. A sad and beautiful sight, and one that made you shudder. At the entrance of the village lay the bodies of about fifty civilians who had been shot for having fired upon our troops from ambush. In the course of the night, many others were shot in the same way, so that we could count more than two hundred. The women and children, lamp in hand, had to watch the horrible scene. We then ate our rice in the midst of the corpses, for we had not tasted food since morning."

Reservist Schlauter of the 4th regiment of field-artillery of the Guard mentions the following incidents of the march through Belgium occurring on August 25th:

"Three hundred of the inhabitants of the town were shot and those who survived the volley were requisitioned as gravediggers. You should have seen the women at this moment! But you can't do otherwise. During our march on Wilot things went better. The inhabitants who wished to leave could do so and go where they liked, but anyone who fired was shot. When we left Owele shots were fired; but there, women and everything were fired upon. At the frontier they have to-day shot a Hussar and destroyed the bridge. The bridge has been rebuilt by the gallant infantry."

Professor Bédier reproduces the text of an order of the day issued by Major-general Stenger, commander of the 58th brigade, on August 26th, as follows:

"From to-day no more prisoners will be made. All prisoners will be put to death. The wounded whether armed or not will be put to death. Even prisoners grouped in larger formations will be put to death. No enemy shall be left alive behind us.

(*Signed*)

"First-lieutenant and Company-chief Stoy, Colonel and Regimental-commander Neubauer, Major-general and Brigade-commander Stenger."

According to Professor Bédier, furthermore, about thirty German soldiers belonging to General Stenger's brigade, who had been taken prisoners by the French, affirmed under oath that this order of the day was transmitted to them on the 26th and five of these declared that they had actually seen instances of its execution. Other evidence seemingly confirmed the tenor of their testimony.

The text of a personal narrative describing the German occupation of St. Dié on August 27th, written by a Bavarian officer, First-lieutenant A. Eberlein, and published in the *Münchner Neueste Nachrichten* on October 7th, is quoted as proof of the frequent but almost unbelievable assertion that the Germans shielded themselves from the enemy's fire by the shameful method of forcing civilians to stand or advance in front of them.

Lieutenant Eberlein's company as advance-guard entered St. Dié in the belief that the town had been entirely evacuated by the enemy, but they were suddenly assailed by French troops, who poured a volley into their ranks from behind a barricade, while rifle-fire blazed from the windows of the neighboring houses. A brilliant idea came to the lieutenant. Bewildered at first and entirely cut off, he and his men barricaded themselves in a house while waiting for reinforcements and by blows from the butt end of a rifle compelled three male inhabitants, whom they had captured, to go into the middle of the street and remain there seated on chairs during the fight. This apparently produced the desired effect, as the firing subsided soon after. Lieutenant Eberlein related further that a German reserve regiment which entered St. Dié by another road had recourse to a similar stratagem. Of the four civilians who were compelled to march in front of this column, two were killed and two were severely wounded, according to a later investigation made by the French authorities.

In a number of instances the diaries offer corroborative evidence of the allegations made in the French Official Report. A striking example of such coincidence is afforded by the accounts of the conduct of the Germans at Nomeny in the Department of Meurthe-et-Moselle. According to the French account the Germans, in a state of terrible excitement, entered this town, where the inhabitants had taken refuge in their cellars to escape the fire of artillery, on August 20th, and gave themselves over to abominable excesses, sacking and burning the town systematically and killing many of the people as they were attempting to escape from the burning houses. The report states further "that an officer arrived at the end of the butchery and ordered the women who were still alive to get up and shouted to them: 'Go to France!'" The town was almost entirely destroyed. The French emphatically denied the charge that the inhabitants had fired on the German troops.

The same incident is described in the diary of a private of the 8th Bavarian regiment of infantry as follows:

"A shell burst near the 11th company and wounded seven men, three of them seriously. At five o'clock the order was (given) to us by the officer commanding the regiment to shoot all the male inhabitants of Nomeny and to raze the whole town to the ground because the people

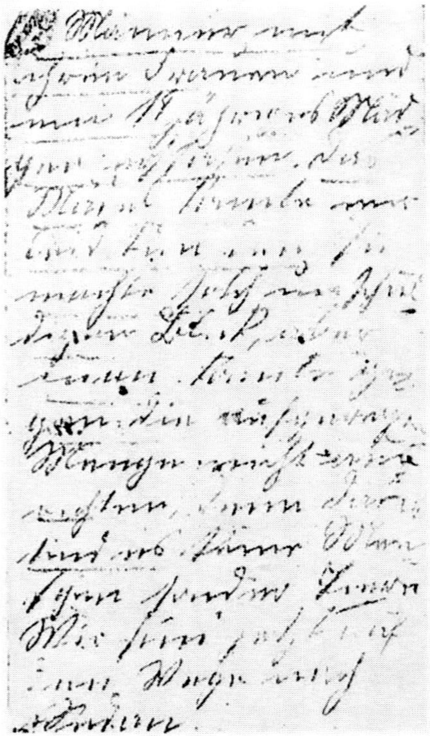

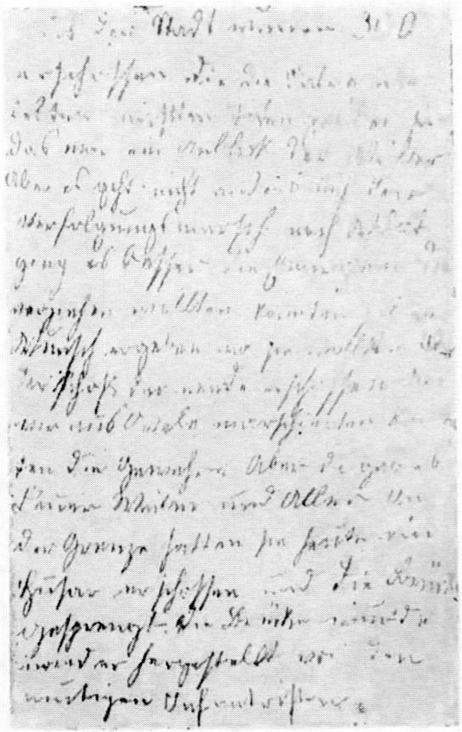

Pages from the note books of German soldiers, published in *German Atrocities from German Evidence*, by Joseph Bédier.

Translation of the above.

"In this way we destroyed eight houses with their inmates. In one of them two men with their wives and a girl of eighteen were bayoneted. The little one almost unnerved me, so innocent was her expression. But it was impossible to check the crowd, so excited were they, for in such moments you are no longer men, but wild beasts."

Translation of the above.

"25th August (in Belgium): Three hundred of the inhabitants were shot and the survivors were requisitioned as grave-diggers. You should have seen the women at this moment! But you can't do otherwise. During our march on Wilot, things went better: the inhabitants who wished to leave could do so and go where they liked. But anyone who fired was shot. When we left Owele, shots were fired: but there, women and everything were fired on . . ."

were making a mad attempt to oppose with arms the advance of the German troops. We broke into the houses and seized all those who resisted, to execute them according to martial law. The houses which either the French or our own artillery had not yet set on fire were burnt by us, and in consequence almost the whole town was reduced to ashes. It is a terrible thing to see women and children, defenseless and henceforth destitute of everything, driven along like a flock of sheep to be shoved off in the direction of France."

A committee was appointed by the British Prime Minister on December 15, 1914, "to consider and advise on the evidence collected on behalf of His Majesty's Government as to outrages alleged to have been committed by German troops during the present war, cases of alleged maltreatment of civilians in the invaded territories, and breaches of the laws and established usages of war; and to prepare a report for His Majesty's Government showing the conclusion at which they arrive on the evidence available." The committee was composed of the former British Ambassador to the United States, the Right Honorable Viscount Bryce, as chairman; of the eminent lawyers and jurists, the Right Honorable Sir Frederick Pollock, the Right Honorable Sir Edward Clarke, and Sir Alfred Hopkinson, with Sir Kenelm E. Digby added later; the Vice-chancellor of the University of Sheffield and distinguished historian, Mr. Herbert A. S. Fisher; and the editor of the *Edinburgh Review*, Mr. Harold Cox.

The highmindedness and sincerity of these gentlemen and their well-founded reputation for discernment and judicial capacity commended the results of their investigation to the thoughtful consideration of neutral nations. Doubt may have subsisted as to the reliability and adequacy of the available evidence, which was mainly derived from the depositions of Belgian refugees, whose testimony might naturally be biased, and as to the facilities for obtaining a correct appreciation of the situation, inasmuch as the inquiry was necessarily conducted in places remote from the scene of the occurrences. The committee endeavored as far as possible to compensate for these unavoidable disadvantages by their zeal in enlarging the range of their evidence and by the thoroughness and critical accuracy with which they sifted it.

More than 1,200 depositions which had already been taken from Belgian witnesses and British officers and soldiers by authority of the Home Office were examined by the committee, and to this mass of evidence they added diaries taken from the German dead and a number of proclamations issued by, or at the bidding of, German military authorities in Belgium and France.

The persons of legal knowledge and experience who had taken the depositions in different parts of the United Kingdom are said to have performed their task in a strictly fair and impartial manner. They reported that the Belgian witnesses exhibited very little vindictiveness or emotional excitement. The committee, for their part, found that the depositions, "though taken at different places and on different dates, and by different lawyers from different witnesses," often corroborated one another in a striking manner.

After a thorough analysis and comparison of the material the committee arrived at the following conclusions:

(1) That there were in many parts of Belgium deliberate and systematically organized massacres of the civil population, accompanied by many isolated murders and other outrages.

(2) That in the conduct of the war generally innocent civilians, both men and women, were murdered in large numbers, women violated, and children murdered.

(3) That looting, house burning, and the wanton destruction of property were ordered and countenanced by the officers of the German army, that elaborate provision had been made for systematic incendiarism at the very outbreak of the war, and that the burnings and destruction were frequent where no military necessity could be alleged, being indeed part of a system of general terrorization.

(4) That the rules and usages of war were frequently broken, particularly by the using of civilians, including women and children, as a shield for advancing forces exposed to fire, to a less degree by killing the wounded and prisoners, and in the frequent abuse of the Red Cross and the White Flag.

Of special interest is the observation that most of the systematic outrages or severities took place between August 4th and 30th, and along geographical lines forming an irregular Y, the trunk extending from the German frontier to Liége, and the arms from Liége to Charleroi and from Liége to Mechlin respectively, corresponding with the invaders' most prominent lines of advance and later of communication. In the words of the report, "for the first fortnight of the war the towns and villages near Liége were the chief sufferers. From the 19th of August to the end of the month, outrages spread in the directions of Charleroi and Malines (Mechlin) and reach their period of greatest intensity. There is a certain significance in the fact that the outrages round Liége coincide with the unexpected resistance of the Belgian army in that district, and that the slaughter which reigned from the 19th of August to the end of the month is contemporaneous with the period when the German army's need for a quick passage through Belgium at all costs was deemed imperative."

The charges elaborated in these reports, as briefly outlined, constituted the most formidable indictment for brutality which had been preferred against any enlightened nation in modern times, and Germany's answer was awaited with feelings of scornful incredulity or intense suspense.

While there was a tendency in some quarters in Germany to regard any apology for the conduct of the army as an indignity, the general outcry of indignation abroad provoked the natural response. Voices were raised both privately and officially to denounce the "campaign of lies," and the civilian population of Belgium was accused of having treated German wounded and prisoners with the most inhuman cruelty.

But in vain the amiable character of the German people and the proverbially inflexible discipline of the German army were invoked as proof of the inability of their soldiers to commit the enormities charged against them. No previous record, however lustrous, could alone avail against the evidence of palpable and gruesome facts, the graves of five or six thousand civilians recently slain, the ruin and desolation of once prosperous communities, and the flight of a million terrified inhabitants.

Wolff's Bureau issued on January 12, 1915, a reply to the report of the French committee with an indignant refutation of all the charges contained in it, but apparently before full knowledge of its contents had been received, for the article declared that "all the general points are without specific particulars as to time, place, the guilty parties, or proofs of these acts of murder, incendiarism, and rape." It stated furthermore that wherever the French government had quoted particular cases, the German government had ordered an investigation, the results of which "could be awaited with calmness and confidence," but that the French allegation that the German army had wantonly burned seventy houses in Lunéville could be immediately refuted, since this severity had been necessitated by the

Vandalism in Louvain, the intellectual capital of the Low Countries since the Middle Ages: ruins of the Hôtel du Nord and other buildings.

Louvain. The Town Hall unscathed, while of the church of Saint Pierre and of the university nothing remains but the walls.

treachery of the civilians who fired on German troops in the Military Hospital on the afternoon of August 25, 1914, and in the streets on the following morning.

A criticism of Professor Bédier's pamphlet appearing in the semi-official *Norddeutsche Allgemeine Zeitung* in seven columns on February 28, 1915, confined itself mainly to the exposition of a few unimportant inaccuracies of translation and seemed to admit by inference the incontrovertibility of the general basis of facts, but at the same time very pertinently emphasized the probability that the episodes narrated in the diaries of the German soldiers were almost exclusively incidents in the *franc-tireur* warfare, an argument which in reality constitutes the only effectual basis for the attempted refutation of the charges.

Finally, on May 15, 1915, the German government published an Official White Book for the purpose of proving by the sworn depositions of a large number of eye-witnesses in the German army that the Belgian people of all classes and ages, and of both sexes, engaged in a contest with the German troops in flagrant violation of international law, making treacherous attacks in parts of the country, and in towns and villages, which had already been occupied and were ostensibly pacific, and that the severities practised by the German troops against the civilian population of Belgium were in all cases absolutely necessary and justifiable measures for the suppression of this unlawful conduct.

The compilers of the report were convinced that this popular outbreak was systematically planned at the instigation, or at least with the complicity of the Belgian authorities. They claimed, furthermore, that the people, blinded with rage or perverted by unscrupulous leaders committed savage acts of cruelty, mutilating and killing the wounded and captives, engaged in every form of treachery, violated the sanctity of the Red Cross, and fired on surgeons and nurses who were engaged in performing their professional services.

The arduous problem of reconciling the German report with the statements made by the Allies, particularly by the Belgians themselves, is strikingly illustrated by the case of Andenne. Here, according to the German account, the population received the German troops in a friendly manner on August 20th; but suddenly, upon the ringing of the church-bells at 6.30 in the evening, the inhabitants barricaded their houses and opened fire on the unsuspecting soldiers from cellar-windows and apertures in the roofs, utilizing hand-bombs, hand-grenades, boiling water, and even machine-guns for their furious onslaught. The German troops upon recovering their presence of mind forced their way into the houses, shot about two hundred persons who resisted or were caught with arms, and burned the buildings from which shots had been fired as rightful retaliation for the treacherous attack. The Belgians, on the contrary, insisted, as we have seen, that the inhabitants of Andenne did nothing to offend the Germans or provoke their resentment. Besides the testimony contained in the official reports, many other accounts by eye-witnesses, especially accounts of the occurrences in Belgium by Hollanders, citizens of a neutral country, are very important, but cannot be discussed here in detail.

One general reservation must be made before we proceed to the weighing of the evidence.

The examination of the legal and moral principles involved in the conduct of Germany toward Belgium in the first and second volumes of the present work revealed ample grounds for the opinion that the Belgian people possessed within their national boundaries, collectively and individually, the unqualified right to life, liberty, and the pursuit of happiness, and the absolute use, disposition, and enjoyment of their property both public and private, unrestricted by any intrusion or encroachment in consequence of an external state of hostilities. And governed by this strict, but theoretically irrefutable, opinion, many persons will emphatically designate any infringement of these incontestable privileges by foreign interference simply and unconditionally as an odious enormity, regarding the insolent treatment of the least of the Belgians as a grave misdemeanor, every curtailment of the use of property as a trespass, every confiscation as a felony, every destruction by fire as arson, every exaction as brigandage, every violation of personal liberty as an outrageous assault, and the taking of human life in Belgium, in any situation whatever, as homicide, murder, or assassination. In accordance with this conception, the invasion of Belgium was tantamount to a filibustering excursion, which justly exposed the participants without any legitimate redress to whatever measures of self-defense or retribution the inhabitants of the country were impelled to take.

This extreme opinion would eliminate all discussion regarding the justifiability of the conduct of Germans in Belgium, where the chief part of the alleged atrocities occurred, by regarding the invasion of that country in itself as the supremely culpable atrocity and consequently all the particular actions of the invaders as inherently outrageous without reference to the circumstances in each individual case.

For the sake of the present argument, therefore, we shall disregard this extreme view altogether and assume that the conditions of belligerency and the usages of warfare applicable to the operations in Belgium were precisely the same as those that prevailed elsewhere. It may be assumed, moreover, as established by the various reports,—although space forbids a detailed exposition of all the steps by which these conclusions have been reached on the basis of the evidence,—on the one hand, that the Germans committed acts of great severity in Belgium and France; on the other, that the participation of civilians or irregular combatants in the fighting, alleged as cause of these severities, actually occurred in many localities.

All destruction of property incidental to the development of the hostile engagements of regular warfare, or unavoidably produced in the course of the actual fighting, must be excluded from the category of atrocities and from every consideration of moral responsibility for lawless or wanton conduct. For within the arena of combat, tactical requirements unquestionably overrule or supersede all other interests and considerations.

In many instances, however, it is impossible to trace the distinction with such accuracy as to avoid every ambiguity. The battle-lines in northern France and Belgium passed very close to a number of cities and towns celebrated for their artistic treasures, whose partial or total destruction, defended by the Germans as a military necessity, has precipitated a series of prolonged and inconclusive controversies. The German military authorities insisted, for example, that the dismantling or destruction of convenient points of observation for directing the fire of artillery, whether cathedral towers or municipal belfries, was an indispensable military measure for the safety of their armies.

As already mentioned, the cathedral at Reims, the ancient coronation place of the kings of France, was first damaged by artillery-fire soon after the French reoccupied the city. First-lieutenant Wengler of the German heavy artillery, who directed the firing of the much-maligned shots, stated that a French observer on the northern tower of the cathedral was first noticed on September 13th, when the firing of the French artillery became very much more

Part of the ruins of the archbishop's house at Reims and of the chapel which connected it with the rear of the cathedral, photographed from one of the towers of the ruined cathedral.

Sand bags being arranged in an attempt to protect some of the remaining sculptures of Reims cathedral after the first German bombardment.

accurate, and that observation from this point continued until the 18th, when a German projectile from a 15-centimeter howitzer struck the observation tower and another from a 21-centimeter mortar fell on the roof of the cathedral and set it on fire.

The Germans, while admitting the bombardment of Reims for the dispersion of hostile artillery, maintained that these were the only shots deliberately aimed at the venerable old building. But Mr. Richard Harding Davis, who arrived at the cathedral about three o'clock on the afternoon of the 18th asserted positively that howitzer shells had already penetrated the wall and killed two of the German wounded who were lying in the nave, that, twenty-four hours after Lieutenant Wengler claimed to have ceased firing, shells set fire to the roof of the cathedral and wrecked the archbishop's palace and the chapel which connected it with the rear of the cathedral, and that the whole quarter around had been largely reduced to ruins by the bombardment. Mr. Davis was again in the cathedral on the 22d, while it was again being shelled, in company with the Abbé Chinot, Mr. Gerald Morgan, Captain Granville Fortescue of Washington, and the Honorable Mr. Robert Bacon, formerly American ambassador to France.

Throughout the remainder of the campaign Reims was repeatedly shelled, the bombardment continuing many days, or even weeks, almost without interruption. Showers of projectiles often fell in the vicinity of the cathedral, especially in the Place du Parvis in front of the western, or main, façade.

Although in a structural sense the cathedral has not been destroyed, the ineffable elegance and wealth of decoration of its interior have been rudely treated. The priceless sculptural adornment of the principal front has suffered the most grievous damage, caused in large part by the burning of the scaffolding covering this entire façade, which had been erected for the work of restoration that had been in progress for several years. This conflagration extended to the interior and destroyed the confessionals, choir-stalls, and other woodwork. Measures were eventually taken to protect the remaining figures on the west façade from being damaged by the fragments of shells exploding in the square in front.

In view of the effective activity of French artillery stationed in Reims, or in its suburbs, and of the cardinal importance of the city as a center of military communications, the pretension that the Germans, by subjecting it to their severe cannonading, violated article 25 of The Hague Convention, which forbids the bombardment of undefended places, scarcely seems tenable.

The same question of responsibility based on a similar situation arises in connection with the monuments of Soissons, particularly the beautiful Gothic cathedral, which was perforated by shells in a number of places.

The graceful, richly decorated belfry of Arras was completely destroyed, and the Hôtel de Ville, a precious monument of the Spanish-Flemish Transition style of the sixteenth century, was mostly reduced to a sorrowful mass of ruins.

Saddest of all are the losses at Ypres, which have already been briefly described. The grand old Cloth Hall, unique example of its kind, stands rent and shattered, a pathetic wreck, the noble belfry reduced to a formless fragment, the exquisite Renaissance addition at the side ground to dust. The defense for the action of the Germans in destroying the monuments of Ypres is undoubtedly weaker than the justification of their conduct in the other instances. For Ypres was at no time within the tactical field of conflict. The defense of military necessity can only be invoked on the ground that the bombardment of the city impeded the passage of troops, or the reinforcing and munitioning of the troops on the actual battle-line.

An unbiased judgment on the basis of a comprehensive knowledge of the circumstances would probably ascribe the havoc wrought at Mechlin to the same category as the damage inflicted upon the places already discussed.

Deplorable as these losses are, it is unthinkable that the German military authorities were so heedless of every consideration of expediency as uselessly to offend the sensibility of the civilized world by the deliberate defacement or destruction of artistic monuments, as is sometimes insinuated, with no practical purpose in view. The idea, moreover, that the destruction of these monuments was part of a calculated policy of terrorism intended to destroy the resolution of the enemy imputes to the German General Staff an unbelievable ignorance of the fundamental impulses of human conduct. For artistic monuments cannot be used with effect like pawns to secure the tranquil behavior of an infuriated people.

A similar problem respecting the limits of legitimate warfare has been created by the bombardment of cities and towns from aëroplanes and dirigibles. German aircraft dropped bombs on Louvain, Namur, Antwerp, and many other places, and for a time paid almost daily hostile visits to Paris. This practice, which brought wounds and death to many harmless non-combatants, excited intense and widespread indignation, and has been regarded as a part of the alleged German policy of calculated frightfulness. Although most of the places were defended by very strong fortifications, the Allies contended that these attacks were a gross violation of the regulations of The Hague Convention of 1907, either because they were made without any previous warning, or else because, as in the case of Paris, the points at which they were directed were so far from the forts that they obviously served no strictly military purpose whatever. As an exhaustive examination of all the questions involved in this complicated problem would detain us too long, we must dismiss it with the general observation that the practice of bombarding indiscriminately from the sky the streets and buildings of cities and towns has not been justified by the familiar argument of military necessity or success.

It is impossible not to recognize that a fundamental distinction for the purposes of the present discussion exists between the isolated acts of brutality committed by individuals and the systematic acts of severity inflicted by command of the regular military authorities. In the gigantic armies of even the most civilized nations in time of war there must be many depraved individuals whose criminal propensities are likely to be inflamed by the privations and excitement of warfare and by separation from any restraining influences of their customary environment. The crimes committed by such men may affect unfavorably the general estimate of the state of morality and refinement among the people whom they so unworthily represent, but they do not involve the government or the military chiefs in the charge of violating the rules of warfare, unless the latter make themselves virtually accomplices in guilt by their culpable laxity.

In spite of the indignant assertions of the German press that the German commanders effected with full success the maintenance of the strictest discipline, we feel constrained to urge in their behalf that many of the acts of brutality which occurred in Belgium and France were probably committed without their sanction. The occurrence in individual cases of such lawless excesses does not seem to be entirely inconsistent with the recorded conditions of criminality.

A comparison of analyses of available German Imperial Statistics for the ten-year period (1897 to 1907) and of the reports of the Home Office of the United Kingdom during the nearest similar ten-year period published (1900 to 1910) shows the following average yearly convictions for crimes:

	In Germany.	In the United Kingdom.
Murders	350	97
Felonious woundings	172,153	1,262
Rapes	9,381	216
Incest	573	56
Malicious damage to property	25,759	358

(These reports show also that during the same period the average yearly number of illegitimate births in Germany was 178,115; and in the United Kingdom 37,041).

It may be thought that the law is more strictly enforced in Germany: this may be true as to the punishment of minor misdemeanors, but certainly not as to serious crimes, for British courts are proverbially strict in respect of crimes such as those in the table given above. Of course some allowance should be made for the difference in population, but Germany has only about half as many million more people than the United Kingdom.

In face of the abundant evidence, especially that furnished by passages in the diaries of the German soldiers themselves, it is difficult to escape the conclusion that during the invasion of Belgium and France the German troops frequently plundered wine-shops and wine-cellars and drank to intoxication.

But for the reasons already adduced our attention will be directed mainly to the severities executed by order of the military authorities. The treatment of these will be based upon the instances in which there is practically no question as to the nature and extent of the severities inflicted.

Some particular cases in the chapter of atrocities have been invested with particular interest and significance by the sensational character of the occurrences themselves, the extent of the losses, and the ardor of the ensuing polemics. A clear statement of the basis of accepted fact and of the chief points of controversy in a few of these instances will help to illuminate the whole field of discussion.

We shall consider first the case of Aerschot, a town of 8,000 inhabitants about ten miles northeast of Louvain, where the 8th German infantry brigade was quartered on August 19th. The municipal authorities of Aerschot received the Germans with every mark of respect, the mayor, M. Tielemans, offered the brigade-commander, Colonel Stenger, the hospitality of his own home which faced the market-place, and the relations between civilians and the military were apparently harmonious until about eight in the evening, when shooting suddenly began in the streets near the market-place.

According to the German account civilians opened fire from the houses upon the soldiers in the streets and upon the supply trains passing through the town. The houses from which firing took place were accordingly burned and a number of guilty or suspected civilians were led off for execution. Soon after the firing began, Colonel Stenger was found lying mortally wounded on the floor of a front room in the mayor's house, where he had been sitting near the open balcony doors. He had probably been hit by a shot from the opposite side of the market-place. The circumstances of the outbreak seemed to show that it had been deliberately prepared in anticipation of the coming of the Germans.

But the Belgians and their allies affirmed that the existence of a plot, the participation of the population of Aerschot in acts of hostility, and the alleged guilt or complicity of the mayor and of his family were pure fabrications. They suggested that the commotion might have arisen from the aimless shooting of drunken soldiers who were engaged in terrorizing the inhabitants, and that the killing of the commander might easily have been explained as caused by a German shot discharged at random in the general excitement and confusion.

However the case may be, the mayor, his son, and his brother were executed by the Germans the next day on purely circumstantial evidence of guilt or else as an act of reprisal. Colonel Jenrich, the post-commander, had warned the mayor that he would be held responsible with his own life for any attack that should be made by the civilian population.

Dinant was of fundamental strategical importance for the Germans in the early weeks of the war, because, in the development of their great plan for outflanking and enveloping the Allies, it was essential that the Twelfth Army Corps should cross the Meuse at that point. The tentative occupation of Dinant on August 15th need not detain us. On the evening of August 21st the 2d battalion of sharpshooters of the 108th (Fusilier) regiment and a detachment of pioneers made a reconnoissance in force as far as this town, which was occupied in part by the troops of the enemy.

The Germans coming from the direction of Ciney were received with a fusillade from the nearest houses as they were entering the town by the Rue St. Jacques. They made their way as far as the bridge, but were assailed from all sides and forced to retreat, setting fire to a number of houses in the streets through which they passed.

On the 23d the Twelfth Corps made a determined effort to gain the left bank of the Meuse and advanced in the direction of Dinant and its vicinity, the Thirty-second Division towards the north and the Twenty-third Division towards the south.

According to German accounts it required a desperate conflict to overcome the resistance of the civilian population in Dinant and the neighboring places. The first battalion of the regiment of Body Guards was subjected to a galling fire as it descended the steep slope into Dinant, and the troops were compelled to storm each house separately. The civilians who were caught with arms in their hands were shot on the spot, while all those who were suspected were held as hostages. The Germans themselves suffered considerable losses. The 108th and 182d regiments approaching Dinant further north had a similar experience. Eventually the troops had to be withdrawn from Dinant so that the obstinate resistance of the population could be crushed by a bombardment.

The German report is singularly reticent about the ultimate fate of the town. But the ruin of this picturesque and prosperous place, the destruction of all but 200 of its 1,400 buildings, and the conversion of an unusually attractive panorama into a mournful scene of waste and desolation can neither be hidden nor disguised.

According to the statements published by the Allies the Germans committed barbarous outrages in Dinant both on the 21st and 23d without any provocation from the civilian population. Finally, they sacked and burned the town systematically and executed groups of from 50 to 120 civilians, massacring in all about 800 persons.

From the extensive testimony presented in the German report and from some of the diaries of the German soldiers, particularly a passage in the diary of a private of the 108th

1. — Jusqu'au 6 Septembre 1914, à 4 heures de relevée, toutes les armes, munitions, explosifs, pièces d'artifices qui sont encore en possession des citoyens, seront remises au Château des Bruyères. Celui qui ne le fera pas sera passible de la peine de mort. Il sera fusillé sur place ou passé par les armes, à moins qu'il prouve qu'il n'est pas fautif.

2. — Tous les habitants des maisons occupées des localités de Beyne-Heusay, Grivegnée, Bois-de-Breux, Fléron, devront rentrer chez eux à partir de la chute du jour (en ce moment, à partir de 7 heures du soir — heure allemande). Les maisons prédésignées seront éclairées aussi longtemps que quelqu'un y sera sur pied. Les portes d'entrée seront fermées. Celui qui ne se conformera pas à ces prescriptions s'exposera à des peines sévères. Toute résistance quelconque contre ces ordres entrainera la mort.

3. — Le Commandant ne doit rencontrer aucune difficulté dans ses visites domiciliaires. On est prié, sans sommation, de montrer toutes les pièces de la maison. Quiconque s'y opposera sera sévèrement puni.

4. — A partir du 7 septembre, à 9 heures du matin, je permetterai l'occupation des habitations de Beyne-Heusay, Grivegnée, Bois-de-Breux, par les personnes qui y demeuraient précédemment, aussi longtemps qu'aucune défense formelle de fréquenter ces lieux n'aura été prononcée pour les habitants susvisés.

5. — Pour avoir la certitude qu'il ne sera pas abusé de cette permission, les Bourgmestres de Beyne-Heusay et de Grivegnée devront dresser immédiatement des listes de personnalités qui seront retenues, par alternance de 24 heures, comme otages au Fort de Fléron. Le 6 septembre 1914, pour la première fois, de 6 heures du soir jusqu'au 7 Septembre, à midi.

Il y va de la vie de ces otages à ce que la population des communes précitées se tienne paisible en toutes circonstances.

Pendant la nuit, il est sévèrement défendu de produire des signaux lumineux quelconques. La circulation des vélocipèdes n'est autorsée que de 7 heures du matin à 5 heures du soir (heure allemande).

6. — Je désignerai, hors des listes qui me seront soumises, les personnalités qui, de midi d'un jour à midi de l'autre jour, ont à séjourner comme otages. Si le remplacement n'a pas lieu en temps utile, l'otage reste de nouveau 24 heures au Fort. Après ces nouvelles 24 heures, l'otage encourt la peine de mort si le remplacement n'est pas fait.

7. — Comme otages, sont placés en première ligne, les Prêtres, les Bourgmestres et les autres Membres de l'Administration.

8. — J'exige que tous les civils qui circulent dans ma circonscription, principalement ceux des localités de Beyne-Heusay, Fléron, Bois-de-Breux, Grivegnée, témoignent de leur déférence envers les officiers allemands, en ôtant leur chapeau ou en portant la main à la tête comme pour le salut militaire. En cas de doute, on doit saluer tout militaire allemand. Celui qui ne s'exécute pas doit s'attendre à ce que les militaires allemands se fassent respecter par tous les moyens.

9. — Il est permis aux militaires allemands de visiter les véhicules, paquets, etc., de tous les habitants des alentours. Toute résistance à ce sujet sera punie sévèrement.

10. — Celui qui a connaissance que des quantités supérieures à 100 litres de pétrole, benzine, benzol et d'autres liquides analogues se trouvent à un endroit déterminé des Communes précitées et qui ne l'a pas annoncé au Commandant militaire qui y siège, lorsqu'il n'y a aucun doute sur le lieu et la quantité, encourt la mort. Les quantités de 100 litres sont seulement visées.

11. — Celui qui n'obtempère pas de suite au commandement « levez les bras » se rend coupable de la peine de mort.

12. — L'entrée du Château des Bruyères, de même que celle des allées du parc, est interdite, sous peine de mort, depuis le crépuscule jusqu'à l'aurore (de 6 heures du soir à 6 heures du matin — heure allemande), à toutes personnes qui ne sont pas des soldats de l'armée allemande.

13. — Pendant le jour, l'entrée du Château des Bruyeres n'est permise que par l'entrée Nord-Ouest, là où se trouve la Garde, et pour autant de personnes qu'il y a de cartes d'entrée distribuées. Tout rassemblement à proximité de la Garde est défendu dans l'intérêt de la population.

14. — Quiconque, par la communication de fausses nouvelles qui seraient de nature à nuire au moral des troupes allemandes, de même celui qui, de n'importe quelle manière, cherche à prendre des dispositions contre l'armée allemande, se rend suspect et encourt le risque d'être fusillé sur-le-champ.

15. — Tandis que, par les dispositions susvisées, les habitants de la région de la forteresse III B. sont menacés de peines sévères lorsqu'ils enfreignent ces dispositions d'une manière quelconque, ces mêmes habitants peuvent, lorsqu'ils se montrent paisibles, compter sur la protection la plus bienveillante et le secours en toutes occasions lorsqu'on leur fait ou pourrait leur faire du tort.

16. — Les demandes de remise de bétail pour une quantité déterminée se font journellement, de 10 à 12 heures avant midi et de 2 à 3 heures après-midi, au Château des Bruyères, auprès de la Commission du bétail.

17. — Celui qui, sous l'égide de l'insigne de la Convention Suisse, nuit ou même cherche à nuire à l'Armée allemande et est découvert, est pendu.

(s.) **DIECKMANN**,

Major-Commandant.

Pour copie conforme :

Le Bourgmestre,

Victor **HODEIGE**.

Grivegnée, le 8 Septembre 1914.

Facsimile of typical proclamation posted by German commandants. *(For translation see tissue.) It will be observed that penalty of death for trivial cause without formality of trial is threatened and that the latter part of paragraph 8 offers to the soldiers great opportunity of license.*

1.—Before the 6th of September, 1914, at 4 o'clock, in the afternoon, all arms, munitions, explosives and fireworks which are still in the hands of the citizens, must be surrendered at the Château des Bruyères. Anyone who does not obey will be liable to the death penalty. Unless he can prove that he is not at fault, he will be shot on the spot, or given military execution.

2.—All inhabitants of houses in Beyne-Heusay, Grivegnée, Bois-de-Breux and Fléron must remain at home after sunset (at present 7 o'clock P.M., German time). The aforesaid houses must be lighted as long as anyone remains up. The entrance doors must be shut. Anyone who does not conform to the regulations exposes himself to severe penalties. Any resistance to these orders will involve death.

3.—The Commandant should meet no opposition whatever in his domiciliary visits. Each inhabitant is asked to open all the rooms of his house without even a summons. Whoever makes any opposition will be severely punished.

4.—Beginning September 7th, at 9 o'clock in the morning, I will permit the houses of Beyne-Heusay, Grivegnée, Bois-de-Breux, to be occupied by persons formerly dwelling in them as long as no formal prohibition to frequent these places shall have been issued against the inhabitants above referred to.

5.—In order to be sure that this permission is not abused, the Burgomasters of Beyne-Heusay and of Grivegnée shall immediately draw up a list of persons who shall be held as hostages, at the fort of Fléron, in 24 hour shifts; on September 6th, for the first time, from 6 o'clock in the evening until mid-day, September 7th.

The lives of these hostages will depend upon the population of the aforesaid communes remaining pacific under all circumstances.

During the night, it is strictly prohibited to make any luminous signal whatever. The use of bicycles is allowed only from 7 A. M. until 5 P.M., German time.

6.—I will designate from the lists submitted to me the persons who will be detained as hostages from noon of one day to noon of the next day. If the substitute does not arrive in time, the hostage will remain at the Fort another 24 hours. After this second period of 24 hours, the hostage incurs the penalty of death if the substitution is not made.

7.—Hostages will be chosen, primarily, from among Priests, Burgomasters and other members of the civic administration.

8.—I demand that all civilians living in my district, especially in Beyne-Heusay, Fléron, Bois-de-Breux and Grivegnée, shall show deference toward the German officers by taking off their hats or by carrying the hand to the head as in military salute. In case of doubt, every German soldier must be saluted. If anyone refuses to do so, he must expect the German soldiers to employ any means to make themselves respected.

9.—The German soldiers have the right to inspect wagons or packages belonging to the inhabitants of the surrounding country. Any opposition in this respect will be severely punished.

10.—Anyone knowing of more than one hundred litres of petroleum, benzine, benzol, or other similar liquids being in any place in the aforesaid communes and who does not report same to the military commander on the spot, incurs the penalty of death, provided there is no doubt about the quantity and the place. Quantities of 100 litres are alone referred to.

11.—Anyone who does not instantly obey the command of "hands up" becomes liable to the death penalty.

12.—Entrance to the Château des Bruyères and to the alleys of the park is prohibited under the penalty of death, from dark till dawn (6 P. M. to 6 A. M., German time) to all but soldiers of the German Army.

13.—During daytime, entrance to the Château des Bruyères is allowed only by the northwest entrance, where there is a guard, and only to as many people as there are cards of admission issued. Any gathering near the guard is prohibited in the interest of the population.

14.—Anyone who by spreading false news, prejudicial to the morale of the German troops, or who, by any means whatsoever, tries to take measures against the German Army, renders himself a suspect and incurs the risk of being shot immediately.

15.—Whereas, by the above regulations, the inhabitants in the vicinity of the fortress III B. are threatened with severe penalties if they violate these regulations in any way, on the other hand, these same inhabitants, if they remain peaceful, may rely upon the most benevolent protection and help on all occasions when wrong is done them.

16.—The requisition of cattle in specified quantities will take place daily, from 10 A. M. until noon and from 2 P. M. to 5 P. M., at the Château des Bruyères, before the Cattle Commission.

17.—Anyone who, under the protection of the insignia of the Swiss Convention, harms, or even tries to harm, the German Army, and is discovered, shall be hanged.

(*Signed*) DIECKMANN, Major in Command.

For certified copy: The Burgomaster.

(*Signed*) VICTOR HODEIGE.

Grivegnée, September 8th, 1914

regiment cited by Professor Bédier, it seems impossible not to conclude that civilians engaged in acts of hostility at Dinant and its vicinity. It is unlikely, moreover, that the German commanders dissipated the strength of the maneuvering wing of their armies at this very critical juncture by detaining considerable forces for the gratuitous spoliation and destruction of an unresisting town, or merely for inculcating by another and rather superfluous example of frightfulness their salutary lesson of submission.

It is of importance to note in this connection that the civilians in Dinant did not wait until the town had been occupied to attack the Germans, but openly contested their approach.

After emphasizing the importance of the aim of the Twelfth Corps to cross the river at Dinant, the German Military Commission for the Investigation of Offenses against the Laws of War comments upon the destruction of the town, the shooting of civilians who engaged in hostilities, and the execution of hostages, as follows:

"It was a military necessity quickly to overcome the resistance of the inhabitants who opposed that aim,—an aim which had to be attained by every means. From that point of view it was certainly justifiable to bombard with artillery the town which had taken active part in the fight, to burn the houses which were occupied by *franc-tireurs*, and to shoot the inhabitants who were caught with arms in hand.

"Likewise in agreement with the law was the shooting of the hostages which took place in various localities. . . . The hostages were secured in order to stop the action of the *franc-tireurs*. As, nevertheless, the people continued to inflict losses on the fighting troops the shooting of the hostages had to be resorted to. Otherwise their seizure would only have meant a vain threat."

The first German troops marched into Louvain on August 19th, but there was no interruption of peaceful intercourse between the people and the troops until the 25th. On that day, as already noted, the Belgian army made a vigorous sortie from the intrenched camp of Antwerp and there was an engagement in the neighborhood of Louvain. The Ninth Reserve Corps, which was coming to support the Third Reserve Corps was detraining at Louvain and the column began to pass through the streets at six in the evening. The staff of General von Boehm, commander of the Ninth Reserve Corps, had made its headquarters in the famous Town Hall.

As the Germans claim, about eight in the evening (German time), soldiers of a Landsturm company (von Sandt), which had just marched from the northwest exit of the city to the railway station in the east, noticed the appearance of a green and then a red rocket, and simultaneously with this signal the inhabitants opened fire upon the German troops at the station, and at the Town Hall and in the intervening section of the city, throwing the transport column into confusion and killing and wounding a number of officers and soldiers. It was natural to assume that this outbreak had been carefully planned to synchronize with the sortie from Antwerp. General von Boehm returned from the battlefield to Louvain about 11.30 and ordered a Landwehr brigade to march into the city for suppressing the uprising, seized the mayor and other leading citizens as hostages, and caused them to be led through the streets to summon the inhabitants to cease hostilities. This measure was apparently without success, for the attacks are said to have continued on the 26th and 27th.

In the meantime the Germans had been breaking into the houses from which shooting was supposed to have occurred, and setting fire to them. The conflagration spread throughout the neighborhood of the Town Hall during the night of the 25th-26th. By morning the University Library, the pride and delight of many generations, had been consumed, and the flames were attacking the Church of St. Peter, which was partially damaged. Dr. Ingermann of the Landwehr is said to have been treacherously wounded while he was saving the pictures from this church. As mentioned in a former chapter, the incomparable Town Hall was saved by the efforts of the Germans themselves.

The alleged discovery of large quantities of arms and ammunition and the detection of soldiers disguised as civilians are used as evidence to prove the existence of a plot officially organized in which important functions were said to have been attributed to the Garde Civique and to the clergy.

The Allies, on the contrary, claim that there is absolutely no evidence to prove that civilians fired on the German troops. The origin of the commotion is attributed by many witnesses to a mistake due to the nervousness of the German troops at the station who fired upon their own comrades as they were returning from the battlefield. On the pretext that civilians had shot at the soldiers the Germans burned and bombarded Louvain and slaughtered many of the inhabitants as a deliberate act of terrorism calculated to shatter the resolution of the Belgians.

Whatever may have been the real situation which furnished the occasion for the commotion, there is scarcely room for doubt that the German leaders decided after deliberate reflection that the entire section of Louvain between the Town Hall and the railway station should be systematically destroyed. This decision was relentlessly executed on the 27th and one-sixth of the city, the most important section, was reduced to a wilderness of blackened ruins.

We have noticed that the severities which were denounced as atrocities by the Allies were represented by the Germans as necessary measures of retribution for repressing the unlawful participation of civilians in acts of hostility. The contradiction leads us to examine, first, the rightfulness of the employment of retributory measures in the given situation, and, secondly, the justice of the measures themselves and of the manner in which they were applied.

The fact that in some instances civilians undoubtedly attacked the German troops does not in itself constitute an offense against the established rules of war. It is necessary to distinguish between the conditions in which the *levée en masse,* or people's warfare, is lawful and those in which it is prohibited, and therefore a question of law arises beside the question of fact in connection with the first subject of discussion just mentioned.

It has been shown that the atrocities practised in Serbia were largely the consequence of a fundamental difference of attitude respecting the extension of the capacity to possess the rights of belligerency. We shall presently discover that a similar discrepancy existed between the views of the antagonists in the western theater.

The question as to how far the rights of active belligerency are to be conceded to irregular combatants was discussed at The Hague Convention of 1907, when the German delegation uniformly advocated a narrow conception of legitimate warfare, which would restrict the relative defensive capacity of all states whose regular military organization is inferior to that of Germany by depriving them, to a very large extent, of the right to defend themselves in an emergency by improvised and informal methods. But the conference adopted a broader conception of legitimate warfare in conformity with the view of the British, Belgian, and Swiss delegations. Recognizing in principle the inalienable right of the inhabitants of an invaded territory to take up arms in their own defense, the conference formulated the legal basis for the belligerent rights of irregular combatants in the first and second articles of the Rules of The Hague Convention dealing with the Conduct of War on Land, as adopted in 1907. These rules were eventually agreed to by the German delegation.

A distinction was made between an organized and an unorganized people's warfare. In the former the militia or bands of volunteer combatants, in order to be recognized as belligerents, are obliged to conform with the following four conditions: they must have responsible leaders, they must wear a distinctive emblem recognizable at a distance, they must carry their arms openly, and they must observe the laws and customs of war. The unorganized people's warfare is free from the first two conditions, but in return is bound by two other particular stipulations: it may only be waged in the territory not yet occupied by the enemy, and there must have been lack of time to prepare the organized people's warfare.

The German and Belgian governments agree, accidentally, in one very important affirmation, namely, that there was no organized people's warfare in Belgium, the first by declaring that the civilians who engaged in acts of hostility had neither responsible leaders nor any distinctive emblem, the second by denying emphatically that any measures were taken which were calculated to induce civilians to fight. It follows, therefore, unless both disputants are wrong, that any hostile action on the part of the civilian population must fall under the head of unorganized people's warfare and ought to conform to the special conditions for the same, which limit it to territory not yet occupied by the enemy and to situations in which the government has not had time to organize the people's warfare. Any civilian attacks on the German troops in Louvain, Aerschot, Andennes, or elsewhere occurring after the occupation by the invaders had been effected must therefore have been violations of the established usages of war, provided, of course, that the hostile action occurred at the people's own initiative, as the Germans claim, and was not provoked by the lawless behavior of the soldiers.

But in other cases the circumstances seem to justify the active resistance of the civilian population according to the spirit of The Hague Convention. In many localities the hostilities in which the civilians were accused of engaging were attempts to repel the invaders as they approached, not treacherous attacks in places already effectively occupied. Dinant is the most striking example of this class. The Germans themselves declare that they were fired upon as they first approached the town, but insist that there the unorganized people's warfare was a contravention because the Belgian authorities had had ample time to organize the people's warfare.

Without pausing to discuss the delicate question of the proper limit for the term of legitimate exemption from the obligation to organize the people's warfare, we shall turn to a more convincing example of the unorganized *levée en masse* in circumstances conformable with article 2 of The Hague Convention.

The Germans invaded Belgium almost before the inhabitants of the country had received any warning of their intentions, and to judge by the invaders' own accounts, which doubtless merit reliance, they encountered from the first the desperate resistance of the population. But, although this unorganized people's warfare was apparently quite legal, since there had been no time for the otherwise prescribed organization and the territory had not yet been occupied by the enemy, the Germans punished the combatants with the same implacable severity which marked their conduct in the treatment of the hostilities that were unquestionably forbidden by The Hague Convention.

The contention that the danger existing for years of a Franco-German war imposed the precautionary obligation of organizing the people's warfare implies a reproach of the Belgians for their misplaced confidence in the German government's own repeated pacific assurances.

A thoughtful consideration creates the impression that the Germans neither made, nor intended to make, in their treatment of civilians engaged in acts of warfare, any distinction corresponding to the principles adopted by The Hague Convention. Their rule was absolutely simple and could be expressed with laconic brevity in the words of one of the diaries: "*der schoss, der wurde erschossen*"—"anyone who fired was shot."

The application of the repressive measures was characterized by the principle of collective responsibility and by the summary infliction of severe reprisals. The first involves vicarious expiation for the offenses of isolated individuals, and it offends the most elementary sense of justice by making the innocent suffer with, or for, the guilty. It treats the lives of the people of whole communities as rightly forfeit, severally or collectively, for local violations of the laws of war, and exposes representative personalities or others chosen at random to execution for the thoughtless conduct of the most worthless or irresponsible persons.

The conduct of the Germans on the basis of this doctrine constitutes the most serious tenable charge against their practice. It is scarcely necessary to accumulate extensive evidence, since the Germans themselves acknowledge it without any hesitation. The *Kölnische Zeitung* declared, for example, in regard to the suppression of the people's warfare:

"We all made one fundamental principle clear: for the fault of the individual the community to which he belonged must suffer. The village in which our troops had been shot at by the civilian population was burned down. If the culprit was not discovered, a few representatives were taken out of the general population and shot. Women and children were not touched, except when they were found with weapons in their hands.

"This principle may seem hard and cruel,—it has been developed from the customs of modern and ancient military history, and, as far as it can be spoken of at all, recognized. It is also justified by the theory of setting an awful example. The innocent must suffer with the guilty; and, when the latter cannot be found, they must suffer for the guilty."

The practice here so clearly described is an unmistakable violation of article 50 of the Convention adopted at the Second Hague Conference, in 1907, which forbids expressly the infliction of any collective penalty, pecuniary or other, upon communities for the action of individuals for which they could not be considered collectively responsible.

In vain the plea of military necessity was solemnly advanced as final argument for every sanguinary, wholesale execution. The world instinctively condemned the hasty, inconsiderate acts of retaliation with their inseparable complement of hideous injustice. Was there time to weigh with any semblance of judicial method the guilt imputed to each victim of the days of wild commotion at Louvain, Andenne, Dinant, and scores of other places? Bloody execution must have overtaken many whose only fault was physical inability to escape the dangerous proximity of their suspected neighbors.

The impetuous incendiaries of Gué d'Hossus were unconstrained by any discriminating scruples. The promiscuous

"The shields of Rousselaere." Civilians compelled to march in front of German troops as a protection from gunfire.

"Culture from the air." Many innocent persons have been killed by bombs dropped from German Zeppelins.

Reproductions from a series of cartoons relative to the war, drawn by the Dutch artist Louis Raemaekers.

shooting of the male inhabitants of Nomeny was manifestly unaccompanied by any judicial process. The little children mutilated or slain in the wild nocturnal raiding of the village near Blamont are silent witnesses to the brutal iniquity of such proceedings.

The practice of taking hostages, a natural corollary of the doctrine of collective or vicarious responsibility, was generally employed by the Germans as a means for restraining the population in localities which had already been occupied. An authentic illustration of this practice, as of the application of the theory of collective responsibility in general, is afforded by the proclamations issued by the German military authorities in different parts of the conquered territory in Belgium and France. The following extract from a proclamation of Field-marshal von der Goltz, then Governor-general of Belgium, dated at Brussels, October 5, 1914, will serve as a general example:

"In future the communities (localités) nearest the place where such acts occur (destruction of railways or telegraph lines), whether accomplices or not, will be relentlessly punished. To this end, hostages have been taken from all the communities near the railways threatened by such attacks and at the first attempt to destroy the railways or the telegraph or telephone lines, they will be immediately shot."

The proclamation issued at Grivegnée on September 8th is an even more specific illustration of these repulsive methods. It confers, moreover, a startling warrant for impulsive acts of violence. In this case the exacting harshness is explained, though scarcely palliated, by the proximity of an important fortress. The gratuitous humiliation of the inhabitants decreed in article 8 is an example of the petty arrogance which makes the conciliation of a conquered people by official Germany seem almost hopeless.

The practice of compelling individuals selected from the civilian population to cover with their own persons the march of the German soldiers through towns and other places where treacherous attacks by the inhabitants were feared is a special application of "hostage-right." For in such cases the apprehended attacks would inflict automatically the retributory penalty upon the hostages. In view of the confusion regarding the limits of legitimate warfare it is not surprising that this practice should involve the Germans in the charge of protecting themselves in battle behind a living shield of civilians. From the German point of view the justice of the conduct of Lieutenant Eberlein at St. Dié depends upon the character of the opponents, whether they were regular troops or *franc-tireurs*.

The foresight of the Germans, who had provided against the hostility of civilians as against every other contingency, does not prove that a systematic policy of terrorism in enemy territory was part of their original design. Our opinion that they did not anticipate serious resistance on the part of Belgium is incompatible with the supposed intentional program of intimidation in that country. The opposition of the Belgians, and especially the stubborn resistance of the civilian population in the first few days of the campaign, was a painful disillusionment.

The German commanders were doubtless exasperated by this forced and unforeseen distraction from warfare on the large and organized scale in which they had been trained, and in which their superiority chiefly lay, to the insidious conflict of irregular combatants whom they professionally despised. Instead of showing any forbearance in their behavior towards a people whom they publicly confessed to have wronged, the invaders, treating every hostile action by civilians, without distinction, as a violation of the law of war, punished such offenses by the infliction of the severest penalties for which there was the remotest semblance of a precedent.

In the conduct of the Belgian people's warfare there were doubtless many instances in which the prescriptions of The Hague Conventions were violated. The sinister impression of these early days endured. The legend of German barbarity growing more frightful by transmission, which drove thousands to panic-stricken flight, probably impelled others to desperate but hopeless acts of opposition. On the other hand, rumors of treacherous attacks, rendered doubly terrible by hints of dark and awful mutilations, quickly spread among the German soldiers. Fearless in the face of visible opponents their fancy shuddered at the hidden pitfalls of the narrow, overshadowed streets of towns and villages, and perhaps at times their harassed nerves gave way to groundless panic at an accidental shot or other threatening circumstance.

In the execution of a maneuver upon which the whole campaign depended, when every faculty was strained, when the destiny of the universe seemed to hang upon the progress accomplished in a single day, it is not surprising that the German leaders took drastic measures to eradicate what they believed to be a vital menace.

But when a people not devoid of virile spirit believe themselves to be the victims of a treacherous attack, when they see their harvests rudely trampled under foot, when they behold the ruthless desecration of places hallowed by the dearest ties, where their forefathers repose and they themselves were born, where their offspring learned to prate their first uncertain accents at their knees, where their tenderest sentiments and deepest feelings cluster, a choking, palpitating passion grips their heart. The normal range of their impulses narrows down to one intense, resistless force of execration. Expediency and calculation are instantly forgotten, and the act of retribution bountifully compensates for every sacrifice and risk. A common fiery zeal coördinates their efforts and their rage supplies the weapons. The legal subtleties and chivalrous affectations of formal warfare vanish before the concentrated fury of their rage. They will fight with wild primeval violence, with the splendid ardor of the charge at Marathon, with the grim persistence of the Minute Men, with the glorious madness of the immortal Five Days' uprising at Milan, with the obdurate ferocity of the populace who dwelt between the German frontier and Liége, where the invaders were compelled, not only to burn the villages, but literally to level them with the dust, before resistance was eradicated. Regardless of all conventions and of the consequences, they will struggle with the fierce love of independence which has thrilled the fancy of the ages, and the sentimental enthusiasm of mankind will applaud their magnificent perversity.

CHAPTER XV

Signs and Expressions of Public Opinion and Sentiment

In France: the hopeful outlook of the French people, M. Viviani's speech on December 22d. In Great Britain: truce between the two great political parties, the Prime Minister's speech at Guildhall, the king's visit to the front, the raiding of the east coast, civilian deaths at Hartlepool, Scarborough, and Whitby. In Germany: the general view of the nation's obligation, hatred for Great Britain, the session of the Reichstag on December 2d, the position of the Socialists. In Austria-Hungary: New Year's address of the Hungarian Premier, Count Tisza. In Turkey: demonstrations and predictions, opening of parliament. In Russia: alleged mistreatment of Mohammedan population, disaffection in the Ukraine, popular impressions and the surviving discords. Bismarck's foreign policy and the later policy of Germany. General tendencies.

The reports described in the last chapter were more than mere collections of alleged historical evidence. Their publication was a historical event in itself, because they helped to confirm the spirit of determination in each of the belligerent countries. The Allies were more than ever convinced that they were fighting for freedom against military despotism, that they were fighting the battle of humanity against barbarism, while at the same time the indignation of the Germans was sustained by the ostensible proof that after Belgium had become the misguided tool in a wicked conspiracy against themselves, their soldiers had been victims of the treachery and savage cruelty of the Belgian people.

Napoleon once declared that moral factors count for three fourths in warfare and the conflict of the physical forces for only the remaining fourth. The meteoric progress of events in the field of physical conflict has thus far enthralled our attention in the present volume, but we ought not to close without casting at least a fleeting glance at the state of public opinion and sentiment in the period about the end of the first campaign.

In the darkest hour, when the menace of a siege hung over Paris, M. Alfred Capus, a talented academician, had written an editorial for *Figaro,* expressing the highest spirit of devotion of the French nation, in which he said:

"What is now necessary,—in fact, it is the sole condition of national salvation,—is an inexhaustible reserve of moral force. . . .

"Let me repeat. The one condition is that the army and its chiefs shall feel back of them a country ready for all sacrifices, with undaunted souls, unflinching wills, and a definite, coherent government.

"A weakening of the will, a feebleness of soul, would be as detestable as desertion. To lack constancy to-day is to desert before the enemy. It is treason. We all have our duty,—the government, the press, public opinion. This duty, in a single word, is firmness, which implies harmony, calmness, the stoic acceptance of events, and ardent confidence in the destiny of our country.

"At certain critical hours a cry of anger is a blasphemy, a doubt may be a crime. The victory is hard to win, but certain. Let the whole nation deserve it. Each man to his post! Let us take for ourselves the simple and sublime words of the great Englishman: 'France expects every man to do his duty.'"

In the period at which we have arrived the French people could draw confidence from the reflection that the political parties had adhered to the sacred union inaugurated at the commencement of hostilities and that the defense of the country had not been paralyzed by internal dissension, that the nation had faced the most critical hours with unshaken firmness, and that, in addition to the traditional qualities of courage and fervor, it had given wonderful proof of patience, tenacity, and stoicism.

In a celebrated speech delivered before the Chamber of Deputies on December 22d, Prime Minister Viviani expressed the aim from which the French government has never swerved, namely, that France must continue the war until she and her allies could dictate the terms of peace.

"Just now," he declared, "there is only one policy, a relentless fight until we attain definite freedom for Europe by gaining a victory which shall guarantee peace." More specifically he asserted that, faithful to her obligations, "France, in accord with her allies, will not lay down her arms until she has avenged outraged right and regained forever the provinces which were torn from her by force, restored heroic Belgium to the fulness of her material prosperity and political independence, and broken Prussian militarism so that the Allies may eventually reconstruct a regenerated Europe founded upon justice and right."

A truce had been imposed upon the strife of the political parties in Great Britain at the beginning of the war, when the support of the Unionists was pledged to the ministry in the following note of their leader on August 2d:

"Dear Mr. Asquith,—Lord Lansdowne and I feel it our duty to inform you that in our opinion, as well as in that of all the colleagues whom we have been able to consult, it would be fatal to the honor and security of the United Kingdom to hesitate in supporting France and Russia at the present juncture; and we offer our unhesitating support to the government in any measures they may consider necessary for that object.

"Yours very truly,

"A. BONAR LAW."

Only a small minority of Labor representatives maintained an attitude of protest against the prosecution of the war.

The Prime Minister, Mr. Asquith, formulated the intentions of Great Britain in a noteworthy speech at the Lord Mayor's banquet at the Guildhall on November 10th, when he declared that they would "not sheathe the sword until Belgium recovered all and more than all that she had sacrificed, until France was adequately secured against the menace of aggression, until the rights of smaller nationalities were placed on an unassailable foundation, until the military dominion of Prussia was fully and finally destroyed."

In this connection allusion may be made to the visit of King George to the British army at the front in Flanders, because it served as a demonstration of the harmonious coöperation of the three allies in the West. It was noteworthy also as the first instance of the presence of royalty with a British army in the field since George II fought at Dettingen in 1743.

The king crossed to France in a warship on November 29th, visited some base-hospitals and reached the British headquarters on the 30th. During his tour of inspection along the front he met President Poincaré, Prime Minister Viviani, and General Joffre on December 1st and was received by King Albert at his headquarters in the last fragment of Belgian territory not occupied by the enemy on the 4th. The return to England was effected on the 5th.

One of the factors which agitated public opinion in Great Britain was the raiding of the east coast by a German naval squadron in December, when 119 civilians were killed by the bombardment at Hartlepool, seventeen at Scarborough, and two at Whitby, many women and small children being included among the slain. The German authorities were probably actuated by the belief that a palpable demonstration of the vulnerability of Great Britain would profoundly influence opinion both in the British Isles and in neutral countries.

The Germans claimed that their bombardment of these open towns was justified by the fact that Hartlepool ranked as a coast defense and had a regular garrison, that there were earthworks and a battery of six 6-inch guns at Scarborough, and that Whitby was a coast-guard station. But the English regarded the affair as simply an application of the German policy of frightfulness, which had recently been enunciated by von Hindenburg in his remark that the most ruthless manner of conducting warfare is in reality the most merciful since it brings hostilities to the speediest termination.

It is hardly too much to say that the raid was a failure, except as an opportunity for practice, and that the results were incommensurate with the risks. The chief effect seems to have been to intensify the belligerent spirit in England.

It would be difficult to exaggerate the importance of gaining a clear, objective conception of the prevailing point of view in Germany, unclouded by preconception or sentimental bias, because it is an absolutely fundamental factor and an appreciation of it makes intelligible many things which at first seem illogical, distorted, or incomprehensible.

There were doubtless many German extremists whose attitude on every particular question of policy had been colored by their instinctive conviction that with the existing unequal division of the earth's surface even abstention from war on the part of Germany was in itself a striking proof of magnanimity in the strongest military power, and that any further demand upon her forbearance must necessarily be regarded as an intolerable effrontery.

It seemed to the majority of the German people, as we have had occasion to observe, that after Austria-Hungary had exhibited exceptional patience in the face of repeated provocations, but had finally been compelled to take up arms for her absolute self-preservation, safety and honor alike made German intervention a bitter but inevitable necessity. To the average German the war was essentially a defensive war.

We have seen how resentment against Great Britain had grown in intensity and become a powerful motive force. The fires of German wrath were fed by the seemingly vapid argument that Great Britain was responsible for the war because she could have prevented it, although it is just as obvious that Germany herself, and in fact any one of the other contestants, except Belgium, could have done the same. For it is always possible to prevent a quarrel when one of the parties is willing to resign his claim. But the grounds for this belief on the part of the Germans were probably less superficial than it might at first appear. Their passionate conviction was proof of the weight attributed in the popular imagination to formal alliances rather than of intellectual obtuseness or a warping of the critical faculty. Germany was pledged to support Austria-Hungary just as France was pledged to help Russia in precisely the situation which had arisen. For them it was a question of national honor, there was no alternative. But Great Britain, according to the specific statements of her responsible ministers, had been free from every formal obligation of this kind. Of course it did not occur to the German public that quite apart from any formal obligation it was just as necessary for the security of Great Britain that the integrity and vitality of France should be maintained as it was for the safety of Germany that Austria-Hungary should be preserved.

For German opinion it was just a step to the assumption that what the British government was unwilling to prevent it had in reality desired and devised. The image of Great

Britain, perverted by envy from the course of her own true interests and with characteristic hypocrisy concealing her treacherous designs under the specious cover of pacific and conciliatory proposals until the favorable moment for striking her rival, when the latter was involved in a desperate struggle on opposite fronts, hovered like an inflaming fury in the German imagination.

Foreign observers, impressed by the strength of popular opposition to the military and foreign policy of the German government before the war and convinced that the exceptional appearance of unanimity since the war began must conceal profound currents of discontent, looked forward to the assembling of the Reichstag early in December with a vague anticipation that something sensational or epochal might occur when an opportunity should be afforded for free discussion and criticism.

The speech delivered by the Imperial Chancellor before the Reichstag on December 2d has already been cited, but another extract may appropriately be quoted in the present connection. In the following words he voiced the yearning hope of many Germans that the spirit of harmony created by the common peril might be preserved after the war had been terminated:

"The wonderful fervor glowing in the hearts of the German people, the unprecedented unity and unconditional self-surrender of one to another, must be and will be victorious. And when a glorious and happy peace is ours, we shall hold this national spirit sacred and regard it as the holiest bequest of this terribly grave and great age. As by magic the walls have fallen which separated for a time, a dull and barren time, the various classes of our people, the walls which we have raised against one another in misunderstanding, envy, and mistrust. It is a liberation and blessing that the rubbish heap of social prejudice has been swept away; that man alone has value now; that all count alike and that each holds out his hand to his fellow-man for a common and holy end. So once more I repeat the Kaiser's words uttered at the outbreak of the war: 'I know no more parties, I only know Germans.' After the war parties will revive; even the freest and most united nation cannot fully live out its political life without parties and political strife. But let us strive—I for my part promise you to do so—that also in this strife there shall be nothing but Germans."

The Social Democrats, who, after vigorously opposing, had in the end complacently acquiesced in the war policy, and had brought upon themselves the denunciation of their comrades abroad by their alleged betrayal of the principles of the International, reiterated their point of view of August 4th, in the following words of their leader, Herr Haase:

"In connection with the words of the Imperial Chancellor regarding Belgium, I wish to declare in the name of my party that no subsequent disclosures warrant in our judgment a departure from our attitude of August 4th. The Social Democratic party adheres to-day to the point of view expressed on August 4th in its declaration concerning the war, the cause of which in the final analysis was economic rivalries. Our confines are still menaced by hostile forces; hence the German people must exert their whole strength for the protection of the country. For this reason the Social Democratic party approves the new credits demanded. . . . As on August 4th we still uphold, in harmony with the International, the imperishable right of every nation to integrity and independence. To deprive foreign nations of these privileges is to sow the seed of fresh war. We stand, therefore, by what we said on August 4th. We demand that, as soon as the goal has been reached in which the enemy is desirous of peace, the war be ended on terms conducive to friendship with other nations."

The only spectacular event of the session was the solitary vote of Karl Liebknecht against the war-budget in violation of the discipline of the Social Democratic party which requires that all members shall vote as a unit. His action was repudiated by the following resolution of the parliamentary group, as published in *Vorwärts:*

"The Social Democratic party strongly condemns Karl Liebknecht's breach of discipline, and it repudiates the misleading information which he has spread concerning proceedings within the party. The party is determined that it shall vote as a unit in the Reichstag. If any deputy is unable conscientiously to participate in the voting he is at liberty to abstain, but he must not give his abstention the character of a demonstration."

The Reichstag, after convening at 4.15 P. M. and sanctioning the second war appropriation of 5,000,000,000 marks, was adjourned about 6 P. M. until March, 1915.

The position, unquestionable prominence, and personal authority of the fiery controversialist, uncompromising nationalist, and conservative statesman, Count Tisza, Prime Minister of Hungary, made him a very significant spokesman for the hopes and aspirations of those who regarded the Dual Monarchy with feelings of loyalty and attachment.

Replying to the New Year's greetings of parliamentary supporters of the government, he delivered an address which contained a number of noteworthy passages. He recalled with satisfaction the chagrin of the enemy upon discovering so many unexpected signs of vigor, harmony, and self-sacrificing devotion in the supposedly decadent monarchy. But they themselves, he admitted, had been surprised by the rapidity and violence of the first attack in the northeast. The bitterest surprise had been the forced evacuation of Serbia after their troops had victoriously advanced into the heart of the country.

"I am not so much disturbed," he said, "by the disadvantages of the present situation in a military sense, for we shall very soon cancel the military consequences of the retreat. But it grieves me that an army which has struggled against the superiority of an altogether very competent enemy and has carried on a heroic conflict with feverish impetuosity for weeks and months in the midst of the greatest physical difficulties, should lose this glory, at least in the eyes of the public, simply because it was expected to perform a superhuman task."

In regard to the harmonious coöperation of the Austro-Hungarian and German troops, Count Tisza declared: "Our troops are permeated with reciprocal feelings of trust, affection, and appreciation. German and Austro-Hungarian troops accomplish marvels together. There is absolute harmony between the leaders." His recent visit to the German headquarters had given him ample proof of the loyalty and candid sympathy of their ally. As to the effect of the war upon the constitution of the Dual Monarchy, he declared: "Dualism, which offers a basis for the preservation of Hungarian independence and nationalism, has stood the fiery test of warfare. The centralizing tendencies which still crop out at times in Austria have lost every justification in the trials of the great war; and in the face of all that the Hungarian nation has accomplished and sacrificed for the lofty common aims of the monarchy, only a pernicious madness could return to an agitation for centralization. History has definitely settled to-day the problem of the constitution of the monarchy. Friction on constitutional questions has no longer any place."

In Turkey the indispensable basis for public opinion in the truer, comprehensive sense, homogeneity of civilization, a common intellectual outlook, and a general feeling of community of interests and responsibilities, is manifestly lacking. Its place is taken by the selfish impulses and designs of the political cliques by whom the superficial manifestation of popular enthusiasms or passions is surreptitiously controlled. The news of Turkey's plunge into the vortex of the world-war was followed in the German press by copious accounts of the demonstrations of satisfaction at Constantinople. But the publication of these glowing narratives at precisely the time when it was expedient to counteract in the imagination of the German people the embarrassing retardation of the seemingly impending victorious consummation of operations in the western campaign is not free from suspicion that the occurrences themselves were adroitly elaborated or that the reports of them were judiciously embellished.

A popular demonstration before the German Embassy on the evening of November 14th was described as an event of special significance. The appearance of the German ambassador at the balcony was greeted by a prolonged ovation, and after the playing of the German national hymn, "*Heil dir im Siegeskranz,*" Nazim Bey, the leader of the Young Turks, delivered a fervent address, to which the ambassador replied in the warmest tones, expressing his conviction that the victory of the three allies would mark the beginning of a new era of prosperity for Turkey and Islam.

A proclamation announcing the Holy War on November 21st recalled that Russia, the inveterate foe of Islam, had associated herself in the present conflict with Great Britain and France, powers that held millions of Mohammedans under their yoke and were driven by their insatiable greed to plot the ruin of the caliphate. This document declared that the powers composing the Triple Entente had robbed many Mohammedan peoples of freedom and independence and it denounced them as instigators of the recent Balkan War and of the present conflagration which threatened the heart of Islam. All Ottoman subjects from twenty to forty-five years of age were summoned to take up arms, and all other Mohammedans, including those who lived under the tyranny of the enemy powers, were commanded either to take part themselves in the Holy War or to contribute to it from their financial resources.

The Turkish Parliament was convened with an impressive ceremony on December 14th by the Sultan in person assisted by an imposing retinue, which included the heir apparent and other princes, the Khedive, and Goltz Pasha, and in the presence of the highest military, religious, and civil dignitaries, foreign ambassadors, and the German military mission. The speech from the throne described how the great European crisis had broken in upon Turkey while she was engaged in the peaceful work of healing the wounds of the Balkan War and removing the remaining sources of friction with her neighbors. General mobilization had been carried out solely for the preservation of Turkish neutrality. But Russia's unprovoked attack on the Turkish fleet in the Black Sea and the hostile acts of Great Britain and France on land had compelled the Sultan to declare war. The subversive designs of Russia, Great Britain, and France against Islam made it a religious duty to invoke the Holy War against these powers. The Sultan was confident that the achievements of the Turkish forces and of the other Mohammedan warriors called out in the Holy War would match the glorious victories of their allies in Europe.

After the withdrawal of the court, Halil Bey, President of the Parliament, opened the session by a speech in which he emphasized his conviction that the present contest did

Longwy after bombardment and capture by the Germans.

The British Lion between the Crown Prince, "Conqueror of Longwy," and von Hindenburg, "the victorious leader of the Army of the East." Statues placed in front of the Eberlein Museum, Berlin, in the early days of the war.

not involve solely an isolated question or the vindication of national honor, that it was not a struggle for the protection of a single province, but for existence itself. It was necessary for the Turks to persevere with unsparing efforts until they had won a durable peace which would guarantee to their descendants the opportunity of pursuing their civilizing task in tranquillity. Formerly they had been compelled to combat the Muscovite tyranny alone. Henceforth they would struggle in defense of civilization in league with Germany whose superiority in industry, administration, and organization was no less marked than in war. "I am convinced," he said, "that after the war the French and English, who will have to acknowledge with sorrow that the progress of Germany cannot be destroyed by violence, will seek a reconciliation with us."

Soon after the collision which inaugurated the hostilities between Russia and Turkey a deputation of Ruthenians from the Ukraine came to Constantinople and issued an address to the Ottoman nation, declaring that Russia had always been the enemy of Turkey, that the treatment of Mohammedans in Russia was inhuman, and that 30,000,000 people in the Ukraine looked for deliverance from oppression to Turkey, the old ally of the Cossacks of that region. The world to its astonishment was informed at the same time that such a question existed and that it had attained the proportions and degree of bitterness thus indicated. It is a well-known fact that there is a very marked temperamental distinction between the Great Russians and the Little Russians, the inhabitants of the Ukraine, and that the intolerant attitude of the government in respect to the dialectical peculiarities which distinguish the speech of the latter has been a source of discontent. But the fantastic tone of the address and its timely utility as a factor contributing to an atmosphere of exhilaration in Turkey create the suspicion that a disagreement of only minor significance was being exaggerated and exploited. The fact remains, nevertheless, that the wave of patriotic fervor which swept over Russia at the outbreak of the war had no magic efficacy to obliterate every form of abuse and dissatisfaction.

The war was commonly regarded as the consummation of the long struggle for the emancipation of the Russian national genius from bondage to foreign, that is German, influence and institutions. The passionate devotion to the native tradition found expression in a great variety of ways, in acts and impulses, great and petty, sane and bigoted. Thus on September 1st an imperial order directed that the Russian designation "Petrograd" should be substituted for the German "St. Petersburg" as the name of the capital of the empire.

There was undoubtedly a tendency on the part of the reactionary elements to turn the effusion of nationalistic enthusiasm to the promotion of their own designs. On the other hand, the Social Democrats, who in a body left the hall of the Duma before the vote was taken in the historic session of August 8th as a demonstration of their abhorrence of the war, were still unreconciled. They declared in a letter to M. Vandervelde their intention of continuing their war against Tsardom with greater energy than ever. "The Russian government," they asserted, "as well as the German government is the enemy of democracy. Even now that it is at war it persecutes the working men and the non-Russian nationalities, and should it be victorious, it would propagate political reaction in all Europe."

Bismarck had always dreaded the possibility of a league of hostile states and succeeded during the twenty years of his chancellorship in preventing such a combination by a foreign policy of circumspection, by cultivating useful friendships, and by isolating the chief potential enemies of Germany. But later, as we have seen, by her policy of gaining a dominating influence in the Balkan peninsula and Turkey, while upholding the pretensions of her ally, Austria-Hungary, by her unswerving effort to create a formidable sea-power, and by her uncompromising determination not to recede from any part of the Reichsland, Germany provoked the enmity of Russia and Great Britain, and kept alive the animosity of France.

The military leadership of Germany undertook to shatter the resulting coalition by crushing one of the partners before the others could effectively intervene. Possessing the initiative on all the fronts throughout the greater part of the first campaign, convinced that the supreme issues would be decided upon the principal battlefields of Europe and that a speedy decision was all-important for themselves, the Germans concentrated their energy for the offensive in the West and repeatedly hurled their tremendous masses against the armies of the French and their immediate supporters. But while these redoubtable efforts failed to reduce France to helplessness, or to eliminate her from the number of Germany's opponents, the unforeseen alacrity of the other foes compelled the Teutonic powers to divert an ever-increasing portion of their energy and strength to other fields, and to make enormous efforts where for the time only subordinate operations had been contemplated in the original plan.

Consequently, in spite of the unprecedented scale of their exertions and an astonishing succession of stupendous performances, no decisive results were anywhere obtained, and the execution of the original German plan reached its culmination without success, although its failure was mitigated by the occupation of the greater part of the invaded territory.

Both sides had been deceived in all the specious expectations upon which their respective hopes of speedy triumph had been founded. The British Empire had not been paralyzed by Irish discontent, the Boer revolt, or Indian disloyalty; France had not become the prey of discouragement or partisan dissension; and Russia had not collapsed from internal disorder or corruption; while at the same time Germany had displayed an almost universal spirit of unanimity and unflinching resolution that belied all the predictions of her enemies, and the supposedly incoherent aggregation of Austria-Hungary had shown unexpected tenacity and harmony. Contrary to the common notion that the maximum warlike strength of autocratic states, by reason of the concentration of authority, is available from the start and therefore necessarily diminishes as a contest proceeds, while that of democracies is only gradually attained in the course of the struggle, the energy and force of Germany were destined still to increase for many months far beyond the utmost prevision of her antagonists.

The palpitating excitement and glamor of mobilization and of the early spectacular operations of warfare had faded into the grim monotony of incessant, mechanical, exhausting toil requited by only microscopic results. But the spirit of the peoples had lost none of its fervor, and each nation deliberately schooled itself in the conviction that an unconquerable determination to win at any cost would make victory absolutely certain.

CHAPTER XVI

The Naval Situation at the Beginning of the War

Indications of the outbreak of war. Strategy of the belligerent powers. Obstacles to a British blockade. British and French fleets in the Mediterranean. British forces in home waters. Mobilization of the British fleet and its review by the king. The Turkish force. The belligerents' forces in the Far East and the Pacific. Seizure of German merchant ships in enemy ports. Relative power of Great Britain and Germany. The North Sea as the great area of naval action. The Baltic and Germany's control of that sea. Strategic value of the Kiel Canal. Safety of the German coast. The German base at Heligoland. The British bases at Scapa Flow and Rosyth. Status of power in the Mediterranean. Mine-laying. The German ships *Goeben* and *Breslau*.

That the great war was not unexpected is clearly shown by published documents as well as by preliminary events known to all. Among the documents one of the most notable is the letter to the French ambassador in London, dated November 22, 1912 (an enclosure to Document 105 of the British Diplomatic Correspondence), beginning with: "From time to time in recent years the French and British naval and military experts have consulted together." Such consultations could have but one meaning: looking to common action in war. The withdrawal of the British fleet from the Mediterranean to home waters, leaving this sea practically entirely to the French was a strong indication of the British attitude which certainly was not lost upon Germany. The German army increase of 1913 and the return of the French to three years as the term of service in the army were also marked indications of the great tension which needed but a spark to produce the conflagration. Nor can the establishment of a great northern dockyard at Rosyth on the Firth of Forth, the first steps towards which were taken in 1903, be omitted as a sign of the times. The general feeling is well expressed in the published dispatches of the Belgian ministers at London, Paris, and Berlin from 1905 to 1914.

The British naval strategy was, of course, of the simplest: to clear the ocean of German shipping, naval and commercial. Her forces were ample for the purpose, as results have shown. In addition, her battle fleet concentrated, as mentioned, in 1913, in the North Sea, was ready to resist any attempt at invasion or attack upon the British coast, or if opportunity afforded, to attack the German fleet at sea.

Of all these matters Germany and Austria were, of course, fully aware. Germany could not expect to do much more on the high seas than, for a time, to raid British commerce. Her foreign squadrons were small but of fast and efficient ships. Two notable squadron actions were to be fought, but the end was visible from the beginning. Naturally, too, effort was to be made in the direction of raiding by employing fast armed merchantmen. Germany, however, had a naval strategy in a large sense, in addition to an offensive by submarines, thus expressed by Baron von Maltzahn: "Our fleet law of 1900 was founded on 'Defense by Battle.' It states in its preamble that 'Germany must possess a battle fleet of such strength that war, even for the most powerful naval adversary, would involve such risks as to endanger the latter's supremacy.' By 'Defense by Battle' is meant to bring the enemy to battle on the high seas. It cannot be hoped to defeat him decisively once for all—the difference of strength which is a presumption of the strategic defensive would indeed prevent this—but it must be able to deprive him of so much of his strength that what remains is not sufficient for his purpose." Events later than the period covered in this volume show that Germany has held to this view. There could be, as von Maltzahn says, no hope on the part of Germany of a decisive victory over the British battle fleet with the great odds against her.

Germany could, too, in addition to the hope of decreasing the British fleet by losses equal to the whole of her own, have an expectation of wearing out in large degree the endurance of men and ships, as in this respect her situation was much more favorable than that of Britain. Her ships were in immediate contact with her dockyards, and her officers and men in touch with their home life in greater degree than were the British.

The conditions in the Baltic gave the Germans much the same superiority that existed for the British in the North Sea. The fortress of Kronstadt, the base of the Russian fleet, was practically unattackable, while the superiority in naval force of the Germans gave them command of the sea. The entrances to the Baltic were soon to be mined and, for reasons to be mentioned later, there was no danger in the Baltic from the British fleet.

The new element of submarines and mines prevented the establishment by the British of a blockade. Any effort at the establishment of one in the sense known to international law would have resulted in losses too serious to contemplate. The command of the Baltic Sea was thus of great advantage to Germany, in enabling her to maintain her commerce with Scandinavia—and would have been immensely greater had the United States enforced its protests against British action.

French strategy in the Mediterranean was akin to that of the British in the North Sea, and that of Austria akin to that of Germany. The impregnable coasts of the Central Powers were a factor of enormous weight in the naval part of the contest. As for Russia, her weakness in both the Baltic and the Black Sea at the beginning of the war precluded any real effort. Japan's action was to be confined to the seizure of German islands and the fortified port of Tsingtau.

The cataclysm of the war found the fleets of the belligerent powers distributed as had been arranged some years before. The British had withdrawn, practically, from the Mediterranean, leaving but 3 battle-cruisers, 4 armored cruisers, 4 light cruisers, 16 destroyers and depot ships, 2 gunboats, 6 submarines (3 at Malta, 3 at Gibraltar), and 16 torpedo boats. Thus, on the outbreak, the general command was taken over by the French admiral, Boué de Lapeyrère; Sir Berkeley Milne, the British commander-in-chief, returning home. The entire French fleet, practically, was now in this sea: 4 dreadnoughts, 18 pre-dreadnoughts (all carrying 12-inch guns), 20 armored and protected cruisers, 11 light cruisers, 84 destroyers, 153 torpedo boats (mostly small and ill-adapted to modern war), and 70 submarines.

Similarly, nearly the whole of the British navy was in home waters, organized as follows:

A. First Fleet (except Fourth Cruiser Squadron).

1. *First Battle Squadron.* Eight dreadnoughts, one carrying ten 13.5-inch guns; seven carrying ten 12-inch.
2. *Second Battle Squadron.* Eight dreadnoughts, all carrying 13.5-inch.
3. *Third Battle Squadron.* Eight pre-dreadnoughts, *King Edward* type, carrying four 12-inch.
4. *Fourth Battle Squadron.* Three dreadnoughts, carrying 12-inch; one pre-dreadnought.
5. *First Battle-Cruiser Squadron.* Four battle-cruisers.

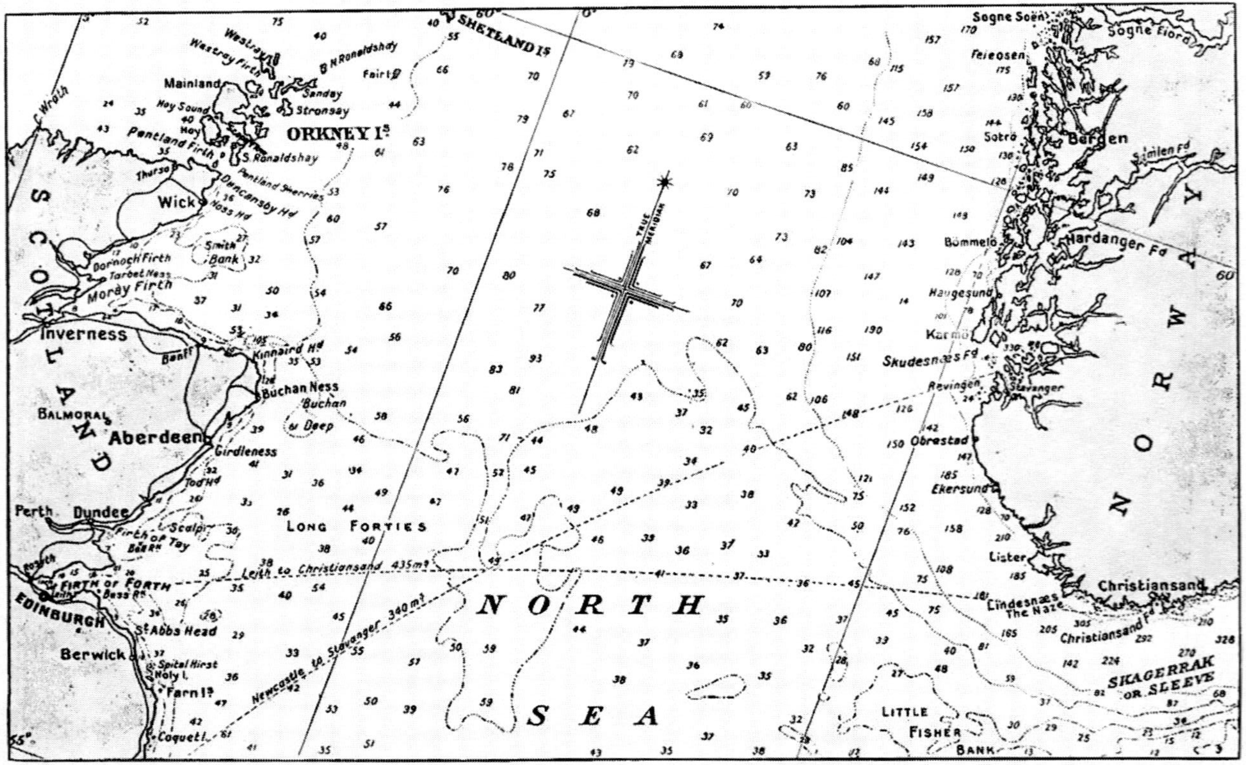

Chart of the North Sea showing positions of British naval bases at Rosyth and Scapa Flow.

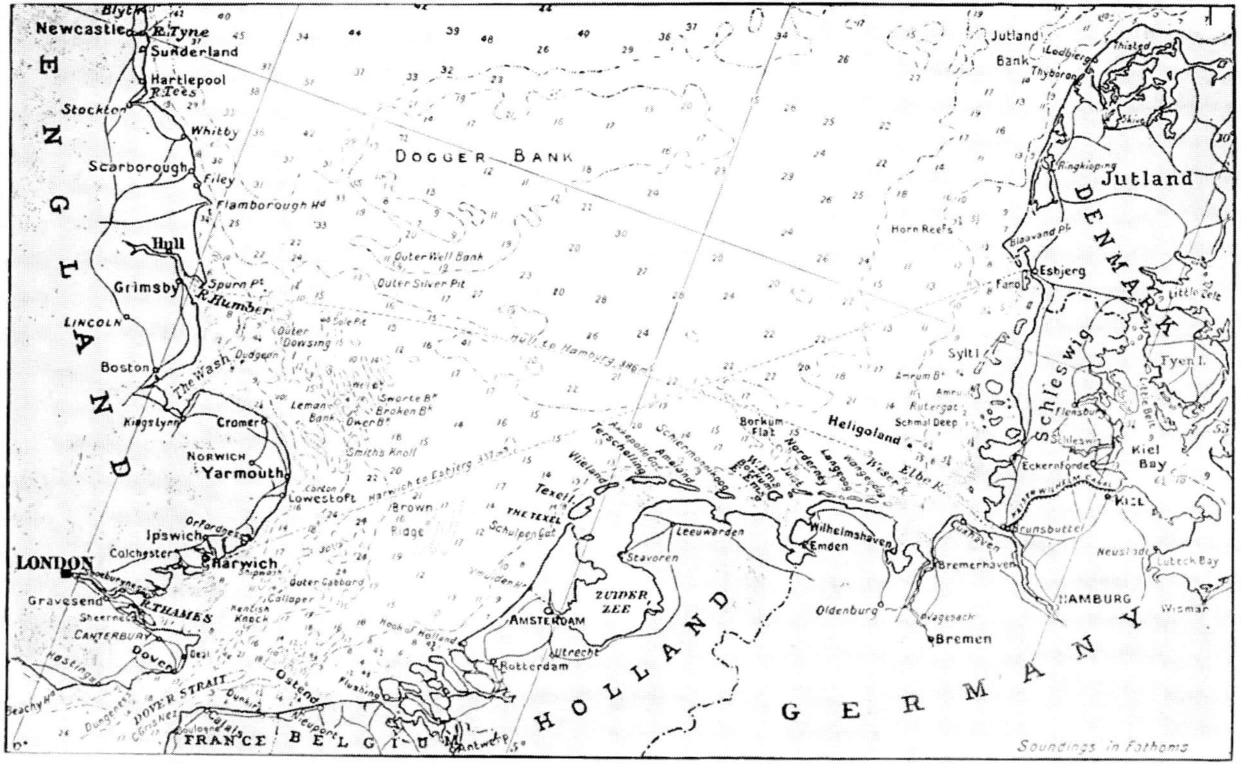

Chart showing North Sea and English Channel, and positions of German naval bases at Heligoland, Wilhelmshaven, and Cuxhaven.

6. *Second Cruiser Squadron.* Four armored cruisers.
7. *Third Cruiser Squadron.* Four armored cruisers.
8. *First Light Cruiser Squadron.* Four light cruisers.

B. Second Fleet.

1. *Fifth Battle Squadron.* (Pre-dreadnoughts, *Bulwark* type.)
2. *Sixth Battle Squadron.* (Pre-dreadnoughts, *Duncan* type.)
3. *Fifth Cruiser Squadron.* (*County* class.)
4. *Sixth Cruiser Squadron.* (*Drake* class.)

C. Third Fleet.

1. *Seventh Battle Squadron.* (Pre-dreadnoughts, *Majestic* type.)
2. *Eighth Battle Squadron.*
3. *Seventh, Eighth, Ninth, Tenth, Eleventh, Twelfth Cruiser Squadrons.* (Cruisers of all sorts.)

D. Mediterranean Fleet.

1. *Second Battle-Cruiser Squadron.* Three battle-cruisers.
2. *First Cruiser Squadron.* Four armored cruisers.
3. *Second Light Cruiser Squadron.* Four light cruisers.

In May, 1914, there was a mobilization of the fleet, thus described by a British Service paper: "The climax of the test mobilization was reached July 19th, when King George inspected his stupendous fleet at Spithead. Steaming at 11 knots an hour, the imposing cavalcade was headed by the first battle-cruiser squadron, consisting of four ships in battle-line ahead, under the command of Rear Admiral Sir David Beatty, who commands the squadron. The honor of leading the way was given to these dreadnought battle-cruisers—the *Lion, Queen Mary, Princess Royal,* and *New Zealand.* The battle-cruisers were followed by 29 ships of the first, second, third, and fourth battle squadrons. These, the flower of the British navy, came in two columns abreast, Admiral Sir George Callaghan (commander-in-chief of the home fleet) leading the way in the flagship *Iron Duke.* In the first battle squadron was the *Marlborough* (flagship of Vice Admiral Sir Lewis Bayly), with a displacement and horse-power equal to those of the *Iron Duke* herself. Then came the *Colossus* and *Hercules,* the *Neptune* and *St. Vincent,* the *Superb* and *Collingwood.*

"The powerful second battle squadron came next, the *King George V* (flagship of Vice Admiral Sir George Warrender) being followed by the *Audacious, Ajax, Centurion, Orion, Conqueror, Monarch,* and *Thunderer,* all tremendous pieces of naval architecture of the most modern type, but showing variety in shape. The third battle squadron brought before the notice of His Majesty ships of lighter tonnage, and, as things move so swiftly in naval matters nowadays, some experts say, almost out of date. These were the *King Edward VII* (flagship of Vice Admiral E. E. Bradford), which was launched at Devonport in 1903, and others built from nine to eleven years ago. Closely in their wake came the fourth battle squadron, consisting of such fine specimens as the *Dreadnought* (flagship of Vice Admiral Sir D. A. Gamble), and the *Téméraire, Bellerophon,* and the *Agamemnon.* Other big ships, small ships, snake-like destroyers, and gunboats, and the almost myriads of tiny craft which go to make up Britain's first line of defense, came along from the westward, submitted themselves to the inspection of the King, and proceeded on their way. These were the vessels of the second and third cruiser squadrons, light cruisers attached to the different fleets, destroyers, and a regular swarm of small craft. Although the big ships for the most part steamed two and three abreast, the line was nearly 14 miles long, reaching from near Osborne to the eastern end of the Isle of Wight. Two hundred of the most complete sea fighters ever known; the greatest congregation of ships ever assembled had passed in view before the King, comprising 24 dreadnoughts, 35 pre-dreadnoughts, 18 armored cruisers, 7 protected cruisers, and 78 destroyers, together with mine-layers, repair ships, and all kinds of auxiliaries."

The London *Times* of May 28, 1914, two months before the outbreak of war, referring to the decision to send the naval cadets from Osborne and Dartmouth to sea when the fleet is mobilized in July, said: "That this is being tried in July indicates that what Mr. Churchill calls the test of mobilization of the Third Fleet is really a mobilization for war, for this step would only be taken in view of the imminence of hostilities." A startling statement in view of later events.

The fleets in the Mediterranean and the Black Sea were to play but a secondary rôle. The Turks had been deprived of their two fine ships, the *Reshadieh* of 23,000 tons, built at Barrow, and carrying ten 13.5-inch and sixteen 6-inch guns, and the *Birinji Osman* of 27,500 tons, built at Elswick, and carrying fourteen 12-inch and twenty 6-inch guns. The two had cost Turkey some £6,000,000. The money had been paid, but on the day it had been arranged to hoist the Turkish flag both were seized by the British government and the Turks were minus both money and ships. The Turks, to be at war on November 5, 1914, were, however, soon to have a reinforcement from Germany in the *Goeben* and the *Breslau*, which arrived at Constantinople on August 11th, after a week of daring escapes, and went through the form of a sale to Turkey, thus creating a difficult and unprecedented situation in international law. The story of these ships on the outbreak of the war will have mention later.

In the Far East the Germans had two armored cruisers, the *Scharnhorst* and the *Gneisenau*, of 11,600 tons, eight 8.2-inch, six 6-inch, 40 caliber guns, and 23 knots, the light cruiser *Emden*, of 3,500 tons, ten 4.1-inch, 40 caliber guns, and 24 knots, three old cruisers, four other small vessels and two destroyers; Austria, a small cruiser, the *Kaiserin Elisabeth*. France had two armored cruisers, a destroyer, a gunboat, and four river gunboats; Great Britain, a battleship (the *Triumph*, of 12,000 tons), two armored cruisers, two light cruisers, eight destroyers, four torpedo boats, three submarines, and a number of small (chiefly river) craft; Japan, of course, her whole navy.

In the Pacific, Germany had in Australian waters, three old cruisers; in East Africa, the light cruiser *Königsberg* and the survey ship *Möwe;* and on the West Coast of North America, the light cruisers *Nürnberg* and *Leipzig*. There were one British battleship and two light cruisers in the East Indies; three cruisers and a gunboat at the Cape of Good Hope; two submarines at Vancouver. There were also the New Zealand navy of three cruisers and a sloop, and the Australian, of one battle-cruiser, three light cruisers, three destroyers, and two submarines. France had one gunboat and a despatch boat in the islands.

Distributed in all seas was a great number of German merchant ships. The war, though long brewing and known to be certain finally to come, broke so suddenly that none of these ships could have sufficient premonition of the danger to reach home ports, except those near to Germany. There was immediate seizure of all such in enemy ports, the procedure going to the extent of detaining some in British ports previous to the actual outbreak. The whole was in marked contrast to American procedure in the Spanish War, the President's proclamation, dated April 26, 1898, allowing twenty-six days to reach a home port, reading as follows:

"4. Spanish merchant vessels, in any ports or places within the United States, shall be allowed till May 21,

1898, inclusive, for loading their cargoes and departing from such ports or places; and such Spanish merchant vessels, if met at sea by any United States ship, shall be permitted to continue their voyage, if, on examination of their papers, it shall appear that their cargoes were taken on board before the expiration of the above term; provided, that nothing herein contained shall apply to Spanish vessels having on board any officer in the military or naval service of the enemy, or any coal (except such as may be necessary for their voyage), or any other article prohibited or contraband of war, or any dispatch of or to the Spanish government.

"5. Any Spanish merchant vessel which, prior to April 21, 1898, shall have sailed from any foreign port bound for any port or place in the United States, shall be permitted to enter such port or place, and to discharge her cargo, and afterward forthwith to depart without molestation; and any such vessel, if met at sea by any United States ships, shall be permitted to continue her voyage to any port not blockaded."

The relative forces actually available at the outbreak of the war, are given best in a table:

	British.	German.	
Super-dreadnoughts	13	—	
Dreadnoughts	11	16	
Dreadnought battle-cruisers	7	4	(Battle-cruisers)
Pre-dreadnought battleships	38	24	
Armored cruisers	30	9	
Light cruisers	22	33	
Protected cruisers	44	—	
Destroyers	198	151	
Torpedo boats	89	47	
Submarines	72	38	
Mine-layers	7	—	
Repair ships	3	—	

It needs no labored analysis to show the very great superiority of the British fleet—a superiority which inclines one to wonder at any fear of Germany's attaining, in a generation at least, an approach to equality. Certainly such fear is not understandable to the writer. Great Britain was very soon to have afloat in ships of 18,000 tons and over, forty 15-inch guns in five ships, ten 14-inch in one ship, a hundred and fifty 13.5-inch in sixteen ships, and a hundred and fifty-two 12-inch in sixteen; a total of thirty-eight dreadnoughts and super-dreadnoughts. Five others, each to carry eight 15-inch, had been laid down before the breaking out of the war. Germany had but twenty-six such ships built and building, carrying sixteen 15-inch, one hundred and sixty-four 12-inch, and eighty-six 11-inch.

Tabulated these are:

	British.	German.
15-inch	40	16
14-inch	10	—
13.5-inch	150	—
12-inch	152	164
11-inch	—	86
	352	266

Leaving aside the 11-inch, which no navy is now using as a primary battery, the relation of heavy guns was 352 to 180. A very careful estimate of the initial energy of the heavy guns of the two fleets (including the earlier but serviceable battleships) shows the power of the German gunfire to have been but 48% that of the British. There was, however, an element of superiority in favor of the Germans, in the greater elevation of 30 degrees which they were able to give their guns, as against 15 degrees of the British, and by a superiority in initial velocity. The range of the German 12-inch is 47% in excess of the British 13.5-inch, the former having an initial velocity of 3,080 foot-seconds, the latter 2,700. The newer British 12-inch has a range of over 1,000 yards in excess of the British 13.5-inch, but it falls short of the German by 7,670 yards, or 38%, chiefly through its lower possible elevation.

Naval battles are now fought at ranges undreamed of in former days, a range of 16,000 yards being, for example, used in the battle off the Falkland Islands. With the eye at an elevation of 20 feet a ship is hull down at such a distance. But the danger zone extends thousands of yards beyond this range, though the angle of fall in the greater ranges becomes so obtuse that the danger of hitting is much diminished. The flatter the trajectory the greater the chances of striking the target.

The great area of naval action was, of course, to be the North Sea, though in the peculiarly impregnable conditions of the German coast, there could be no attack upon the coast itself. Naval action in this sea was long to be confined to minor operations, chiefly submarines. The sea thus known may be taken as a rectangle, one end of which is bounded by a line 315 miles long, extending from the northernmost Orkneys to Bergen in Norway, the south end, taken from the Wash to the Elbe, is of equal breadth. The length of the northwest and southeast axis between these lines is 480 miles. Between Texel and Yarmouth, at the southwest corner, is a great teat with a general breadth of about 110 miles narrowing quickly at the Straits of Dover to about 25. The area of the sea is about twice that of our five Great Lakes combined. In general it is shoal enough to anchor at almost any point south of a line drawn from Moray Firth to the north end of Denmark (known as the Skaw), but north of this the water deepens, until at the Skagerrack, which extends along southern Norway to the west coast of Sweden, there is a depth of several hundred fathoms. Over the whole area of the shallower (and much greater) part of the sea, mines can be anchored without difficulty, but not in the Skagerrack or in the more northerly parts except on the borders of the Scotch coast, a depth beyond 360 feet precluding such action.

Leaving the Skagerrack, which extends northeast 120 miles with a breadth of 60, one turns suddenly south by east into the Cattegat, another 120 miles in length with a breadth varying from 30 to 60 miles. At the southwest corner begin the Great Belt and Little Belt, circuitous passages for 90 miles among numerous islands and sandbanks and the only entrances to the Baltic besides the still more narrow and difficult passage of "the Sound," which separates from Sweden the large island of Zealand, in which is Copenhagen. For centuries, Denmark claimed jurisdiction over these passages, and charged tolls which all nations paid until they were abolished in 1857 by a payment, in which all seafaring powers shared, of a *quid pro quo* to Denmark. Excluding the Gulf of Bothnia (itself 400 miles long), the Baltic is about half the size of the North Sea. The distance from Kiel to the entrance of the Gulf of Finland, at the eastern end of which lies the great fortress of Kronstadt, is about 500 miles, the breadth, south of the Gulf of Finland, is from 150 to 200 miles. This sea, like the North Sea, is shallow and ships can anchor in most parts. Navigation is, of course, greatly obstructed in winter by ice. One can readily see that operations in these two seas involve much greater distances (given here in statute miles) than are generally supposed.

The strategic conditions almost wholly favored Germany. The Russian fleet as compared with the German was weak. The British fleet could not aid the Russians

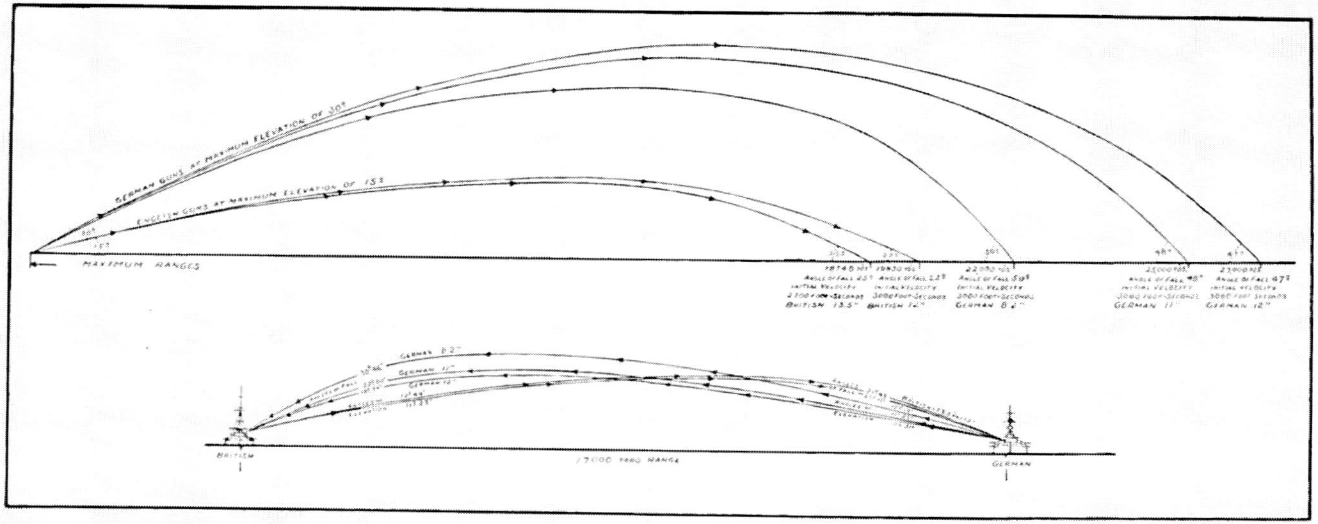

DATA FOR MAXIMUM POSSIBLE ELEVATION, 30-DEGREE GERMAN, 15-DEGREE ENGLISH*

BRITISH 13.5" GUN
Fifteen Degrees
Initial velocity, 2,700 f.s.
Weight of projectile, 1,250 lbs.
Angle of elevation, 15°.
Range, 18,745 yds.
Angle of fall, 25°.

BRITISH 12" GUN
Fifteen Degrees
Initial velocity, 3,000 f.s.
Weight of projectile, 850 lbs.
Angle of elevation, 15°.
Range, 19,830 yds.
Angle of fall, 22°.

GERMAN 12" GUN
Thirty Degrees
Initial velocity, 3,080 f.s.
Weight of projectile, 860 lbs.
Angle of elevation, 30°.
Range, 27,500 yds.
Angle of fall, 47°.

GERMAN 11" GUN
Thirty Degrees
Initial velocity, 3,080 f.s.
Weight of projectile, 660 lbs.
Angle of elevation, 30°.
Range, 25,000 yds.
Angle of fall, 48°.

GERMAN 8.2" GUN
Thirty Degrees
Initial velocity, 3,080 f.s.
Weight of projectile, 275 lbs.
Angle of elevation, 30°.
Range, 22,000 yds.
Angle of fall, 50°.

DATA FOR 17,000-YARD RANGE

BRITISH 13.5" GUN
Initial velocity, 2,700 f.s.
Weight of projectile, 1,250 lbs.
Range, 17,000 yds.
Angle of elevation, 12° 14'.
Angle of fall, 21° 43'.
Striking energy, 12,806 ft. tons.

BRITISH 12" GUN
Initial velocity, 3,000 f.s.
Weight of projectile, 850 lbs.
Range, 17,000 yds.
Angle of elevation, 11° 25'.
Angle of fall, 21° 10'.
Striking energy, 8,353 ft. tons.

GERMAN 12" GUN
Initial velocity, 3,080 f.s.
Weight of projectile, 860 lbs.
Range, 17,000 yds.
Angle of elevation, 10° 31'.
Angle of fall, 19° 34'.
Striking energy, 8,690 ft. tons.

GERMAN 11" GUN
Initial velocity, 3,080 f.s.
Weight of projectile, 660 lbs.
Range, 17,000 yds.
Angle of elevation, 11° 26'.
Angle of fall, 20°.
Striking energy, 6,136 ft. tons.

GERMAN 8.2" GUN
Initial velocity, 3,080 f.s.
Weight of projectile, 275 lbs.
Range, 17,000 yds.
Angle of elevation, 15° 19'.
Angle of fall, 30° 40'.
Striking energy, 2,115 ft. tons.

* From the *United States Naval Institute Proceedings*.

materially for two powerful reasons. They could not weaken their main fleet in the North Sea lest the remainder should be attacked by the combined German battle fleet, and any force sent into the Baltic would be subject to like attack through the command by the Germans of the Kiel Canal, officially the Kaiser Wilhelm Canal. Further, the Baltic ports of Germany were so heavily fortified (the coast fortifications, as in most European countries, being under control of the navy) as to be invulnerable to naval attack. The Kiel Canal was thus an invaluable asset to Germany both in war and peace. It had just been deepened and widened to meet the needs of the largest ships. It now has a surface breadth of 350 feet, instead of the former 130, and a bottom breadth of 130, instead of 60. At each end are locks, two of which (at Brünsbuttel), are larger even than those of Panama, as they are 1,083 feet long with a breadth at the entrance of 148; these are available for ordinary use in docking. The canal is 56 miles long, and ends in the Elbe at Brünsbuttel, 50 miles below Hamburg and 22 miles above Cuxhaven, the port on the south side, 16 miles within the lightship marking the entrance of the Elbe. There is no need to set forth at length the immense advantage to the Germans of this interior connection with the two seas. In a day or so the fleet can transfer from one sea to the other, avoiding the dangerous navigation (in fact impossible in present conditions) of 680 miles from the mouth of the Elbe round Denmark by the Cattegat to the Baltic. The two great war ports, Kiel and Wilhelmshaven (the latter on the estuary of the Jade and which during the war has been the headquarters of the German fleet, of which Admiral von Ingenohl was in chief command), are thus in close connection. The mouth of the Jade is but 14 miles from the Elbe lightship, 38 miles above which is the entrance to the Kiel Canal.

But the great safeguard and what makes the German coast on the North Sea practically unattackable are the sands, the deposits of the German rivers, which extend, in a fringe of shallows varying in breadth from 10 to 20 miles. The intricate channels through these are unusable except by careful buoyage. Naturally the buoys have been removed and only such marks are used as can be recognized by the Germans themselves. Nature herself has thus done more for the Germans than any degree of ordinary fortification could do. She has, in fact, made the German coast impregnable.

Fifty miles at sea, WNW of Cuxhaven and 40 miles from the nearest mainland, lies the island of Heligoland, a watering place in peace, a fortress and naval base in war. It is but a small plateau, a mile in length and a third of a mile in greatest breadth, with steep red cliffs some 200 feet high. On this plateau is a village. At the southeast end is a low beach on which, as also on the sand island known as the Dünen-Insel to the eastward and parallel with the main island and once connected with it, are villages made up of hotels and bathing establishments. The island population is about 2,300. The conformation affords a harbor for the largest ships, protected from attacks by very strong fortifications. It is thus a very powerful and practically unattackable advanced naval base. Until 1807 the island was Danish. It was seized in that year by the British and remained under their control until 1890 when, Lord Salisbury being British Premier, it was ceded to Germany in return for release by the latter of any claim in Zanzibar, and now forms part of Schleswig-Holstein.

On the outbreak of the war the British Battle Fleet, with Admiral John R. Jellicoe as commander-in-chief, was moved north to the Scapa Flow, a harbor formed by the many islands at the southern end of the Orkneys, with five deep channels for entrance and exit. The harbor is a quadrangle with sides of eight and ten miles with deep water through its whole extent; it thus affords 80 square miles of thoroughly protected anchorage. Here the large ships were in comparatively easy reach of the great docks of Belfast and the Mersey and were able to go to sea for gun practice with comparatively small danger of attack from submarines.

The battle-cruiser squadron, with many adjuncts of armored and light cruisers, destroyers and submarines, used the newly established naval station of Rosyth, three miles above the Forth Bridge, and nearer by 100 miles to the German coast. Every precaution of nets and other obstacles was established against torpedo attack. At the opening of the war a beginning only had been made on the large docks which were to form a part of the Rosyth equipment. There were, however, many available on the east coast.

The British fleets were thus, the one about 400, the other some 500 nautical miles from the mouths of the Elbe and the Jade, or a 24-hours' run at 17 and 21 knots for the one and the other.

The situation in the Mediterranean was so entirely secondary in the beginning that it needs but short attention. The Italian fleet was neutral: the French fleet so outclassed the Austro-Hungarian, both in numbers and power (reckoned by experts as three to one), that there could be no question of any great sea action; the Turkish sea power, when Turkey entered the war on November 5, 1914, was practically represented only in the *Goeben*, purchased from Germany. But the Austrian coast, stretching some 360 miles along the Adriatic, was practically almost as unattackable as that of Germany. This coast, south of Istria, is fringed with a series of narrow islands, some as extensive as 40 miles in length, which are parallel with the coast, and inside of which are deep, narrow passages, easily mined, the important points of which are heavily fortified. Practically no effort has been made against these.

The question of mine-laying rapidly assumed great importance. The Germans at once mined the waters of their own coast, and British reports of German mine-laying on the British coast were communicated to the Washington government, stating that "on or about August 26th an Iceland trawler is reported to have struck a mine 25 miles off the Tyne and at least one foreign newspaper has stated that the mine was English. Although the German action in laying mines has forced the Admiralty to reserve to itself the right to do likewise, the statement already made of His Majesty's Government that no British mines have been laid remains absolutely true at this moment." The question of precedence is of small moment, as all the nations at war were sure to use so effective a weapon. Thus the London *Times* of the 3d October published an official map of a British mine-field covering an area from latitude 51° 15′ N. and 51° 40′ N. and longitude 1° 35′ E. and 3° E., an area 25 nautical miles broad and 43 nautical miles long. As this area reached within three miles of the British coast, it enabled the British to exercise control over all traffic to and from Holland, as all ships bound to or from Holland had to come into British waters. This fact was used to claim the right to take off and examine mails, to which the United States has made strong but, as yet, ineffective protest.

The extent to which mines were laid may be judged by a dispatch of the International News Service *via* Sayville, May 14, 1916: "A dispatch from Amsterdam says that during April ninety mines drifted up on the Dutch coast. Fifty were British, three French, thirteen German, and twenty-four of unestablished nationality.

"Since the beginning of the war, continues the dispatch, 1,014 mines have landed on the Dutch coast, of which 535 were British, 61 French, 193 German, and 225 of unknown nationality."

All the nations involved in the war were equally sinners in mine-laying, though judged by the foregoing all were not equally efficient in anchoring their mines, which often drifted with fatal effect. The extreme depth at which mines can be securely anchored is regarded as three hundred and sixty feet, but it is evident that much less may cause mines to be insecure.

The first important incident of the war on the sea was the escape of the German battle-cruiser *Goeben* and her companion the light cruiser *Breslau* and their taking refuge in Turkish waters. On August 2d, news having been received at Messina of the declaration of war against Russia, the *Goeben* and *Breslau* left, and on the evening of August 4th the *Goeben* was off Philippeville, the *Breslau* having parted company with the purpose of bombarding Bona. The two ports are on the Algerian coast, Bona being about 175 miles west of the city of Tunis, the other some 40 miles further. War with France being now known to have been declared, the *Goeben* entered the port and opened a bombardment which inflicted much injury in the harbor, but being met with a heavy fire she withdrew. She, now joined by the *Breslau*, met the British battle-cruisers *Indefatigable* and *Inflexible* and the light cruisers *Gloucester* and *Weymouth*, which closing in on them were asked what was wanted. Reply was made that war was threatening between Great Britain and Germany. The Germans separated, putting on their best speed, and though followed ran the British out of sight. The same night they heard by wireless of the British declaration of war.

On August 5th they were again at Messina, at which, as it was a neutral port, they could stay but twenty-four hours. The time was spent in coaling. A British squadron was known to be on the watch for them in the strait, but in the evening of Thursday, August 6th, they left. Just what occurred in the strait is unknown, but the result was the escape of the two German ships and a court-martial of the senior British commander, who was finally exonerated of the charges made against him by the Admiralty. The German cruisers reached the Dardanelles on August 10th, boarded several British and French ships, but did nothing beyond destroying the wireless apparatus of the French steamer *Saghalien*, facts in themselves which should have thrown light on the Turkish situation. Shortly after reaching Constantinople they passed under the Turkish flag, the *Goeben* receiving the name *Sultan Yawuz Selim* and the *Breslau*, *Midellu*. The situation was unprecedented in international law. The *Goeben* was to the Turks but a fair offset in equity, though not in value, for the seizure of their ships built in England.

CHAPTER XVII

Operations in European Waters

In the North Sea: August 5, 1914, the German mine-layer *Königin Luise* destroyed. The British cruiser *Amphion* mined. August 9th, German submarine attack. August 28th, battle off Heligoland, German losses. The *Pathfinder* torpedoed. Victims of the *U-9* on September 22d and October 15th. Loss of four German destroyers. October 26th, sinking of the British battleship *Audacious*. British monitors in Belgian defense. Isolated casualties. German cruisers raid the east coast of England. Great battle on January 24, 1915, loss of the German cruiser *Blücher*. In the Baltic: German casualties. A German submarine sinks a Russian cruiser. In the Mediterranean. Austrian casualties. British submarine success in the Dardanelles. Russian losses in the Black Sea.

The naval war opened promptly. On August 5, 1914, a German mine-layer, the *Königin Luise*, a Hamburg liner of 2,163 gross tons, converted for the purpose, was sighted off the Suffolk coast by the light cruiser *Amphion* of 3,500 tons, accompanied by three destroyers. The German ship was chased some thirty miles and was sunk by gunfire when nearing the Scheldt. The high speed of all four of the British ships made destruction certain. The German had a quick revenge, for the *Amphion* when returning over the region of the former's operations struck one of the mines.

The official account says that a sheet of flame instantly enveloped the bridge, rendered the captain insensible, and he fell onto the fore and aft bridge. As soon as he recovered consciousness he ran to the engine-room to stop the engines, which were still going at revolutions for twenty knots. As all the fore part was on fire, it proved impossible to reach the bridge or to flood the fore magazine. The ship's back appeared to be broken, and she was already settling down by the bows. All efforts were therefore directed towards placing the wounded in a place of safety in case of explosion, and towards getting her in tow by the stern. Twenty minutes after the mine was struck the men, officers, and captain left the ship. Three minutes after the captain left his ship another explosion occurred, which enveloped and blew up the whole fore part of the vessel. The effects showed that she must have struck a second mine, which exploded the fore magazine. The after part now began to settle quickly until its foremost part was on the bottom and the whole after part tilted up at an angle of 45 degrees. In another quarter of an hour this, too, had disappeared.

On August 9th the First Light Cruiser Squadron of the main British fleet was attacked by submarines. They approached submerged, only the periscopes showing above water. The *Birmingham* by a lucky shot struck the periscope of the nearest, and later on her rising above water, rammed and sank her. The other got away. The lost vessel was supposedly of the earlier type, of 300 tons displacement. On August 19th the Press Bureau issued the following statement: "Some desultory fighting has taken place during the day between the British patrolling squadron and flotillas and German reconnoitering cruisers. No losses are reported or claimed. A certain liveliness is apparent in the southern area of the North Sea." "A reassuring statement was made by the *Daily Chronicle's* correspondent in Hull to the effect that the mine-sweeping fleet of trawlers had almost cleared the areas of the North Sea that were strewn with mines by the Germans. Meanwhile, trading and passenger steamers have been resuming their regular sailings. Cargoes of foodstuffs have been arriving at several east coast ports both in England and Scotland. As the *Times* naval correspondent remarks, 'British fishing boats are putting out, coastwise traffic has been resumed, mail and passenger boats are running to and fro between Britain and Northern Europe, and a Norwegian bark, the *Ingrid*, is said to have arrived at Dover on August 13th from the Baltic, having crossed the waterway without seeing any signs of war.'"

It was not until August 28th that the first real clash of the war on sea came, the scene of this being the vicinity of Heligoland. The British preliminary movements are described by Commodore Roger Keyes, commanding the submarine flotilla. The whole of the British force was under the command of Admiral David Beatty, who had his flag in the battle-cruiser *Lion*, which, with three others of her class, the *Queen Mary*, *Princess Royal*, and *Tiger*, formed the First Battle-Cruiser Squadron, under his immediate command. All carried eight 13.5-inch guns and several exceeded 30 knots speed, the *Princess Royal* on her trials having reached 34.7. The *Invincible* and *New Zealand*, battle-cruisers of 17,000 and 18,000 tons and 26 knots, joined him on August 28th. Accompanying were the Light Cruiser Squadron, with the *Euryalus* (flagship of Rear Admiral A. H. Christian), of 12,000 tons, and 21.5 knots, as flagship, with the First and Third Destroyer Flotillas and the submarines.

The movements of the British are best described by Commodore R. Y. Tyrwhitt, whose broad-pennant was in the protected cruiser *Arethusa*, of 3,560 tons, 29 knots, and two 6-inch and six 4-inch guns.

He sailed August 27th with the First and Third Destroyer Flotillas to carry out the prearranged operations. Four destroyers were absent, but he was joined by the *Fearless* (of 3,450 tons and ten 4-inch guns) in the afternoon.

At 6.53 A.M., on August 28th, a German destroyer was sighted and chased. From 7.20 to 7.57 A.M. the *Arethusa* and the Third Flotilla were engaged with numerous destroyers and torpedo boats making for Heligoland; course was altered to port to cut them off. Two cruisers, one with four and the other with two funnels, were sighted at 7.57, the nearest of which was engaged. The *Arethusa* received a heavy fire from both cruisers and several destroyers until 8.15 (18 minutes), when the *Magdeburg* transferred her fire to the *Fearless*. Close action was continued with the two funnelled cruiser (the *Ariadne*) until 8.25, when a 6-inch projectile from the *Arethusa* wrecked the *Ariadne's* fore bridge. The latter turned for Heligoland, now slightly visible on the starboard bow. All ships were now ordered to turn westward and shortly after, speed was reduced to 20 knots.

The following are vividly descriptive paragraphs from the commodore's report:

"During this action the *Arethusa* had been hit many times and was considerably damaged; only one 6-inch gun [of which, as mentioned, she carried two and six 4-inch] remained in action, all other guns and torpedo tubes having been temporarily disabled. . . .

"A fire occurred opposite No. 2 gun-port side, caused by a shell exploding some ammunition, resulting in a terrific blaze for a short period and leaving the deck burning. This was promptly dealt with. . . .

"The flotillas were reformed in divisions and proceeded [westward] at 20 knots. It was now noticed that the *Arethusa's* speed had been reduced.

"*Fearless* reported that the Third and Fifth Divisions of the First Flotilla had sunk the German commodore's destroyer and that two boats belonging to the *Defender* had been left behind, as our destroyers had been fired upon by a German cruiser during their act of mercy in saving the survivors of the German destroyer. At 10 A.M. hearing that the *Lurcher* and *Firedrake* were being chased by light cruisers, the *Fearless* and First Flotilla went to their assistance until 10.37 A.M., when having no news, and being in the vicinity of Heligoland, I ordered the ships in company to turn to the westward.

"All guns except two 4-inch were again in working order and the upper deck supply of ammunition was replenished.

"At 10.55 A.M. a four-funnelled German cruiser was sighted, and opened a very heavy fire at about 11 o'clock.

"Our position being somewhat critical, I ordered *Fearless* to attack, and the First Flotilla to attack with torpedoes, which they proceeded to do with great spirit. The cruiser at once turned away, disappeared in the haze, and evaded the attack.

"About ten minutes later the same cruiser appeared on our starboard quarter. Opened fire on her with both 6-inch guns; *Fearless* also engaged her, and one division of destroyers attacked her with torpedoes without success. . . .

"The cruiser was badly damaged by *Arethusa's* 6-inch guns, and a splendidly directed fire from *Fearless*, and she shortly afterwards turned away in the direction of Heligoland.

"Proceeded, and four minutes later sighted the three-funnelled cruiser *Mainz*. She endured a heavy fire from *Arethusa* and *Fearless* and many destroyers. After an action of approximately 25 minutes, she was seen to be sinking by the head, her engines stopped, besides being on fire.

"At this moment the Light Cruiser Squadron appeared, and they very speedily reduced the *Mainz* to a condition which must have been indescribable.

"I then recalled *Fearless* and the destroyers, and ordered cease fire.

"We then exchanged broadsides with a large four-funnelled cruiser on the starboard quarter at long range, without visible effect.

"The Battle-Cruiser Squadron now arrived and I pointed out this cruiser to the admiral commanding, and was shortly afterwards informed by him that the cruiser in question had been sunk and another set on fire. . . .

"*Arethusa's* speed was about six knots until 7 P.M., when it was impossible to proceed any further, and fires were drawn in all boilers except two, and assistance called for. . . .

"His Majesty's ship under my command was then towed to the Nore, arriving at 4 P.M. on August 29th. Steam was then available for slow speed, and the ship was able to proceed to Chatham under her own steam."

Of the part played by the destroyers and submarines in this first serious naval encounter some important details may be quoted from the report of Commodore Keyes, commanding the Submarine Flotilla (himself in the destroyer *Lurcher*). He says:

"At midnight on August 26th I embarked in the *Lurcher*, and, in company with *Firedrake* and submarines *D 2*, *D 8*, *E 4*, *E 5*, *E 6*, *E 7*, *E 8*, and *E 9*, of the Eighth Submarine Flotilla, proceeded to take part in the operations in the Heligoland Bight arranged for August 28th. The destroyers scouted for the submarines until nightfall on the 27th, when the latter proceeded independently to take up various positions from which they could coöperate with the destroyer flotillas on the following morning.

"At daybreak on August 28th the *Lurcher* and *Firedrake* searched the area, through which the battle-cruisers were to advance, for hostile submarines, and then proceeded towards Heligoland in the wake of submarines *E 6*, *E 7*, and *E 8*, which were exposing themselves with the object of inducing the enemy to chase them to the westward. . . .

"Lieutenant Commander Ernest W. Leir, commanding submarine *E 4*, witnessed the sinking of the German torpedo boat destroyer *V 187* through his periscope and, observing a cruiser of the *Stettin* class close, and open fire on the British destroyers which had lowered their boats to pick up the survivors, he proceeded to attack the cruiser, but she altered her course before he could get within range. After covering the retirement of our destroyers, which had had to abandon their boats, he returned to the latter, and embarked a lieutenant and nine men of the *Defender*, who had been left behind. The boats also contained two officers and eight men of *V 187*, who were unwounded, and 18 men who were badly wounded. As he could not embark the latter, Lieutenant Commander Leir left one of the officers and six unwounded men to navigate the British boats to Heligoland. Before leaving he saw that they were provided with water, biscuit, and a compass. One German officer and two men were made prisoners of war.

"Lieutenant Commander Leir's action in remaining on the surface in the vicinity of the enemy, and in a visibility which would have placed his vessel within easy gun range of an enemy appearing out of the mist, was altogether admirable."

The action of Lieutenant Commander Leir, which Commodore Keyes very justly describes as admirable, soon brought the former well-merited promotion to the rank of commander.

Commodore Keyes in dealing in general with the operations of the submarines speaks of the sinking, six miles south of Heligoland, of the German light cruiser *Hela* on September 13th by submarine *E 9* and on October 6th, by the same submarine (the commander of which was Lieutenant Commander Max K. Horton), the sinking of the German destroyer *S 126*. He also mentions the "short steep seas which accompany westerly gales in the Heligoland Bight [which] made it difficult to keep the conning-tower open. There was no rest to be obtained, and even when cruising at a depth of 60 feet the submarines were rolling considerably and pumping, *i.e.*, vertically moving about 20 feet."

The following is Admiral Beatty's report:

"I have the honor to report that on Thursday, August 27th, at 5 A.M., I proceeded with the First Battle-Cruiser Squadron and First Light Cruiser Squadron in company, to rendezvous with the Rear Admiral, *Invincible*.

"At 4 A.M., August 28th, the movements of the flotillas commenced, as previously arranged, the Battle-Cruiser Squadron and Light Cruiser Squadron supporting. The Rear Admiral, *Invincible*, with *New Zealand* and four destroyers, having joined my flag, the squadron passed through the prearranged rendezvous.

"At 8.10 A.M. I received a signal from the Commodore (T) [Tyrwhitt], informing me that the flotilla was in action with the enemy. This was presumably in the vicinity of their prearranged rendezvous. From this time until 11 A.M. I remained about the vicinity ready to support as necessary, intercepting various signals, which contained no information on which I could act.

"At 11 A.M. the squadron was attacked by three submarines. The attack was frustrated by rapid maneuvering, and the four destroyers were ordered to attack them. Shortly after 11 A.M. various signals having been received indicating that the Commodore (T) and Commodore (S) [Keyes] were both in need of assistance, I ordered the Light Cruiser Squadron to support the torpedo flotillas.

"Later I received a signal from the Commodore (T), stating that he was being attacked by a large cruiser, and a further signal informing me that he was being hard pressed, and asking for assistance. The Captain (D) [name not published], First Flotilla, also signalled that he was in need of help.

"From the foregoing the situation appeared to me critical. The flotillas had advanced only 10 miles since 8 A.M., and were only about 25 miles from two enemy bases on their flank and rear respectively. Commodore Goodenough had detached two of his light cruisers to assist some destroyers earlier in the day, and these had not yet rejoined. (They rejoined at 2.30 P.M.) As the reports indicated the presence of many enemy ships—one a large cruiser—I considered that his force might not be strong enough to deal with the situation sufficiently rapidly, so at 11.30 A.M. the battle-cruisers turned to ESE and worked up to full speed. It was evident that to be of any value the support must be overwhelming, and carried out at the highest speed possible.

"I had not lost sight of the risk of submarines, and possible sortie in force from the enemy's base, especially in view of the mist to the southeast.

"Our high speed, however, made submarine attack difficult, and the smoothness of the sea made their detection comparatively easy. I considered that we were powerful enough to deal with any sorties except by a battle squadron, which was unlikely to come out in time, provided our stroke was sufficiently rapid.

"At 12.15 P.M. *Fearless* and First Flotilla were sighted retiring west. At the same time the Light Cruiser Squadron was observed to be engaging an enemy ship ahead. They appeared to have her beat.

"I then steered NE to sounds of firing ahead, and at 12.30 P.M. sighted *Arethusa* and Third Flotilla retiring to the westward engaging a cruiser of the *Kolberg* class on our port bow. I steered to cut her off from Heligoland, and at 12.37 P.M. opened fire. At 12.42 the enemy turned to NE, and we chased at 27 knots.

"At 12.56 P.M. sighted and engaged a two-funnelled cruiser ahead. *Lion* fired two salvos at her, which took effect, and she disappeared into the mist, burning furiously and in a sinking condition. In view of the mist and that she was steering at high speed at right angles to *Lion*, who was herself steaming at 28 knots, the *Lion's* firing was very creditable.

"Our destroyers had reported the presence of floating mines to the eastward, and I considered it inadvisable to pursue her. It was also essential that the squadron should remain concentrated, and I accordingly ordered a withdrawal. The battle-cruisers turned north and circled to port to complete the destruction of the vessel first engaged. She was sighted again at 1.25 P.M. steaming SE, with colors still flying. *Lion* opened fire with two turrets, and at 1.35 P.M., after receiving two salvos, she sank.

"The four attached destroyers were sent to pick up survivors, but I deeply regret that they subsequently reported that they searched the area but found none.

"At 1.40 P.M. the battle-cruisers turned to the northward, and *Queen Mary* was again attacked by a submarine. The attack was avoided by the use of the helm. *Lowestoft* was also unsuccessfully attacked. The battle-cruisers covered the retirement until nightfall. By 6 P.M. the retirement having been well executed and all destroyers accounted for, I altered course, spread the light cruisers, and swept northwards in accordance with the commander-in-chief's orders. At 7.45 P.M. I detached *Liverpool* to Rosyth with German prisoners, seven officers and 79 men, survivors from *Mainz*. No further incident occurred."

It is clear that the Germans, despite their loss of the protected cruisers, the *Köln* and *Mainz*, both of 4,280 tons, 26 knots and twelve 4.1-inch guns, and the *Ariadne*, of

2,618 tons, 22 knots and ten 4.1-inch guns, and a destroyer, had carried off the honors of the fight. Their opponents (the *Arethusa* carrying the Commodore's broad pennant) had withdrawn badly injured and, leaving the field of action at 20 knots, were going westward when the battle-cruiser squadron, of overpowering force, came to the rescue. Says Mr. Jane, the well-known British naval critic: "Our popular press feeds us on apparent results. . . . But from the naval war standard the fact remains that if Admiral Beatty had not taken abnormal risks we should have been badly beaten in the Bight of Heligoland on August 28th." The best that can be said is that both sides fought bravely and that no flag was hauled down.

On September 5th the light cruiser *Pathfinder*, of 2,940 tons, 25 knots and nine 4-inch guns, was torpedoed off St. Abb's Head, East Scotland, by the German *U-21*, which had been previously reported as that sunk by the *Birmingham* on August 9th. But on September 22, 1914, the Germans scored a much more marked success in the sinking of the three armored cruisers, *Aboukir*, *Hogue*, and *Cressy*, which were together on patrol duty off the Dutch coast. These ships were 454 feet on the water line, 69½ feet broad, and had an armor belt 11½ feet wide and 230 feet long. They carried each two 9.2-inch, 40 caliber guns, twelve 6-inch, 45 caliber, thirteen 12-pounders, three 3-pounders, and two 18-inch submerged torpedo tubes. The guns were mounted in 6-inch turrets and barbettes. They had 5-inch casemates. Altogether, though built about 1900, they were powerful ships with complements of 700 and, in the flagship, 745 men. Their foe was the German submarine *U-9*, commanded by Lieutenant Commander Weddigen, with a crew of 20 men. The submarine had just been to the Shetland Islands, a journey there and back of 1,200 miles. The commanders of the

Cressy and *Hogue* give in their reports excellent descriptions of what occurred, which are of special interest as illustrating the present dangers of naval life in war. These officers were the second in command, both of their captains being lost.

Commander Nicholson of the *Cressy* says:

"*Aboukir* was struck at about 6.25 A.M. [September 22d] on starboard beam. *Hogue* and *Cressy* closed, and took up position—*Hogue* ahead of *Aboukir* and *Cressy* about 400 yards on port beam. As soon as it was seen that *Aboukir* was in danger of sinking, all boats were sent away from *Cressy* and picket boat was hoisted out without steam up. When cutters full of *Aboukir's* men were returning to *Cressy*, *Hogue* was struck apparently under aft 9.2 magazine, as a very heavy explosion took place immediately after the first explosion.

"Almost directly after *Hogue* was hit we observed a periscope on our port bow about 300 yards off. Fire was immediately opened, and engines put full speed ahead with intention of running her down. Our gunner, Mr. Dogherty, positively asserts that he hit the periscope, and that the submarine then showed her conning-tower, which he struck, and the submarine sank. An officer standing alongside the gunner thinks that the shell struck only floating timber, of which there was much about, but it was evidently the impression of the men on deck, who cheered and clapped heartily, that the submarine had been hit. This submarine did not fire a torpedo at *Cressy*.

"Captain Johnson then maneuvered the ship so as to render assistance to crews of the *Hogue* and *Aboukir*. About five minutes later another periscope was seen on our starboard quarter. Fire was opened. The track of the torpedo she fired at a range of 500 to 600 yards was plainly visible, and it struck us starboard side just before the after bridge. The ship listed about 10 degrees to starboard and remained steady. Time, 7.15 A.M. All water-tight doors, dead lights, and scuttles had been securely closed before the torpedo struck ship. All mess tools, and tables, shores, and all available timber below and on deck had been previously got up and thrown over the side for saving of life.

"A second torpedo fired by the same submarine missed and passed about 20 feet astern. About a quarter of an hour after the first torpedo had hit, a third torpedo, fired from a submarine just before starboard beam, hit us in No. 5 boiler-room. Time, 7.30 A.M. The ship then began to heel rapidly, and finally turned keel up, remaining so for about 20 minutes before she finally sank at 7.55 A.M. A large number of men were saved by the casting adrift of a pattern three target. The steam pinnace floated out of her crutches, but filled and sank.

"The second torpedo which struck *Cressy* passed over sinking hull of *Aboukir*, narrowly missing it. It is possible that the same submarine fired all three torpedoes at *Cressy*.

"The conduct of the crew was excellent throughout. I have already reported the splendid service rendered by Captain Phillips, master of the trawler *L. T. Coriander*, and his crew, who picked up 156 officers and men."

Commander Norton of the *Hogue* reports:

"Between 6.15 and 6.30 A.M. H. M. S. *Aboukir* was struck by a torpedo. The *Hogue* closed the *Aboukir*, and I received orders to hoist out the launch, turn out and prepare all boats, and unlash all timber on the upper deck. The two lifeboats were sent to the *Aboukir*, but before the launch could get away the *Hogue* was struck on the starboard side amidships by two torpedoes at intervals of 10 to 20 seconds. The ship at once began to heel to starboard.

"After ordering the men to provide themselves with wood, hammocks, etc., and to get into the boats on the booms and take off their clothes, I went by Captain Nicholson's directions to ascertain the damage in the engine-rooms. An artificer-engineer informed me that the water was over the engine-room gratings. While endeavoring to return to the bridge the water burst open the starboard entry-port doors, and the ship heeled rapidly.

"I told the men in the port battery to jump overboard as the launch was close alongside, and soon afterwards the ship lurched heavily to starboard. . . .

"A Dutch sailing trawler sailed close by, but went off without rendering any assistance, though we signalled to her from the *Hogue* to close after we were struck.

"The *Aboukir* appeared to me to take about 35 minutes to sink, floating bottom up for about five minutes. The *Hogue* turned turtle very quickly in about five minutes, and floated bottom up for some minutes. A dense black smoke was seen in the starboard battery, whether from coal or torpedo cordite I could not say. The upper deck was not blown up, and only one other small explosion occurred as we heeled over.

"The *Cressy* I watched heel over from the cutter. She heeled over to starboard very slowly, a dense black smoke issuing from her when she attained an angle of about 90 degrees. She took a long time from this angle until she floated bottom up, with the starboard screw slightly out of the water. I consider that it was 35 to 45 minutes from the time when she was struck until she was bottom up."

Somewhat more than half of the 2,200 officers and men, who were aboard the three ships, were lost.

The scene was dramatically described by a Dutch skipper to a temporary member of the American Legation in Holland: "I was called on deck by my mate, who said there were three British men of war in sight. I went up and saw two, then one and then none," so rapid and effective was the action of the submarine. The British naval expert, mentioned above, the late Fred T. Jane, said it was impossible that the exploit could have been the work of one vessel, but this was merely a case of dogmatism too frequent with this writer, as there is no doubt that there was but one submarine present.

The fate of the *Hogue* and the *Cressy*, due to the instinctive and very laudable desire to render assistance, brought an order from the British Admiralty that ships thereafter should not be risked by approaching a vessel so wounded and stopping to make a rescue—a hard necessity.

Three weeks later (October 15th) the cruiser *Hawke*, of 7,350 tons, was torpedoed by the same *U-9*, all but 46 men and three officers of the 600 aboard being lost. Two days after this, October 17th, the light cruiser *Undaunted*, of 3,750 tons and 29 knots, accompanied by three destroyers, sank by gunfire four German destroyers, *S 115*, *S 117*, *S 118*, and *S 119*, each of 420 tons and of a class built at Elbing in 1902-1903. Over 200 of the German crews, of some 240, perished.

On October 27th, however, the British suffered the severest loss of the war up to that time, in the sinking by a mine or torpedo (by which, is unknown) off the north coast of Ireland of the *Audacious*, one of their latest and heaviest battleships, of 23,000 tons and ten 13.5-inch guns. She was at the time, in company with four other ships, carrying on target practice and was just turning to make a run past the target. The White Star steamer *Olympic*, homeward bound, was in reach and lent valuable aid in rescue of the crew, but two of whom were lost. The severity of the blow was shown by the endeavor of the British Admiralty to suppress for a long time the fact, the *Olympic* being detained for a week at Lough Swilly, and the passengers, who were released at Belfast, where the ship docked, instead of Liverpool, warned to keep silence as to what they had seen. There has never, so far as is known to the writer, been any official acknowledgment of the disaster.

On October 31st the *Hermes*, a cruiser of 5,600 tons, 20 knots and eleven 6-inch guns, was sunk by an unknown German submarine off Dover. Nearly 400 of the crew out of some 450 were saved. The disaster was at 9 A.M. The ship was struck twice, the first blow putting her propellers out of action, the second striking her in the vicinity of the engine room, tearing a great rent in the bottom. Notwithstanding, the ship floated, according to some reports, nearly an hour, others, however, stating the time as much less. The ability of a ship of so moderate a size to remain afloat so wounded for at least a considerable time is, in view of later events, of much importance.

Three monitors at this period were actively employed on the Belgian coast, but with what success is not accurately known, the accounts of the British Service papers being so extremely lurid as to damage done by vessels whose gunfire, on account of their quick raft-like motion, is so notably inaccurate, that the reports must be taken with caution. These ships were building for Brazil, but were taken over by Great Britain on the outbreak of the war and named the *Severn*, *Humber*, and *Mersey*. They are 265 feet long with a draft of 8½ feet and a displacement of 1,200 tons, a speed of 11½ knots, and an armament of two 6-inch guns, two 4.7-inch howitzers, and four 3-pounders. As instruments of war, they are of doubtful value in any waters but those of smooth harbors and rivers.

Affairs in the North Sea ran on with varying fortune. The Germans on November 4th lost the armored cruiser *Yorck*, 9,350 tons, 21.4 knots and four 8.2-inch guns in turrets, by the ship's striking one of their own mines at the entrance of Jade Bay. Of her complement of some 700, over 300 were lost. The captain was sentenced on December 23d by a court-martial held at Wilhelmshaven to two years' detention in a fortress and the officer next in rank to one year, the charges being disobedience of orders and negligence. On November 26th the British battleship *Bulwark*, of 15,000 tons, blew up in Sheerness Harbor, nearly her whole complement of 750 men being lost, 14 only being saved. The misfortune is attributed to careless handling in taking on board shells, one falling from such a height as to cause an explosion which extended to the magazines.

That the spirit of venture shown by the reconnaissance in force off Heligoland on August 28th was not singular to the British was seen in the appearance on the English coast, on November 13th, of a German squadron of eight ships which included the battle-cruisers *Seydlitz*, *Moltke*, and *Von der Tann*, and the armored cruiser *Blücher*. These ships bombarded Yarmouth at long range, but with small damage. The raid was probably more for moral than material effect. This was followed on December 16th by an attack, in weather described at an inquest on civilians who were killed as "very thick and hazy," on Scarborough, Whitby, and Hartlepool, with a considerable loss of life at all three places. These attacks, of course, caused much comment, as being unjustifiable. As the subject is of great importance and interest it is not amiss to give space to some authorities. Thus the London *Economist*, of December 26, 1914, said: "We have assumed the burden and responsibilities of war and if the enemy is successful in piercing

The sinking of the German cruiser *Mainz*, off Heligoland, during the action of August 28, 1914. British destroyer standing by to pick up survivors.

Boats from the British battleships rescuing survivors from the *Gneisenau*, after the action off the Falkland Islands, December 8, 1914.

our defenses, it behooves us to face the fact with calm and fortitude. . . . The bombardment of undefended towns is forbidden by the recognized Conventions of naval warfare. Unhappily, no agreement as to the definition of an undefended town has ever been achieved. The term 'undefended' is certainly much wider than 'unfortified.' For instance, in ratifying the Conventions of The Hague Conference on this subject, the British government, supported by those of Germany and France, insisted that the laying of contact mines off a harbor should be sufficient to expose the same to bombardment.

"The Conventions further direct that even in the case of defended towns, the commander of an attacking force must give due notice of bombardment, but only when military exigencies allow, which clearly they do not in the case of a sudden raid; that the enemy must do his best to spare churches, civic buildings, hospitals and the like, where these are distinguishable by the exhibition of large, rigid panels divided into black and white triangles. And here it may be remarked that the official *British Manual of Military Law* lays it down that a town and its defenses constitute an indivisible whole.

"Finally, the immunity of undefended towns does not extend to military works, establishments or depots, or to any workshops capable of supplying military needs, exceptions that would probably be held to cover railway stations, bridges, and coal stores, whether in public or private ownership.

"Now, Hartlepool is clearly a defended town. The War Office reported that the German vessels engaged the fortress, which replied and drove the enemy off. The Germans, on the other hand, pretend to have silenced its guns. There were, presumably, other military targets as well, for shells are reported to have been dropped on the royal engineers' and infantry lines. It is clear, however, that the town suffered far more than the military works. This may have been the result of malice, or carelessness, or incompetence, or it may have been inevitable in the case of a bombardment at considerable range on a misty morning."

Passing, for want of space, the *Economist's* remarks of somewhat like tenor as to Whitby and Scarborough, the paper continues: "There is one consideration which seems to have been lost sight of by some people. It is this—that as indicated above, several proposed restrictions upon the freedom from bombardment have been resisted by the British government in the past. In view of our position as a paramount naval power the decision was very likely a wise one, at least from a military point of view, but we must be prepared to take the consequences, and we shall cut a very poor figure before the world if we complain when others turn to our disadvantage the freedom we have ourselves reserved. . . . Before we give vent to an excess of fury, certainly before we indict a whole nation, let us remember that our own navy has been engaged in similar operations fraught with possible loss to the life and property of non-combatants . . . the fact remains that coast towns have to take their chances. The killing of women and children, and of civilians generally, is an abomination, but war itself is an abomination and will always be so."

In the British naval maneuvers of 1888 there were simulated attacks on several undefended coast towns which brought protests denouncing the action as a breach of international law. Replies were made by naval officers and by several authorities on international law. Among them was Admiral Lord Charles Beresford (now Lord Beresford), who wrote in the London *Times*, of August 18, 1888: "I say boldly and openly that if an officer could damage his enemy and procure panic and demoralization in the enemy's country, he would be wrong to demur a moment in exacting a ransom or in bombarding a seaport town if the opportunity occurs."

As a retaliation for this raid, an attack at daybreak on Christmas Day, 1914, was made on Cuxhaven by seven seaplanes, escorted by a light cruiser, destroyers, and submarines. Seen from Heligoland, two Zeppelins, some seaplanes and submarines were sent out. Apparently nothing was accomplished on either side. Four of the British aviators lost their machines, but themselves returned safe.

On January 1st the British *Formidable,* of the *Bulwark* type, of 15,511 tons, 18.1 knots with four 12-inch and twelve 6-inch guns, was sunk in the English Channel by an unknown submarine. Seventy-one of the crew were picked up by a British light cruiser, 70 had been taken from the water by a trawler, and 40 others, the survivors of 60, who were in a cutter, reached Lyme Regis after tossing about in a heavy sea, without food or water, for twenty hours.

A graphic account of the disaster, from an officer, appeared in the *Daily Telegraph.* The following is taken from an abridgment of this in the *Army and Navy Gazette.* It is valuable not only as illustrating a terrible and dramatic situation, but as throwing some light upon questions involved, in the sinking, later, of much larger ships as to the time they should have remained afloat.

"I was sleeping in my hammock," the officer observed, "when about 2.20 [A.M.] I was awakened by a tremendous crash. I jumped out of my hammock and ran to the upper deck. I noticed that there was already a great list on the ship. At the same time we turned head on the wind.

The explosion occurred on the starboard side, abreast of the foremost funnel, and I should say that the resulting inrush of water flooded the boiler-rooms, because immediately afterwards the electric light and steam power failed on all the engines, and we came to a standstill. . . .

"Just about this time—I should think a quarter of an hour after the first explosion—a second occurred. This proved to be a blessing in disguise, at least temporarily, because the great inrush of water which ensued helped to right the vessel. We got nearly on to an even keel, and this made it much easier for us to get about the deck. The second explosion seemed to me to burst the boilers. All the men eventually got on to the upper deck, each with some piece of woodwork in his possession or near him. Each man wore an Admiralty swimming collar, which, while good enough in a way, simply keeps a fellow's head on a level with the water, with the result that if there is anything of a sea his mouth is nearly always 'awash'—a very unpleasant experience. The officers were wearing a Gieve waistcoat, which is a much better idea than the Admiralty collar. It has a tube on it, and when this is blown up it supports the wearer higher out of the water. There was not the slightest panic. I think this was very creditable. They had been standing for a long time, too, very scantily clothed, in a biting wind, and it was a great test of their courage.

"Everybody seemed to think the ship would hold out and float to dawn, and she did actually float for about two hours and a quarter. She developed a terrible list, however, in spite of the good effect of the second explosion, and in order to correct this the turret crews tried to train their guns on the beam, but there being no hydraulic pressure available, they were unable to do this, and the effort had to be abandoned. Verrey's lights were now sent up. The wireless apparatus was, of course, out of order, there being no current. Then followed an uncomfortable three-quarters of an hour while we waited. All our water-tight doors were closed, and everything done that could be done to keep the ship afloat, but as time went on it was evident she was going under, and her list was increasing terribly. The crews of the starboard side—the side which was in the water—had been down and closed the gun-ports, but it was easily noticeable in one of the gun casements that the water was rising rapidly inside, and coming up the ammunition hoists. During the last 10 minutes that the vessel was afloat—from about 4.20 to 4.30—the list appreciably increased, and matters had reached a climax.

"The captain came down from the bridge on the port boat-deck, and sang out, 'Into the water with you; she's going.' Then it was a question of each man for himself. You must understand that the ship was now nearly flat on her side. Hundreds of the men had climbed over the rails on the upper side, which was out of the water, and stood there in two ranks waiting for orders, and on hearing the captain shout they all slid down the vessel's side into the sea. Many fell with some force against the bilge keel, which was showing above the water, and got some nasty injuries, but eventually swam off. . . . As to myself, I managed to climb over the top rail with great difficulty and slipped down the ship's side with the others. . . . It was with heartfelt gladness that I noticed a cruiser which had seen the end of the *Formidable* come up. I struck out for her. A rope ladder was lowered to us, and I was just able to climb up it with some others."

A large number of officers were saved but the captain went down with the ship, which sank by the head.

It was not until January 24, 1915, that a battle of real importance came in the North Sea, in which the principal forces on the British side, under Vice Admiral Beatty, were the five battle-cruisers *Lion* (flagship), *Tiger, Princess Royal, New Zealand,* and *Indomitable,* and on the German, three battle-cruisers, *Seydlitz* (flagship of Rear Admiral Hipper), *Derfflinger,* and *Moltke,* and the armored cruiser *Blücher*. In the British squadron were twenty-four 13.5-inch guns, and sixteen 12-inch; in the German, eight 12-inch, twenty 11-inch, and twelve 8.2-inch. The weight of British fire to the German was 23 to 14, or about 60% greater. The British battle-cruisers were of 28 and 28.5 knots, except the *New Zealand* and *Indomitable*, which were of 25 and 26. The Germans, in the order named, were 27, 29.2, 28.4, 25.3. The lower speed of the last named, the *Blücher*, was largely the cause of her destruction.

The British Battle-Cruiser Squadron was accompanied by four cruisers of the *Southampton* class, the *Southampton* (carrying Commodore Goodenough's broad pennant), the *Nottingham, Birmingham,* and *Lowestoft*. These ships were of 5,440 tons, 25.5 knots, and each carried nine 6-inch guns. A second cruiser squadron was the *Arethusa, Aurora,* and *Undaunted* (with Commodore Tyrwhitt in the *Arethusa*). These were all of 4,000 tons, 25.5 knots and two 6-inch and four 4-inch guns. The Germans were accompanied by the four cruisers *Rostock, Stralsund, Graudenz,* and *Kolberg,* of from 4,280 to 4,832 tons and from 27 to 28 knots, each carrying twelve 4.1-inch guns. The *Graudenz* and *Rostock* were not engaged. There was the usual accompaniment of destroyers and some submarines (on the German side at least), the number of which is not mentioned.

At daybreak on January 24, 1915, says Admiral Beatty in his report, his whole force was patrolling in company; he continues later:

"At 7.25 A.M. the flash of guns was observed SSE. Shortly after a report reached me from the *Aurora* that she was engaged with the enemy's ships. I immediately altered course to SSE, increased to 22 knots, and ordered the light cruisers and flotillas to chase SSE to get in touch and report movements of enemy.

"This order was acted upon with great promptitude; indeed, my wishes had already been forestalled by the respective senior officers, and reports almost immediately followed from *Southampton*, *Arethusa*, and *Aurora* as to the position and composition of the enemy, which consisted of three battle-cruisers and *Blücher*, six light cruisers, and a number of destroyers, steering NW. The enemy had altered course to SE. From now onwards the light cruisers maintained touch with the enemy, and kept me fully informed as to their movements.

"The battle-cruisers worked up to full speed, steering to the southward. The wind at the time was NE, light, with extreme visibility. At 7.30 A.M. the enemy were sighted on the port bow steaming fast, steering approximately SE distant 14 miles.

"Owing to the prompt reports received we had attained our position on the quarter of the enemy, and so altered course to SE parallel to them, and settled down to a long stern chase, gradually increasing our speed until we reached 28.5 knots. Great credit is due to the engineer staffs of *New Zealand* and *Indomitable*—these ships greatly exceeded their normal speed.

"At 8.52 A.M., as we had closed to within 20,000 yards of the rear ship, the battle-cruisers maneuvered to keep on a line of bearing so that guns would bear, and *Lion* fired a single shot, which fell short. The enemy at this time were in single line ahead, with light cruisers ahead and a large number of destroyers on their starboard beam.

"Single shots were fired at intervals to test the range, and at 9.09 A.M. *Lion* made her first hit on the *Blücher*, No. 4 in the line. The *Tiger* opened fire at 9.20 A.M. on the rear ship, the *Lion* shifted to No. 3 in the line, at 18,000 yards, this ship being hit by several salvos. The enemy returned our fire at 9.14 A.M. *Princess Royal* on coming into range opened fire on *Blücher*, the range of the leading ship being 17,500 yards, at 9.35 A.M. *New Zealand* was within range of *Blücher*, which had dropped somewhat astern, and opened fire on her. *Princess Royal* shifted to the third ship in the line, inflicting considerable damage on her.

"Our flotilla cruisers and destroyers had gradually dropped from a position broad on our beam to our port quarter, so as not to foul our range with their smoke; but the enemy's destroyers threatening attack, the *Meteor* and 'M' Division passed ahead of us, Captain the Hon. H. Meade, D.S.O., handling the division with conspicuous ability.

"About 9.45 A.M. the situation was as follows: *Blücher*, the fourth in their line, already showed signs of having suffered severely from gunfire; their leading ship and No. 3 were also on fire. *Lion* was engaging No. 1, *Princess Royal* No. 3, *New Zealand* No. 4, while the *Tiger*, which was second in our line, fired first at their No. 1, and when interfered with by smoke, at their No. 4.

"The enemy's destroyers emitted vast columns of smoke to screen their battle-cruisers, and under cover of this the latter now appeared to have altered course to the northward to increase their distance, and certainly the rear ships hauled out on the port quarter of their leader, thereby increasing their distance from our line. The battle-cruisers, therefore, were ordered to form a line of bearing NNW, and proceed at their utmost speed.

"Their destroyers then showed evident signs of an attempt to attack. *Lion* and *Tiger* opened fire on them, and caused them to retire and resume their original course.

"The light cruisers maintained an excellent position on the port quarter of the enemy's line, enabling them to observe and keep touch, or attack any vessel that might fall out of the line.

"At 10.48 A.M. the *Blücher*, which had dropped considerably astern of enemy's line, hauled out to port, steering north with a heavy list, on fire, and apparently in a defeated condition. I consequently ordered *Indomitable* to attack enemy breaking forward.

"At 10.54 A.M. submarines were reported on the starboard bow, and I personally observed the wash of a periscope, two points on our starboard bow. I immediately turned to port.

"At 11.03 A.M. an injury to the *Lion* being reported as incapable of immediate repair, I directed *Lion* to shape course NW. At 11.20 A.M. I called the *Attack* [a destroyer] alongside, shifting my flag to her at about 11.35 A.M. I proceeded at utmost speed to rejoin the squadron and met them at noon retiring NNW.

"I boarded and hoisted my flag in *Princess Royal* at about 12.20 P.M., when Captain Brock acquainted me of what had occurred since the *Lion* fell out of the line, namely, that *Blücher* had been sunk and that the enemy battle-cruisers had continued their course to the eastward in a considerably damaged condition. He also informed me that a Zeppelin and a seaplane had endeavored to drop bombs on the vessels which went to the rescue of the survivors of *Blücher*. . . .

"At 2.00 P.M. I closed *Lion*, and received a report that her starboard engine was giving trouble owing to priming, and at 3.38 P.M. I ordered *Indomitable* to take her in tow, which was accomplished by 5 P.M."

No official reports have been published by the Germans, but the account by a correspondent of an interview with participants, mostly officers, gives what seems a fair statement of the German side. This says:

"The *Kolberg* was the first to sight the enemy, a small British cruiser, accompanied by destroyers. The remainder of the British fleet was still below the horizon. The *Kolberg* immediately opened fire. After several minutes the British ship opened with one of her forward guns, and then began an artillery duel between the two smaller cruisers.

"The *Kolberg* steamed ahead and was planning to close with the enemy. However, her sister cruiser, the *Stralsund*, steaming further to the right and a long distance ahead of the squadron, had sighted the main body of the British fleet coming up, and signalled to the admiral on board the *Seydlitz*, 'Eight large hostile ships sighted on starboard bow.'

"The flagship thereupon signalled a command which swung the great German cruisers around and closed the umbrella screen of destroyers. The fleet now headed southeast.

"The British ships again had dropped out of sight, and did not reappear until some time later, when the pilot of the *Moltke* called attention to five big ships on the starboard quarter, that is, to the westward on the opposite side from those seen before.

"The commander and the pilot were still studying through their glasses the five scarcely visible shadows on the gray waves when a big shell struck the water 500 yards away, throwing up a high pillar of water. The enemy had unmasked himself. Either five hitherto unreported big ships had been lurking undiscovered behind our ships, or else five of the eight previously sighted had made a wide circuit around us.

"The German ships immediately answered the fire of the enemy. In order to bring more guns into action, first the enemy, then the German squadron took the familiar echelon formation, like a flight of steps, and steamed along 13 miles apart, each ship trying by constant turning to bring as many guns as possible to bear. The British concentrated their fire on our rearmost ship, the *Blücher*, and shortly landed a severe hit over the

The German cruiser *Blücher* as she capsized after being put out of action by the British in the North Sea, January 24, 1915.

engine-room. This forced the *Blücher* to drop back slowly even before she hoisted her last signal after a second shot reached the engine-room: 'All engines useless.'

"The *Blücher* was a mass of flame from fore bridge to stern, the pillar of fire above her towering to the sky.

"Forty-five minutes later the quarter-deck of the *Seydlitz* also began to blaze.

"The *Seydlitz*, of all the ships which returned, was the only one on which the two-hour bombardment inflicted any real injury. The British were shooting at a very extreme range in order to keep out of reach of our middle artillery. This is probably the reason for the slight damage done to the *Derfflinger* and *Seydlitz*, which were each hit squarely. They each show the mark of a shell which struck their armor, but so weakly that it has not even been necessary to replace the damaged plate.

"The shell which caused the fire on the *Seydlitz* pierced the foundation of a turret and set off some ammunition, causing a fire and some loss of life within the turret. Otherwise the *Seydlitz* was undamaged. Her fighting ability was completely restored as soon as the fire had been extinguished.

"The damage to the whole squadron, in fact, was so slight that the admiral did not need to dock a single ship. They are all at this moment ready to run out against the enemy. The patching of the *Seydlitz's* turret is being done rapidly, and will take at the most only a few days.

"But let us consider the effect of our artillery on the enemy during the two hours of combat. The second ship in the British echelon was the first to waver under the severe fire of the German guns. It sheered out of line and the third ship closed up, leaving a gap between it and the fourth ship. The lame duck was not seen again; presumably she was the one which sank later. After a little more fighting the two ships in the foremost group of English cruisers dropped astern or turned about. Five shells had struck them causing fires.

"The British battle line was now in confusion and its fighting power was broken. This was the reason why its admiral broke off the fight and decided to limp home. He was nowhere near the German mine fields or submarines of which the British report speaks. The fact is the British were finished. They could not follow further. Three of their biggest cruisers were out of action."

The German claim, at first made, of the sinking of a British battle-cruiser, though very specific and evidently believed by them, has never been substantiated and must be regarded as an error. The only large ship lost was the

Blücher and her destruction was due to want of speed; to-day, a main element of success, which demands as much sacrifice of armor and of the smaller caliber armament as can possibly be spared. The *Blücher* (besides her main battery of twelve 8-inch) carried eight 5.9-inch and sixteen 3.4-inch guns, none of which were of any value whatever in a long-range battle. Some of these could well have been omitted to add a knot or so to speed. It would in this instance, at least, have saved a ship and a ship's company.

Whatever the claims of the two sides, the fact is that the weaker force withdrew unpursued. Against such overpowering odds the loss of the Germans should have been far greater. They came out of the fight, not victorious, for victory was impossible in such circumstances, but certainly with untarnished reputation.

Few events of importance in naval operations occurred in Europe in the minor field, as the Baltic, Mediterranean, and Black Sea may, for the period treated, be termed, however interesting and momentous were to be the later operations in the Eastern Mediterranean.

In the Baltic the Germans lost on August 28th the light cruiser *Magdeburg*, of 4,478 tons, 26 knots and an armament of twelve 4-inch guns, by stranding on the Island of Odensholm, one of the Aland Archipelago in the entrance to the Gulf of Bothnia. Attacked, after grounding, by a very superior Russian force, the ship was blown up by the captain's orders, 85 of her crew, including the captain himself, being lost. On December 12th the armored cruiser *Friedrich Karl*, of 8,858 tons, 20 knots and carrying four 8.2-inch and twelve 5.9-inch guns, was sunk by a mine off the Russian coast, only 200 of her crew of 550 being saved. An almost equally important disaster had occurred to the Russians on October 11th in the loss of the armored cruiser *Pallada*, of 7,775 tons, 22.5 knots and armed with two 8-inch and eight 6-inch guns, which was sunk in the Gulf of Finland by the German submarine *U-26*.

In the Mediterranean the Austrian torpedo boat *No. 19* was sunk by a mine near Pola on August 17th, and next day the small cruiser *Zenta* was sunk by the gunfire of French armored cruisers at Castellastua. A small monitor, the *Temes*, of 433 tons, carrying three 4.7-inch guns, was mined and lost in the Danube on November 23d.

The old Turkish armored cruiser *Messudieh*, built in 1874, but reconstructed at Genoa in 1902, of 10,000 tons, 16 knots, two 9.2-inch in turrets and twelve 6-inch guns, was torpedoed and sunk in the Dardanelles by the British submarine *B 11*, which "in spite of the difficult current dived under five rows of mines. . . . Although pursued by gunboats and torpedo boats," she "returned safely, after being submerged on one occasion for nine hours."

The Russians had lost in the Black Sea two gunboats of 1,200 tons, scuttled or torpedoed at Odessa on October 29, 1914.

CHAPTER XVIII

The War on the Ocean

Teutonic ships in the Pacific and in German East Africa. Entente forces in eastern waters. Early movements of the German squadron in the East. Destruction of the British Cable Station at Fanning Island by the *Nürnberg*. Capture of the German Samoan Islands. German attack on Papeete, Tahiti. Battle off Chile, November 1, 1914, destruction of the British ships *Good Hope* and *Monmouth*. Combat at the Falkland Islands, December 8th, destruction of the German ships *Scharnhorst, Gneisenau, Leipzig*, and *Nürnberg*. Destruction of the *Dresden*. Exploits of the cruiser *Emden* in the Indian Ocean; escape and adventures of part of her crew. Operations of the *Karlsruhe* in the Atlantic. Career and destruction of the *Königsberg*. Surrender of the German protectorate of Kiau-Chau. Loss of Germany's oversea possessions. German auxiliary cruisers. British auxiliary cruisers.

Germany, in 1914, had in the Pacific the armored cruisers *Scharnhorst* and *Gneisenau*, the light cruisers *Emden, Dresden*, and *Nürnberg*, the river gunboats *Iltis, Luchs, Tiger*, and *Jaguar*, three other small craft and two destroyers. In the Atlantic was the *Karlsruhe*.

The *Scharnhorst* and *Gneisenau* were sister ships of 11,420 tons and 23 knots. Each carried eight 40 caliber, 8.2-inch guns, four of which were in pairs in turrets of 6.7-inch armor and four in broadside battery. They had as secondary battery six 5.9-inch and twenty 3.4-inch. Each carried four 18-inch submerged torpedo tubes, bow, stern, and broadside. They had complete armor belts 5.9-inch amidships, tapering at the ends to 3.2-inches. The complement of each was 765 men. Their fuel capacity was 1,968 tons of coal and 200 tons of oil.

The *Emden* and *Dresden* were also sister ships of 3,592 tons and 24 knots, with armaments of ten 4.1-inch guns. Their coal capacity was 836 tons, their complements 361. The *Nürnberg* was 200 tons smaller, with the same armament, a half knot less speed, the same coal capacity and a complement of 322. The *Karlsruhe* was of 4,832 tons and 27 knots, with twelve 4.1-inch guns. The squadron in the Pacific was to record a great page in naval history.

In German East Africa were the *Königsberg*, practically the same as another *Nürnberg*, and the *Möwe*, a small surveying vessel, with three 1-pounders.

Austria had in China the *Kaiserin Elisabeth*, an old cruiser of 3,937 tons, 19 knots and eight 5.9-inch guns. Neither in speed nor power was she able to aid the Germans, except as an accessory at Kiau-Chau.

The British force in eastern waters was made up in China of the battleship *Triumph*, of 12,000 tons, with 20 knots speed and a battery of four 10-inch and fourteen 7.5-inch guns, two armored cruisers, two light cruisers, eight destroyers, four torpedo boats, three submarines, and a number of river craft, the last of no war value. In the East Indies were a battleship and two light cruisers; at the Cape of Good Hope, three cruisers and a gunboat; in New Zealand, three cruisers and a sloop (belonging to New Zealand); and in Australia the Royal Australian Navy, of one battle-cruiser, three light cruisers, three destroyers, and two submarines. On the west coast of Canada were two submarines.

France had in her Asiatic possessions, two armored cruisers, a destroyer, a gunboat, and four river gunboats. Japan, of course, had practically her whole fleet in Japanese waters.

The *Scharnhorst* and *Gneisenau* had been in the east, the one since 1909, the other since 1910. They had, of course,

Tsingtau, the free port of Kiau-Chau, the German protectorate, in China, as a base of repairs and supply. Rear Admiral Count von Spee had been in command since December 4, 1912. The *Leipzig* had been on the station for eight years; the *Nürnberg* and *Emden* for four.

The two armored cruisers had left Tsingtau at the end of June for a cruise in Australasia. When the German orders of mobilization reached them on August 1st, they were at Ponape in the Caroline group, along with their supply ship, the *Titania.* The *Nürnberg, Dresden,* and *Emden* were called to join, the two last doing so at Ponape. The *Leipzig* was ordered for the time being to remain on the coast of Mexico, where in July she had relieved the *Nürnberg.* All burnable material was put ashore. On August 6th the squadron left the Carolines for Pagan in the Marianna group (of which Guam is the southernmost island), where it was joined on August 15th by the *Nürnberg* and *Prinz Eitel Friedrich,* the latter now an auxiliary cruiser. The *Emden* here received special orders and left for the Asiatic coast. Having coaled and provisioned the squadron left for the Marshall Islands, 2,600 nautical miles ESE of the Mariannas, and 2,200 WSW of the Hawaiian Islands. The distances give one an idea of the vast spaces of the Pacific.

Coaling again, the *Nürnberg* was sent eastward to call at Honolulu, where she found several German ships. The crews of all wanted places aboard, but only thirty-seven could be taken. On September 6th the *Nürnberg* rejoined the squadron and received orders to leave at once for Fanning Island, a British cable station 550 miles to the eastward. This island, 800 miles south of Honolulu, is almost a desert rock, its highest point is but nine feet above high water. Twenty-six white men and four white women, all connected with the cable service, live amid a population of 260 natives. The *Nürnberg* and her collier consort arrived on September 7th. Their visit is thus described by the commanding officer of the repair vessel *Kestrel*, which arrived September 25th:

"The cable employees were hard at work and were paralyzed to see a German officer at the door of the operating room with a revolver. 'Take your hands off the keys, all of you,' he demanded.

"The men were made to line up against the wall while the sailors with axes smashed the delicate and costly instruments. . . . .

"Another party was engaged near the shore end of the cable, trying to locate it. Failing in this, heavy charges of dynamite were planted and the cable blown to atoms. A crew from the collier grappled for the cable further out to sea with the intention of doing additional damage. Still another party planted dynamite and gun-cotton in the engine-rooms, the boiler-rooms, refrigerating plant, and in the dynamo-rooms. The explosion from these charges was terrific, but no one was hurt. A search was then made of the offices and a number of valuable papers were taken. These papers were taken aboard the *Nürnberg*, and a few hours later an officer returned and hastily summoned a detachment of men. The papers had revealed that several valuable instruments were buried—in reserve for just such contingencies; that a quantity of hidden arms and ammunition existed, and that there was 600*l.* in the office safe. The latter was blown open and the money taken. The officer in charge of this section of the expedition apologized, and said that this was the first time in his life that he had acted the part of a burglar.

"The officers appeared to have a complete knowledge of what was going on in the outside world, and seemed to be in possession of as much information as those who had been in daily cable communication with the mainland. The collier was carefully disguised, and there was nothing which would reveal her identity. She is about 2,200 tons register, and had an elaborate grappling outfit aboard her, whilst her men seemed to be experts in this class of work."

The *Nürnberg* learned from the station records of the seizure on August 30th of German Samoa (a stretch of 1,700 miles from the Marshalls) by a combined British and French force: the *Australia*, a battle cruiser of 18,800 tons, 25 knots and eight 12-inch guns (flagship of the Australian navy), the *Melbourne*, of 5,400 tons, 24.7 knots and eight 6-inch guns, three small cruisers, the *Philomel, Psyche,* and *Pyramus*, of 2,200 to 2,500 tons and eight 4-inch and 4.7-inch guns, and the French armored cruiser *Montcalm*, of 9,367 tons, 21 knots and two 7.6-inch and eight 6.4-inch guns. The British vessels had carried from New Zealand a force of 53 officers and 1,351 men in two troopships. No resistance could be offered by the Germans against so powerful a force; the British flag was hoisted, the troops landed and the naval force withdrawn.

On September 16th the *Scharnhorst* and *Gneisenau* arrived off Apia. The *Scharnhorst* ran into the harbor entrance and lay quietly some considerable time. There was a scattering from the town and into cellars by the inhabitants, the beach was cleared and the exits from the town were crowded. The British account mentioning these details adds: "At this tense moment a squad of about 120 young volunteers came out of the side streets and, marching in fours, swung into the open and deserted roadway. Fully equipped with all their marching swag, heads erect and with martial tread, they proceeded on their course." What the course was is not explained. "But nothing happened. It is impossible to say what motives restrained the German admiral. At any rate, the relief that was felt when the ship steamed away can better be imagined than described." One should at least recognize in the admiral the motive of humanity. His withholding fire in such circumstances does him honor.

On September 21st the squadron arrived at the Society Islands (1,000 miles E by S from Apia), coaled and provisioned and next day were off the French harbor of Papeete in Tahiti. Here the Germans sank by gunfire the French gunboat *Zelée*, of 650 tons, silenced the batteries and destroyed the wharves and coal supplies, the loss being estimated at $400,000.

The squadron then left for the Marquesas, another French group, 750 miles NE of Tahiti, where it remained

Vice-Admiral Beatty, commander of the British First Battle-Cruiser Squadron in the action in the North Sea, January 24, 1915.

Rear-Admiral Cradock, commander of the British squadron in the action off Coronel, November 1, 1914.

Vice-Admiral Sturdee, commander of the British squadron in the action off the Falkland Islands, December 8, 1914.

eight days in Anna Maria Harbor. On October 2d it was at Easter Island, 1,800 miles southeast of the Marquesas. Here the *Dresden,* temporarily detached, joined, and a little later the *Leipzig* from the west coast of Mexico; the latter had destroyed British shipping of an estimated value of $650,000. Having coaled and provisioned, the whole squadron sailed on October 18th and again coaled on October 26th at Juan Fernandez, 1,500 miles ESE of Easter Island and 340 a little south of west from Valparaiso. There they were joined by the *Prinz Eitel Friedrich.* On October 27th von Spee left for the coast of Chile. The events which followed are best described by himself. Writing November 3d, he says:

"My squadron consisting of the large cruisers *Scharnhorst* and *Gneisenau* and the smaller cruisers *Nürnberg*, *Leipzig*, and *Dresden* ran south at 14 knots, distant 20 miles from the Chilean coast to intercept a small English cruiser off Coronel [between Conception and Arauco]. On the way the small cruisers were sent to stop some merchant ships. At 4.15 the *Nürnberg* was out of sight to the northeast and the *Dresden* 12 miles north of Arauco. At 4.17 we sighted to the W by S two ships, and at 4.25 a third ship about 15 miles off. The first two were the *Monmouth* and *Glasgow*, the third was the *Otranto.* They were steering south. The squadron chased at full speed keeping them four points on the starboard bow. The wind was from the south with a strength of 6, with a heavy sea. I had to try not to lose the weather gauge, and also to cut them off from the coast.

"At 4.35 the enemy held more to the west [*i.e.* off the land]. I followed until our course was WSW, so that the *Scharnhorst* with a 22 mile curve slowly drew up, while the *Gneisenau* and *Leipzig* closed up. We interfered as much as possible with their wireless. At 5.20 the joining of an enemy ship was signalled and at 5.30 she took the lead, so we judged her to be the *Good Hope*, Rear Admiral Cradock's flagship. At 5.35, I held a southwest course, later a southerly, and slowed down to let my other ships come up. At 6.7 both lines were parallel and on a southerly course. The distance was 13.5 km. [kilometers = 8.5 miles, 7.3 nautical miles]. The *Nürnberg* was a long way off, the *Dresden*, one sea mile. At 6.25 the distance was 12.4 km. I made a turn toward the enemy and at 6.34 P.M. fired at 10.4 km. [5.6 nautical miles, 11,350 yards]. Wind and sea were ahead; the ships, especially the smaller, labored heavily. The lookouts and those at range-finders suffered much from seas that came over the bows so that they could not always see the enemy. The firing was good on both our large ships and many hits were observed by 6.39 on the *Good Hope.*

"At this time my line was reëstablished. The English then fired. I suppose the high sea troubled them more than it did us. Their two big cruisers were covered with our fire, while the *Scharnhorst* was only struck twice, and the *Gneisenau* four times.

"At 6.35 I turned one point from the enemy: they fired less, while we saw many hits. We saw the turret cover of the forward double turret [of the *Good Hope*] removed and that she was on fire. The *Scharnhorst* thinks she made 35 hits on the *Good Hope.* As the distance, notwithstanding the turning away, diminished, we expected a torpedo attack. The moon had risen at 6; this would help them, so at 7.45 I drew away. It had grown dark. The range-finder aboard the *Scharnhorst* used the fire on the *Good Hope* to measure by. At last all measurements were so uncertain that at 7.26 we ceased firing. At 7.23 we had noticed aboard *Good Hope* an explosion between her smoke-pipes. We think she never fired again. *Monmouth* ceased about 7.20.

[Captain Luce of the *Glasgow* in his report to the Admiralty says: "Enemy firing salvos got the range quickly and their third salvo caused fire to break out on the forepart of both ships, which were constantly on fire till 7.45 P.M. At 7.50 an immense explosion occurred on *Good Hope* amidships, flames reaching 200 feet high. Total destruction must have followed. It was now quite dark."]

"The small cruisers, including *Nürnberg* that had come up, received at 7.30 a wireless to close in and torpedo. It was difficult to see at this time and they could not find the *Good Hope*, but the *Nürnberg* at 8.58 hit the *Monmouth*, running close to her and firing at close quarters. She disabled her before she had fired. Her flag was still flying. No rescue work was possible in the high sea. As the *Nürnberg* thought she saw another ship through the smoke, she pushed ahead. The *Otranto* [a converted cruiser] turned at the beginning of the battle and withdrew at full speed. *Glasgow* sustained her fire longest and then escaped in the dark. *Leipzig* and *Dresden* thought they had scored many hits on her. The small cruisers had no loss or injury. *Gneisenau* two slightly wounded. The men went to battle with spirit and all did their duty."

The British view of this notable action is well given in the following letter, published in the London *Times,* November 12th, from an officer of the *Glasgow,* which escaped. He gives also some interesting details antecedent to the action.

"We were joined by the *Good Hope*, with Sir Christopher Cradock in command, and the *Monmouth* (Captain Brandt) off the Brazilian coast. We then cruised south together—*Good Hope*, *Glasgow*, *Monmouth*, and the armed liner *Otranto*—down to the cold Terra del Fuego and Straits of Magellan. . . .

"Well, after passing and repassing Cape Horn, sometimes twice in one day, we were glad to get orders to proceed north on the Pacific coast and to warmer weather. By this time we found that the two armored enemy's cruisers, *Gneisenau* and *Scharnhorst*, were probably coming over from the Pacific Islands to join up with the cruisers *Leipzig*, *Dresden*, and *Nürnberg*, as they had escaped the Australian and China squadron. We made a rendezvous farther north for our colliers, and went into Coronel and on to Valparaiso to pick up news and receive letters, etc., then back to rendezvous, coaled, and then got orders to go to Coronel alone to send cables, etc. We left Coronel, Chile, on the second occasion about 9 o'clock on the morning of November 1st, and at about 4 P.M. sighted the enemy in force. We put on speed and approached them until we made out four cruisers in line ahead, the two big armored cruisers leading and two 3-funnelled cruisers (about our class) following in open order. They immediately gave chase, so we 'hopped it' in the direction of our own ships and the Flag. We advised the Flag by wireless, but the enemy continually used their wireless in order to jam our signals. We first picked up the *Monmouth* and *Otranto* and ran a line ahead, *Glasgow*, senior, leading.

"In an hour or so the *Good Hope* (Cradock's ship) came up and we wheeled into line behind her, and again approached the enemy, coming round to south when about 7 miles off. The sun by this time was getting low on our starboard beam; the enemy were to the east of us, all proceeding south, they having the advantage both in guns and the light; we being silhouetted against the horizon. Their strategical speed being equal to ours, it was impossible to improve the lights before dark. I did not think he would engage until next day. However, we were now gradually closing. About 6.40 P.M. or so, the foremost armored enemy's cruiser opened fire with her 8-inch, and shells shrieked over and short of us, some falling about 500 yards short, giving the impression of excellent shooting. Soon after the *Otranto* began to haul out of line and edge away to the southwest, she not being fitted to fight men-o'-war. We appeared to close a point or two, and at 7 P.M. opened fire. The enemy replied in rapid salvos, making good and deadly shooting, mostly directed against our Flag, and the *Monmouth*, our next ahead. There was not much doubt as to the result. Shells continued to straddle us, some bursting overhead, throwing pieces of broken shell in all directions. About 10 minutes or so after this the poor *Monmouth* sheered off the line to the westward a hundred yards or so, when I saw her being hit heavily. She appeared to heel a bit and shake, her foremast turret (the 6-inch gun shield) in flames. She fell back again into line and out again to the eastward, still firing her 6-inch intermittently.

"Shortly after the *Good Hope* was seen to be on fire, also about the fore turret, and seemed to steer or fall away to the eastward or towards the enemy. During this time we kept up a continual fire from our two 6-inch and port battery 4-inch guns in the direction of the foremost light cruisers of the enemy's line, the third and fourth ship, of the lines, but owing to the big sea, our rolling, and the gathering darkness it was impossible to spot the fall of our shells. We could only fire at the flash of their guns, and when our heavy rolling allowed our gunlayers to see the flashes at all. About 7.30 P.M. I was standing near the after 6-inch hand up when I felt a shell strike us below deck. It seemed to pass out through the other side, but didn't, and I awaited the explosion, expecting the deck planks to rise up; but nothing visibly occurred at the moment. I was second in command of the starboard battery and, as that was the unengaged side, superintended the supply of ammunition to the port guns and generally kept an eye for casualties, so was able to use my binoculars to see what could be seen. Hills, a marine, carrying ammunition to P5, was struck behind the ear by a fragment of shell and was temporarily out of action, lying down near S5 hand up.

"The *Good Hope* fell more and more out of line to eastward, burning brightly forward, when suddenly an explosion occurred about her after funnel, blowing up *débris* and flames and sparks some 200 feet high or so, quite distinctly to be heard from our deck. Some of our men thought it was the enemy's flagship, so near had she drifted towards them. Soon after I could see nothing of her, and she never fired her guns again. Our speed during the action must have varied from 7 or 8 to 17 knots or so, and when the *Monmouth* dropped back in her distress we had to ease in order not to meet the doses meant for her. The enemy now dropped slowly back, and the armored cruisers directed their fire at us; we continued alone to reply when possible, now at about 4,500 yards. Everybody was remarkably cool, as if at practice. Another shell struck our No. 2 funnel, showing large holes around the casing, and it was this or these shells which wounded three more of our men slightly.

"I cannot understand the miracle of our deliverance; none will ever. We were struck at the water-line by in all five shells out of about 600 directed at us, but strangely not in vulnerable places, our coal saving us on three occasions—as we are not armored and should not be in battle line against armored vessels. We only had two guns that would pierce their armor—the *Good Hope's* old two 9.2's, one of which was out of action 10 minutes after the start. A shell entered the captain's pantry and continued on, bursting in a passage, the fragments going through the steel wall of the captain's cabin, wrecking it completely. Again no fire resulted.

"The *Monmouth*, no longer firing, steamed off to the northwest, and we stood by her signalling. She fell off to northeast, then we asked her if she could steer northwest. She replied, 'I want to get stern to sea as I am making water badly forward.' We followed close by. Shortly after I was on the flying bridge when I spotted the enemy approaching in line abreast, the ship to the right or southward morsing with an oil lamp to the others. They were then about 6,000 yards off or so in the rain, mist, and darkness. I told the captain, who gave me orders to bring them astern, and put on full speed. We drew out of range. The *Monmouth* was silent and hidden by our smoke. . . . Luckily our engines and boilers were intact, and we were able to push through the heavy seas at 24 knots and get away to give an account of the action, and warn the *Canopus*, who, although she no doubt would have fought gallantly, could hardly hope to successfully fight five ships."

The British losses were the armored cruiser *Good Hope*, of 14,000 tons, 23 knots and two 45 caliber 9.2-inch guns in turrets and sixteen 45 caliber 6-inch, with her crew of 900 men, and the *Monmouth*, of 9,800 tons, 23 knots and fourteen 45 caliber guns, four of which were in turrets of 5-inch armor. She carried a complement of 655 men.

As the two German armored cruisers carried together sixteen 40 caliber 8.2-inch guns and twelve 40 caliber 5.9-inch, their weight of fire was distinctly heavier in the heavy guns, but equally distinctly weaker in the secondary battery, as this was but twelve 5.9-inch of 40 caliber against the British thirty 6-inch of 45 caliber. The muzzle energy of each of the 9.2-inch guns of the *Good Hope* was 20,660 foot tons against 14,500 tons of the German 8.2-inch, and that of the British 6-inch 5,830 against 5,340 of the German 5.9-inch. As the battle was well within range of the British 6-inch (11,000 yards diminishing according to the account from the *Glasgow* to 6,000 yards), it is fair to suppose, in view of the almost entire freedom from injury of the Germans that the result was almost wholly a question of superiority in gunnery, and of getting the advantage of delivering the first injury.

Von Spee took his squadron into Valparaiso, where, of course, he could remain but twenty-four hours, and thence to his doom off the Falklands, an error of judgment. Instead of risking such an adventure for the sake of, at most, destroying a telegraph station and a coal supply, both easily replaced, it would have been far better to return to some one of the hundreds of points of refuge in the Pacific not touched by cables, which he could use as a further base of operations. The wireless telegraph, while so wonderful a means of communication between the ships of a squadron, is also an informant to an enemy of the other's vicinity and von Spee might have carried on his operations in the Pacific (coaled, as he apparently was, without difficulty), almost indefinitely. The spaces of the Pacific are so vast, the points of refuge, with no communication with the world, so numerous that in such a region nothing but remote chance would have brought him in contact with an enemy force which he felt he had to avoid. In such a region he could have continued long a thorn in the flesh to his foe. As it was he was to run into the very jaws of death. Accompanied by the colliers *Baden* and the *Santa Isabel* (the latter a new Hamburg freighter of 7,500 tons), the German squadron started for the Falklands.

The only British battleship in the South Atlantic at this time was the *Canopus*, of 12,950 tons, 18.5 knots, carrying four 35 caliber 12-inch guns in turrets of 8-inch armor. On November 10th a squadron of seven ships left Plymouth, England, under Vice Admiral F. C. D. Sturdee, composed of:

	TONS.	SPEED IN KNOTS.	MAIN ARMAMENT.
Invincible (flag)	17,250	28	Eight 45 caliber 12-inch.
Inflexible		27.2	
Carnarvon	10,850	22.2	Four 45 caliber 7.5-inch; six 6-inch.
Cornwall	9,800	23.6	Fourteen 45 caliber 6-inch.
Kent			
Bristol	4,829	25	Two 45 caliber 6-inch; ten 4-inch.
Macedonia	A P. & O. converted cruiser.		

The *Glasgow* (sister to the *Bristol* and escaped from the fight in the Pacific), was in company, having been picked up off Brazil.

The squadron arrived at the Falklands (claimed also by Argentina and called by them the Malvina Islands, but held by the British continuously since 1833) on the morning of December 7th and at once began to coal, with the expectancy of leaving the next day in search of the German squadron. The *Macedonia* was anchored at the entrance as a lookout; the *Invincible, Inflexible, Carnarvon,* and *Cornwall* were in Port William; the *Glasgow* and *Bristol* in Port Stanley.

The following is a paraphrase of Admiral Sturdee's report: At 8 A.M. a signal was received from the station known as Sapper Hill that two steamers were in sight, one with four, the other with two smoke-pipes. The *Kent* was at once ordered to weigh and at 8.45 she passed down the harbor and took station at the entrance. At 8.20 the smoke of another vessel was reported. At 8.47 the *Canopus* reported that the first two ships were 8 miles off and that the smoke reported appeared to be that of two ships 20 miles distant. At 8.50 another column of smoke was reported to the south.

At 9.20 the *Canopus* opened fire on the two leading ships (*Gneisenau* and *Nürnberg*), at a range of 11,000 yards. They at once hoisted their colors and turned away. Their masts were now visible from the upper bridge of the *Invincible* across the low land south of Port William at a distance of about 17,000 yards. They altered course to port apparently to attack the *Kent,* now at the mouth of the harbor, but it would seem that they now saw the battle-cruisers over the land and they altered course easterly and increased their speed. The *Glasgow* was ordered outside at 9.40 and at 9.45 all the others except the *Canopus, Bristol,* and *Macedonia* weighed and stood out in the order, *Carnarvon, Inflexible, Invincible,* and *Cornwall.* On passing Pembroke Light (at the south side of the entrance of Port William), the five German ships were clearly visible to the southeast, hull down. The visibility was perfect, the sea calm with a bright sun, a clear sky and a light breeze from the northwest. At 10.20 was made the signal for a general chase. The battle-cruisers, having the higher speed, took the advance. The *Glasgow* was ordered to keep two miles from the *Invincible* (flag) and the *Inflexible* was stationed on the latter's starboard quarter. Speed was eased to 20 knots at 11.15 A. M. to allow each to get into station. At this time only the smoke-pipes and bridges of the Germans showed above the horizon. Information now came (11.27) from the *Bristol* that three enemy ships, probably colliers or transports, had appeared, and the *Bristol* with the *Macedonia* was ordered to destroy them. As the Germans maintained their distance it was decided to attack and at 12.47 signal was made to open fire; the first shot was fired at 12.55 from the forward turret of the *Inflexible,* at the right hand ship, a light cruiser (the *Leipzig*), with a range of 16,500 yards; the *Invincible* opened a few minutes later at the same ship. The fire from 16,500 to 15,000 yards caused the three rear ships (*Nürnberg, Dresden,* and *Leipzig*) to turn to the southwest; the *Kent, Glasgow,* and *Cornwall* were ordered to follow.

The fire of the battle-cruisers was now directed on the *Scharnhorst* and *Gneisenau.* These now (1.15 P. M.) turned in column seven points to port (with the idea no doubt of closing for a better range for their 8.2-inch guns) and at 1.30 opened fire. The British battle-cruisers then turned together keeping a parallel course with the Germans, eased speed to 24 knots and opened fire at 13,500 yards, increasing to 16,450 (at 2). The Germans then (at 2.10) turned

Admiral Count von Spee, commander of the German squadron defeated by the British off the Falkland Islands, December 8, 1914

Commander Karl von Muller of the German cruiser *Emden* which was destroyed by the Australian cruiser *Sydney* off Cocos, or Keeling Islands, November 9, 1914

about ten points to starboard, the British following the movement and at 2.45 again opening fire. The Germans at 2.53 turned into line ahead (in column, to use the American technical phrase) and at 2.55 again opened fire. The *Scharnhorst* caught fire forward but not seriously, her fire slackened perceptibly; the *Gneisenau* was badly hit by the *Inflexible*.

Admiral Sturdee's report proceeds as follows:

"At 3.30 P.M. the *Scharnhorst* led round about 10 points to starboard; just previously her fire had slackened perceptibly and one shell had shot away her third funnel; some guns were not firing, and it would appear that the turn was dictated by a desire to bring her starboard guns into action. The effect of the fire on the *Scharnhorst* became more and more apparent in consequence of smoke from fires and also escaping steam; at times a shell would cause a large hole to appear in her side through which could be seen a dull red glow of flame. At 4.04 P.M. the *Scharnhorst*, whose flag remained flying to the last, listed heavily to port, and within a minute it became clear that she was a doomed ship; for the list increased very rapidly until she lay on her beam ends and at 4.17 P.M. she disappeared. The *Gneisenau* passed on the far side of her flagship and continued a determined but ineffectual effort to fight the two battle-cruisers. At 5.08 the forward funnel [smoke-pipe] was knocked over and remained resting against the second funnel. She was evidently in serious straits and her fire slackened very much. At 5.15 one of the *Gneisenau's* shells struck the *Invincible*; this was her last effective effort. At 5.30 P.M. she turned towards the flagship with a heavy list to starboard, and appeared stopped, with steam pouring from her escape pipes and smoke from shell and fires rising everywhere. About this time I ordered the signal 'Cease fire,' but before it was hoisted the *Gneisenau* opened fire again, and continued to fire with a single gun.

"At 5.40 P.M. the three ships closed in on the *Gneisenau* and at this time the flag at her fore truck was apparently hauled down, but the flag at the peak continued flying. At 5.50 P.M. 'Cease fire' was made.

"At 6 P.M. the *Gneisenau* heeled over very suddenly, showing the men gathered on her decks and then walking on her side as she lay on her beam ends before sinking."

It is a fine story, one of unsurpassed heroism.

The British admiral made every effort to save life, but the cold water quickly drowned many of the 200 who were estimated to have been in the water unwounded, "life buoys were thrown and ropes lowered but only a proportion could be rescued. The *Invincible* alone rescued 108 men, 14 of whom were found to be dead after being brought on board; these men were buried at sea the following day with full military honors."

To return to the small cruisers which had turned away about 1.00 P.M. with the *Glasgow, Kent,* and *Cornwall* in chase. It was not until 3.00 P.M. that the *Glasgow,* well ahead of her two consorts, exchanged shots with the *Leipzig* at 12,000 yards (6 nautical miles). At 4.17 the *Cornwall* also opened fire, but it was not until 7.17 that with the *Leipzig* on fire fore and aft that the *Cornwall* and *Glasgow* ceased fire. At 9.00 the *Leipzig* turned over to port and disappeared. Seven officers and 11 men were saved.

The *Kent* had been ordered to pursue the *Nürnberg,* the nearest to her. She was in range at 5.00 P.M. and at 6.35 the *Nürnberg* was on fire forward and had ceased firing. The *Kent* closed to 3,300 yards and also ceased fire. As the colors were still flying the *Kent* opened again. "Fire was stopped five minutes later on the colors being hauled down and every preparation was made to save life. The *Nürnberg* sank at 7.27 P.M., and as she sank, a group of men were waving a German ensign attached to a staff. Twelve men were rescued but only seven survived. The *Kent* had four killed and 12 wounded, mostly caused by one shell."

The *Dresden,* with superior speed, escaped. The *Glasgow* was the only one of the three British ships which had any chance in pursuit, but she had become engaged with the *Leipzig* for over an hour before the *Cornwall* and *Kent* could get within range. A change of weather with less visibility also came to the *Dresden's* aid.

The three transports at first reported to the commander-in-chief had been reduced to the two previously mentioned as accompanying the squadron, the *Baden* and *Santa Isabel*. Both ships were sunk after removing the crews, apparently a useless throwing away of two good ships with good cargoes of coal.

There could, of course, be no other result in the circumstances, even in the cases of the *Scharnhorst* and *Gneisenau*. They were overmatched in speed, the *Invincible* making on her trials 28 knots; the *Scharnhorst* 22.7; in muzzle energy of guns (in foot tons) they were as 47,875 to 14,500, or more than three times the power. Notwithstanding, the so much weaker ships made an exhibition of courage and resource than which history mentions nothing finer or more heroic. Men of the sea, of all races, can be proud of such a showing.

The *Dresden*, it may be said here, was destroyed under circumstances akin to the destruction of the American *Essex* in 1814 by the British *Phœbe* and *Cherub*, 101 years before to the month. Both ships were destroyed in neutral waters; both were but a half mile from the shore, and in each case the waters were Chilean.

The *Dresden*, escaping from the Falkland fight, was next heard of at Juan Fernandez, where asking time for repairs and this being refused, her captain decided to intern. On March 14, 1915, at 9 A. M., there appeared off the roadstead (there is no harbor in the true sense) the British cruisers *Glasgow* and *Kent* and the auxiliary cruiser *Orama*. Before the Chilean governor, who had put off to inform the senior officer, Captain Luce of the *Glasgow*, of existing conditions, could reach the latter's ship, fire was opened on the *Dresden*. The captain set her on fire and she was blown up, the crew, of whom 15 were wounded, being saved.

The story of the hostile squadrons ends with the destruction of the *Dresden;* any German ships that remained were of the weaker sort, which could only act as raiders. Of these, the *Emden* was to be the chief. Her adventures make an almost equally dramatic story.

As early as August 4th the *Emden* captured in Tsushima strait the Russian merchantman *Riasan*, which was brought into Tsingtau and receiving the guns of the *Kormoran* was renamed the *Kormoran II* and became an auxiliary cruiser. The *Emden* left Tsingtau on August 6th accompanied by the auxiliary cruiser *Markomannia* and joined von Spee in the Carolines, but on August 13th was detached on special service in the Indian Ocean. Her first capture was on September 10th, after which her prizes came in plenty, many of them of great value. Most of them were sunk, but some were reserved to receive aboard for transportation to safety the crews and passengers of the ships sunk. The captured colliers followed until they were emptied of their coal, when they were sunk. Women and children seem to have been carefully looked after. On the night of September 22-23 the oil tanks at Madras, containing some 600,000 barrels of oil, were set on fire and destroyed by the *Emden's* gunfire, which was ineffectively replied to from the forts. She then cruised, with great destruction to shipping, in the neighborhood of Ceylon, coaled at anchor off Pondicherry on September 29th, and then continued her career of destruction off the Laccadives (south of Malabar).

On October 28th the *Emden* crowned her daring career by entering the harbor of Penang, destroying the Russian cruiser *Jemtchug* (of 3,130 tons and eight 4.7-inch guns), the French destroyer *Mousquet*, and leaving the harbor unharmed. Her visit is excellently described by a correspondent of the *New York Times*, a description worthy of being given in some detail. Writing next day, he says:

"It was probably with the idea of crippling this base from which her pursuers were radiating that the *Emden* made her raid here. Had she

ASE.

os turn to port and after running N.E. 37 minutes
ie German ships, now suffering heavily, slackens rapidly.
asses her to port and at 6.00 sinks; colors flying.

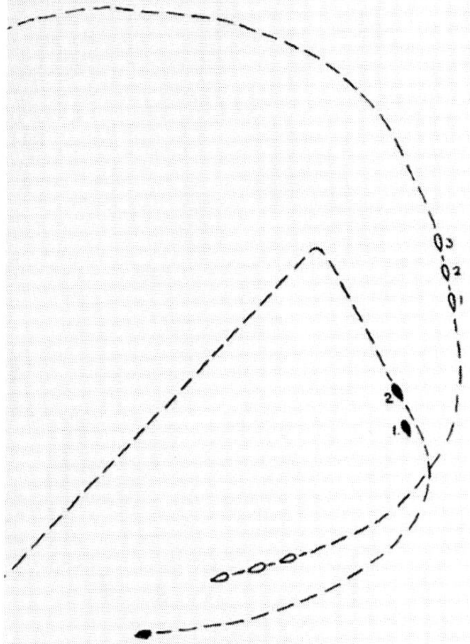

as sunk by the Cornwall and Glasgow at 9 p.m. 7 officers
at 5.00, the latter sank at 7:27 being then about 160 miles
e Dresden only escaped, by superior speed

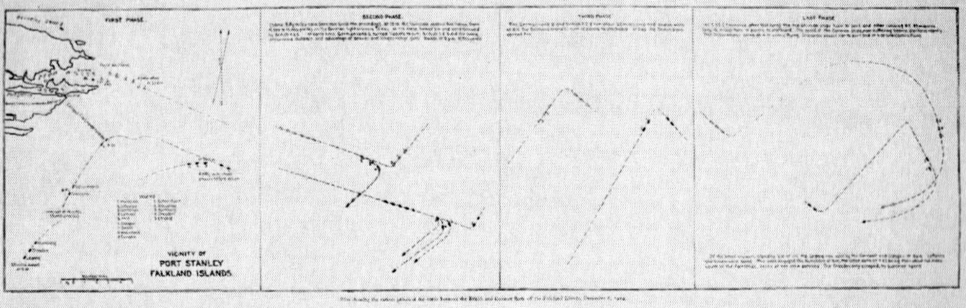
FIRST PHASE
SECOND PHASE
THIRD PHASE
LAST PHASE
VICINITY OF
PORT STANLEY
FALKLAND ISLANDS

found it temporarily undefended she could, at one blow, seriously have embarrassed the English cruisers patrolling these waters and at the same time caused a terrific loss to English commerce by sinking the many merchantmen at anchor in the harbor.

"It was on Wednesday morning that the *Emden*, with a dummy fourth funnel and flying the British ensign, in some inexplicable fashion sneaked past the French torpedo boat *Mousquet*, which was on patrol duty outside, and entered the outer harbor of Penang. Across the channel leading to the inner harbor lay the Russian cruiser *Jemtchug*. Inside were the French torpedo boats *Fronde* and *Pistolet*, and the torpedo boat destroyer *D'Iberville*. The torpedo boats lay beside the long government wharf, while the *D'Iberville* rode at anchor between two tramp steamers.

"At full speed the *Emden* steamed straight for the *Jemtchug* and the inner harbor. In the semi-darkness of the early morning the Russian took her for the British cruiser *Yarmouth*, which had been in and out two or three times during the previous week and did not even 'query' her. Suddenly, when less than 400 yards away, the *Emden* emptied her bow guns into the *Jemtchug* and came on at a terrific pace, with all the guns she could bring to bear in action. When she had come within 250 yards she changed her course slightly, and as she passed the *Jemtchug*, poured two broadsides into her, as well as a torpedo, which entered the engine-room, but did comparatively little damage.

"The Russian cruiser was taken completely by surprise and was badly crippled before she realized what was happening. The fact that her captain was spending the night ashore, and that there was no one on board who seemed capable of acting energetically, completed the demoralization. She was defeated before the battle began. However, her men finally manned the light guns and brought them into action.

"In the meantime the *Emden* was well inside the inner harbor and among the shipping. She saw the French torpedo boats there, and apparently realized at once that unless she could get out before they joined in the action her fate was sealed. At such close quarters (the range was never more than 450 yards) their torpedoes would have proved deadly. Accordingly, she turned sharply and made for the *Jemtchug* once more.

"All the time she had been in the harbor the Russian had been bombarding her with shrapnel, but owing to the notoriously bad marksmanship prevalent in the Czar's navy had succeeded for the most part only in peppering every merchant ship within range. As the *Emden* neared the *Jemtchug* again both ships were actually spitting fire. The range was practically point-blank. Less than 150 yards away the *Emden* passed the Russian, and as she did so torpedoed her amidships, striking the magazine. There was a tremendous detonation, paling into insignificance by its volume all the previous din; a heavy black column of smoke arose and the *Jemtchug* sank in less than 10 seconds, while the *Emden* steamed behind the point to safety.

"No sooner had she done so, however, than she sighted the torpedo boat *Mousquet*, which had heard the firing and was coming in at top speed. The *Emden* immediately opened up on her thereby causing her to turn around in an endeavor to escape. It was too late. After a running fight of 20 minutes the *Mousquet* seemed to be hit by three shells simultaneously and sank very rapidly. The German had got a second victim.

"It was here that the chivalrous bravery of the *Emden's* captain, which has been many times in evidence throughout her meteoric career, was again shown. If the French boats were coming out, every moment was of priceless value to him. Nevertheless, utterly disregarding this, he stopped, lowered boats, and picked up the survivors from the *Mousquet* before steaming on his way.

"The English here now say of him admiringly, 'He played the game.'

"Meantime, boats of all descriptions had started towards the place where the Russian cruiser had last been seen. The water was covered with débris of all sorts, to which the survivors were clinging. They presented a horrible sight when they were landed on Victoria Pier, which the ambulance corps of the Sikh garrison turned into a temporary hospital. Almost all of them had wounds of one sort or another. Many were covered with them. Their blood-stained and, for the most part, naked bodies, were enough to send shivers through even the most cold-blooded person. It was a sight I shall not forget for many a day. Out of a crew of 334 men, 142 were picked up wounded. Only 94 were found practically untouched. Ninety-eight were 'missing.' It is not yet known how many of the crew of 78 of the *Mousquet* were rescued by the *Emden*."

But the *Emden's* career was about to close. On November 9th the cruiser appeared off the Cocos Islands, a small group in the Bay of Bengal, west of British Burma, where there was a cable station connecting Ceylon and Australia. The visit was for the purpose of destroying this station. The *Emden* arrived at sunrise and sent 3 officers and 49 men with 4 machine-guns under command of Lieutenant von Mücke. But the station was quick enough to cable for aid before the party landed, and after two hours' work at cable-cutting the siren of the *Emden* sounded a recall. But even before they could reach the landing stage, the *Emden* had got underway and had begun firing. The

Australian cruiser *Sydney*, of 5,400 tons, 24.7 knots and eight 6-inch 50 caliber guns, which was serving as an escort for a convoy of troops from Australia, bound to Egypt, had appeared. The story of the encounter is told best by an officer of the *Sydney* in a letter to his father:

"On November 9th we were steaming about 50 miles to the eastward of the Cocos Islands (southwest of Java), heading for Colombo, when at 7 A.M. we took in a very interrupted wireless message from the Cocos wireless station—'Strange warship . . . off entrance.' The *Melbourne*, as senior officer, ordered us to raise steam for full speed and go and investigate.

"At 9.15 A.M. the tops of the cocoanut trees of Keeling Islands were sighted. At 9.20 we sighted the *Emden*, or rather the tops of her funnels, 12 or 15 miles away. At 9.40 A.M. she opened fire at a very big range, and shortly after that we started in on her.

"Throughout the action I was almost constantly engaged running backwards and forwards between the ammunition hoist and the forecastle gun, or between the hoist and No. 1 starboard or No. 1 port.

"Once I heard a crash and looking aft saw that a shell had hit near gun No. 2 starboard. But owing to the screen being in the way, I did not know it had knocked out practically the whole of that gun's crew. . . .

"All the time we were going 25 and sometimes as much as 26 knots. We had the speed on the *Emden* and fought as suited ourselves. . . .

"Coming aft the port side from the forecastle gun I was met by a lot of men cheering and waving their caps. I said, 'What's happened?' 'She's gone, Sir, she's gone.' I ran to the ship's side, and no sign of a ship could I see. If one could have seen a dark cloud of smoke it would have been different. But I could see no sign of anything. So I called out, 'All hands turn out the lifeboats, there will be men in the water.' They were just starting to do this when someone called out, 'She's still firing, Sir,' and every one ran back to the guns. What had happened was a cloud of yellow or very light colored smoke had obscured her from view, so that looking in her direction one's impression was that she had totally disappeared. Later we turned again and engaged her on the other broadside.

"By now her three funnels and her foremast had been shot away, and she was on fire aft. We turned again, and after giving her a salvo or two with the starboard guns saw her run ashore on North Keeling Island. So at 11.20 A.M. we ceased firing, the action having lasted 1 hour and 40 minutes. . . .

"We started chasing a collier which had been in attendance on the *Emden*, and when we boarded her we found they had opened the sea cocks and the ship was sinking fast, so we took every one off her and returned to the *Emden*, getting back there at about 4 P.M.

"They sent a man aloft to cut down the colors, and waved a big white flag from forward. It was getting dark and we did not know for certain that the cruiser *Königsberg* might not be near, so we could do no rescue work that night and had to steam away. A cry in the darkness, and we stopped, and lifeboats were lowered to pick up a nearly exhausted but very lucky German sailor. The fourth rescued from the water that day.

"November 10th.—Early in the morning we made for the cable station, to find that the party landed by the Germans to destroy the station had seized a schooner and departed. The poor devils aren't likely to go far with a leaking ship and the leathers removed from all the pumps. Although they had broken up all the instruments, the cable people had a duplicate set buried, so that was satisfactory.

"At 11.10 A.M. we arrived off the *Emden* again. I was sent over to her in one of the cutters. . . . I was received by the captain of the *Emden*. I told him from our captain that if he would give his parole the captain was prepared to take all his crew on board the *Sydney* and take them straight up to Colombo. He stuck a little over the word 'parole,' but readily agreed when I explained the exact scope of it. And now came the dreadful job of getting the badly wounded into the boats. There were 15 of these. Luckily we have a very good pattern of light stretcher into which men can be strapped. We got three badly wounded in each boat. The Germans were all suffering badly from thirst, so we hauled the boats' water casks up on deck, and they eagerly broached them, giving the wounded some first.

"I took an early opportunity of saluting the captain of the *Emden* and saying, 'You fought very well, Sir." He seemed taken aback, and said 'No.' I went away, but presently he came up to me and said, 'Thank you very much for saying that, but I was not satisfied. We should have done better. You were very lucky in shooting away all my voice-pipes at the beginning.'

"When I got a chance, with all the boats away, I went to have a look round the ship. I have no intention of describing what I saw. With the exception of the forecastle, which is hardly touched from forebridge to stern post, she is nothing but a shambles, and the whole thing was most shocking. The German doctor asked me to signal for some morphia, sent me aft, and I never came forward again."

The *Sydney* had 4 killed and 8 wounded. The captain of the *Emden* in his report states his loss as 6 officers, 4 warrant officers, 26 petty officers, and 93 men dead; one commissioned officer and 7 men severely wounded.

Map showing the scene of actions between the British and German fleets on November 1 and December 8, 1914.

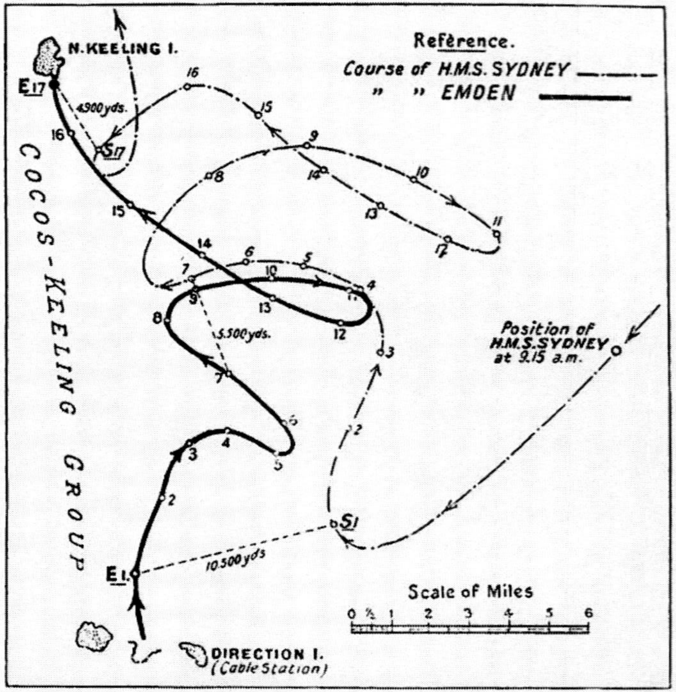

Plan showing the positions of the *Sydney* and the *Emden* during the action off Cocos Island, November 9, 1914.

The captured British collier *Buresk* was recaptured, but her valves had been opened and she sank. Her people, with the German prize crew, were taken on board the *Sydney*. The captain of the *Sydney* sent boats manned by a German crew from the *Buresk* to state that he would return to the assistance of the *Emden* next morning.

The *Emden* was, of course, certain of destruction in such a contest. She had been long at sea and her bottom was necessarily much fouled and her speed reduced. She was smaller than her antagonist by 1,800 tons; her guns were but 4.1-inch of 40 caliber against 6-inch of 50 caliber. She had no chance. The *Sydney* had but to choose her distance and destroy at leisure, all of which was perfectly correct tactics.

The 3 officers and 49 men left ashore by the *Emden*, determined not to fall into the hands of the British, seized the schooner *Ayesha*, of 123 tons, lying at the island, but without sails or running rigging. She had aboard a cargo of rice and cocoa. With the assistance of the inhabitants she was fitted out and at nightfall she was towed off the land, and they set sail without charts or nautical instruments. They had with them their four machine-guns, 28 rifles, and several thousand cartridges. According to Lieutenant von Mücke's account, they headed for Padang, at the middle of the west coast of Sumatra, which, as a Dutch possession, was a neutral port. It was a thousand miles south by east from the Cocos, but they arrived at Emma Bay, a nearby port, on November 27th, a voyage of 18 days. They had suffered much for want of water. They found many German ships in the port, from the crews of which they had an enthusiastic welcome. The *Ayesha*, being claimed by Lieutenant von Mücke as an imperial naval vessel, was allowed but 24 hours' stay. They thus left

November 28th well provided with provisions and clothing and with gifts of tobacco and beer. They had added a Lieutenant Wellman to their number. Light winds kept them on the Sumatra coast 14 days, after which they met a severe gale. The *Choisung*, a German collier of the North German Lloyds, which had been destined for the *Emden*, was met, to which the party aboard the *Ayesha* was transferred and the latter sunk. They renamed their new ship *Ayesha II*. The African coast was sighted on January 4th, and in the night of January 7-8, Lieutenant von Mücke's party, using four boats of their collier, passed safely the Straits of Perim. The *Choisung* reached Massowa on January 13th, but when Italy declared war she fell into the hands of the enemy. The little flotilla was nine weeks skirting the eastern shore of the Red Sea and then only reached a port a little north of Hodeida, which itself is but an eighth of the way to Suez. They stayed eight days in the highlands of Sana to rest, and on March 15th took to two small sailing vessels. They lost one, though apparently there was no loss of life, and reached Lid, 120 miles south of Djidda, the port for Mecca which is 60 miles inland. They marched from Lid to Djidda and were attacked on their way by some 300 Bedouins against whom they defended themselves for three days, when they were rescued by the Emir of Mecca. Lieutenant Schmidt and a stoker were killed. As they could not go to Mecca, they were obliged to take boats again at Djidda and after nineteen days landed at El Weg (April 27th), 260 miles south of the entrance to the Gulf of Suez. Here they met military protection and were escorted inland 150 miles to the station El Ula on the Damascus-Mecca railway. They took train on May 7th and reached Damascus on May 9th. There, says Lieutenant von Mücke, "they were joyfully greeted, but not to compare with their reception in

Constantinople, May 24th, where they marched carrying the *Emden's* flag amid wild cheering." It was truly a brave and adventurous journey, well deserving its happy ending.

In the Atlantic, the *Karlsruhe*, a much larger and speedier ship than the *Emden*, being of 4,832 tons with a speed of 27 knots, and a battery of twelve 4.1-inch 40 caliber guns, was to have no such sensational career, though she was by no means unsuccessful. Very little that is definite is known of her doings beyond the fact that in the early stages of the war she was reported sighted off Sandy Hook by a number of vessels, which was sufficiently believed to cause for a time a marked "hold-up" in British shipping and that she was given coal enough (as international law allowed) at San Juan, Puerto Rico, to carry her to her nearest home port. On October 24th, a cable despatch to the *New York Herald* gave the names of thirteen vessels captured and sunk, twelve of which were British and one Dutch, the captures being made chiefly in the South Atlantic. On October 26th she captured and sank, 500 miles east of Pará, the *Van Dyck*, of 9,800 tons, one of the Lamport and Holt Line between New York and Buenos Ayres. The total of her captures seems to have been 17. The *British Navy League Manual* states that it is believed that she was destroyed by internal explosion at the end of October or early in November. In any case, since then, she has not been heard of.

The only remaining German cruisers abroad were the *Geier*, of 1,604 tons, interned at Honolulu, November 8, 1914, and the *Königsberg*, slightly smaller than the *Dresden* and *Emden*, but with the same armament of ten 4.1-inch guns, in German East Africa. After a considerable destruction of British shipping, she was obliged to take refuge in the Rufiji River in German East Africa, where she remained for some time undiscovered. She was shelled by British cruisers unsuccessfully and it was determined to block the river, for which purpose a merchant steamer, the *Newbridge*, was sunk in the channel, two of the British being killed and eight wounded in the operation. The *Duplex*, a cable steamer which accompanied the *Newbridge*, had five Lascars killed. Five days later it was reported that "the *Königsberg* was finally destroyed and sunk. This is how her end came. The German cruiser had so effectively concealed herself, not only among the palms, but actually covering the ship with foliage, that it was impossible to locate her exact position. To get over this difficulty, the *Kinfauns Castle* arrived on the scene with an aëroplane. This was soon soaring over the river and the position of the hidden cruiser conveyed to the British by means of smoke bombs. Very quickly the big guns of our ships got the range and battered the *Königsberg* till she was sunk."

This is the account appearing in the *Army and Navy Gazette*, of January 16, 1915, which is now given as an example of means employed in the very unusual circumstances, and as an instance of the unreliability of news, apparently definite. The *Königsberg* was not destroyed until some six months after the exploit thus detailed, when, in July, 1915, two monitors, the *Severn* and *Mersey*, were used. The monitors were able to make a nearer approach, the *Weymouth*, 5,250 tons, eight 6-inch guns, flagship of Vice Admiral H. King Hall, and the gunboat *Pioneer* attacking the batteries at the river mouth. Aëroplanes were used to establish accurately the position of the *Königsberg* and after a six hours' bombardment from the monitors, during which they lost four men killed and four wounded, the *Königsberg* was finally destroyed.

On August 16, 1914, the Japanese government addressed an ultimatum to Germany of an extraordinary character declaring that:

"We consider it highly important and necessary in the present situation to take measures to remove the causes of all disturbance of the peace in the Far East and to safeguard general interests as contemplated in the agreement of alliance between Japan and Great Britain. In order to secure firm and enduring peace in Eastern Asia the establishment of which is the aim of the said agreement, the Imperial Japanese Government sincerely believe it to be its duty to give advice to the Imperial German Government to carry out the following two propositions:

"1. To withdraw immediately from Japanese and Chinese waters the German men-of-war and armed vessels of all kinds, and to disarm those which cannot be withdrawn.

"2. To deliver on a date not later than September 15th to the Imperial Japanese authorities, without condition or compensation, the entire leased territory of Kiau-Chau, with a view to the eventual restoration of the same to China."

August 23d was set as the date requiring an unconditional acceptance of these demands.

Kiau-Chau, directly west across the Yellow Sea from Korea, was a territory of 200 square miles, leased by Germany in November, 1897, from China for 99 years, and occupied in March, 1898. On the eastern side of the narrow entrance to a great bay some thirty miles in diameter and badly silted from the mud of the Yellow Sea, the Germans built the city and fortress of Tsingtau. "They excavated," says the *Boston Herald*, "at an expense exceeding $7,000,000, an outer and an inner harbor. They erected great granite piers, so arranged that ships alongside could receive cargo direct from railway trains. The docks and railway terminals at Tsingtau are models of convenience. . . . Some six miles back from the sea a typical German city was built. The scale on which things were done is shown by the fact that the Casino . . . cost more than $1,500,000. Included in the improvements undertaken is the extensive afforestation of the erstwhile treeless hills.

"There are said to be 12 forts; there were barracks built in 1905 for 5,000 men; there is a floating dock 410 feet long and 100 feet wide which will lift 16,000 tons. Exclusive of the Chinese [3,000] . . . the Tsingtau garrison, strengthened by the German and Austrian guards withdrawn from Peking, is about 8,000 men. To this force may be added 1,000 reservists, for every able-bodied German civilian in the colony will be called upon to perform military duty."

Japan declared war on August 24th. The operations on land do not come within the scope of this chapter, but both British and Japanese ships took part in the general attack, though it has been reported that the Japanese looked with disfavor upon the assistance of their allies. About the middle of October the battleship *Triumph* was heavily damaged by gunfire and compelled to withdraw; on October 17th the Japanese cruiser *Takachiho*, of 3,700 tons and eight 6-inch guns, was torpedoed and sunk by the German destroyer *S-90*, with the loss of all but twelve men of her crew of 283 aboard. The same destroyer was stranded and after stranding was destroyed by her own crew three days later in attempting to escape from Tsingtau. On November 6th, Tsingtau being on the eve of a surrender, which occurred November 7, 1914, the Germans sank all the vessels still in the harbor: the gunboats *Jaguar*, *Iltis*, *Luchs*, *Kormoran*, *Tiger*, the destroyer *Taku*, and the minelayer *Rachin*.

Nearly all the oversea possessions of Germany were now in the hands of her enemies: the rest were soon to follow. Togoland, on the gold coast of West Africa, had fallen on

August 15th; Apia and the nine Samoan Islands which Germany owned of the group of fourteen, on August 30th; on September 11th the Bismarck Archipelago protectorate; and on September 24th, Kaiser Wilhelm Land in New Guinea. In regard to the last two, it was remarked that: "their [the Germans] presence . . . has ever been regarded as a menace by Australian public opinion," a curious commentary on the ever-existing greed which is envious of possession by the other man, and exemplifying the need of doing away with the monstrous system of "spheres of influence," by whatever nation exercised. An Anglo-French expedition seized Kamerun on September 27th; the Japanese announced on October 21st their seizure of the Marianna and Marshall Islands. The whole of the German possessions in the Far East were thus in British or in Japanese hands.

The few auxiliary cruisers the Germans were able to get afloat did not remain long uncaptured. They made the great error of not mounting heavier and longer-range guns. Instead of the 6-inch, which was a possible gun for such vessels, they used the 4.1-inch. Thus the *Kaiser Wilhelm der Grosse*, of the North German Lloyds, of 13,952 tons, armed with ten 4.1-inch, fell a victim to the British cruiser *Highflyer*, of equal speed and carrying eleven 6-inch guns. The German ship was sunk off the Rio de Oro coast, a Spanish possession south of Morocco. She is reported to have held up several vessels on the Cape route. One of the most notable was the Union Castle liner *Galician*, which was discovered by the German vessel on August 15th, off Ferro, Canary Islands. The *Galician* was stopped, inspected, had her wireless destroyed, and the following morning she was informed that "on account of the women and children on board we will not destroy your vessel; you are released."

The *Cap Trafalgar*, of 18,710 tons, a new ship of the Hamburg Line in the South American trade, was at Buenos Ayres at the outbreak of the war. Here she coaled and landed everything not necessary for war and went off the Island of Trinidad, where she met the small gunboat *Eber*, from which she took the whole of her armament, two 4.1-inch and six 1-pounders. The *Eber* was sent into Bahia with one officer, one engineer, two warrant officers, and eleven men, where she interned. While the *Cap Trafalgar* was coaling on September 14th, 300 miles east of Rio, she was met by the British auxiliary cruiser *Carmania*, of 19,500 tons and carrying eight 6-inch guns. Notwithstanding, the *Cap Trafalgar* made a defense for two hours (extraordinary in the adverse conditions), when, having listed some 30 degrees, the crew was ordered into the boats. The *Carmania*, according to German statement, withdrew. A last use of the wireless attracted the German collier *Eleonore Woerman*, which picked up the survivors of the *Cap Trafalgar*, numbering 291 men, and landed them at Buenos Ayres, where they were interned. The captain and 12 seamen perished.

There were afloat also as auxiliary cruisers the *Prinz Eitel Friedrich*, the *Kronprinz Wilhelm*, and the *Kormoran II*, formerly the Russian *Riasan*, which was captured on August 6, 1914, by the *Emden*, taken to Tsingtau and armed. There is but little recorded of what they accomplished. All were to be later interned, the two former at Norfolk, Virginia, in April, 1915, the last, at Guam, on December 15, 1914. The names of some fifteen others, of less importance, will be found recorded in the appended list of losses.

A very much greater number of such vessels, naturally, was used by Great Britain. The number and names have not been published, but enough is known to show the

The *Emden* ashore on North Keeling Island, November 10, 1914. Boats from the *Sydney* taking off the survivors.

Landing party from the *Emden*, after having broken the instruments at the cable station on Direction Island, leaving to board the schooner *Ayesha* in which they made their escape.

great extent to which the British merchant marine has thus been utilized, perhaps more than the well-being of her ocean carrying trade justified.

The great disparity in naval power between Germany and Great Britain (the other powers involved being so secondary that they need not be considered) had its natural and inevitable sequence in the early disappearance of German ships, naval and commercial, from the ocean. Though Great Britain declared no blockade of German ports, the former used what was fully the equivalent, in result, of a legal blockade. Ships were stopped on the high seas and examined or detained at will. Naturally all the neutral powers concerned protested. This protest, in the American notes of earlier as well as later periods, was expressed with much force, the later note of October 21, 1915, declaring (paragraph 33) the procedure "ineffective, illegal, and indefensible" and that the United States "can not with complacence suffer further subordination of its rights and interests to the plea that the exceptional geographic position of the enemies of Great Britain requires or justifies oppressive and illegal practices."

Notwithstanding, the offensive action continued with the declared purpose of starving Germany into submission. It is but an illustration of the great fact which has come down to us in undiminished vigor through the ages: *Inter arma silent leges.* It cannot be otherwise. War in itself is the very negation of all law. In every exigency, the great question of national existence always has taken, and always will take, precedence, and Great Britain being preëminent on the water but follows the procedure of all time, and will do so until pressure, economic or other, may make it better for her to yield her methods. Any discussion of this part of the question, beyond the mere statement of what seem to the writer existing facts, is naturally beyond our province.

Germany was thrown on her own resources, which were supplemented by her trade with the Scandinavian countries and with such trade as could enter through Holland and Italy. She was able, with the labor in her fields undiminished by reason of the employment of 2,000,000 or more prisoners of war, to advance her agriculture to a degree which made her independent of exterior supplies of the more important elements of food. Her occupancy of Belgium and of the chief industrial provinces of France, and her unequalled chemistry rendered her likewise independent in the field of munitions of war. And though she has lost her sea-borne commerce, she has a large compensation in escaping the incurring of a heavy foreign indebtedness which must, in time, hang heavily upon those who have the world as a purchasing field. Nor can we yet say how great a factor in such a war, if long continued, may be the submarine or the airship. Thus, with a powerful battle-fleet yet in being, Germany is still a menace to Britain upon the water. To prophesy to what extent, is still too venturesome.

TABLE OF NAVAL LOSSES OF THE SEVERAL BELLIGERENTS SINCE THE OUTBREAK OF THE WAR TO FEBRUARY 1, 1915

From the *United States Institute Proceedings*

Special acknowledgment is here made of the courtesy of the United States Naval Institute in granting the writer the use of the valuable compilations appearing in its Proceedings.

The following table is compiled from various articles appearing in the press of the several belligerent nations. These reports are so contradictory that it is impossible to furnish an absolutely correct list of vessels lost to date. In many cases the belligerent powers have failed to acknowledge the losses of all the vessels. No attempt has been made to include herein a list of the various trawlers, mine-sweepers and such odd small craft which may have been destroyed while mine-sweeping or on submarine patrol duty.

ABBREVIATIONS FOR TYPE OF VESSELS

Dreadnought battleships	d. b.	Submarine	sm.
Battleship	b.	Auxiliary cruiser	ax. c.
Armored cruiser	a. c.	Converted cruiser	o. c.
Protected cruiser	p. c.	Transport	tr.
Cruiser	c.	Mine-layer	m. l.
Light cruiser	l. c	Hospital ship	h. s.
Gunboat	g. b.	Training ship	tr. s.
Torpedo gunboat	t. g.	Armed merchant ship	a. m. v.
Destroyer	d.	Naval tender	n. t.
Torpedo boat	t. b.		

LOSSES OF GREAT BRITAIN AND HER ALLIES

BRITISH WARSHIP LOSSES

NAME	TYPE	TONNAGE	REMARKS	DATE
Audacious	d. b.	24,000	Reported sunk off Irish coast. Cause unknown. British Admiralty non-committal	29-10-14
Bulwark	b.	15,000	Internal explosion at anchor in the Thames	25-11-14
Formidable	b.	15,000	Sunk by German sm., North Sea	1- 1-15
Warrior	a. c.	13,550	By mine	5- 9-14
Hogue	a. c.	12,000	Sunk by German sm. U-29, North Sea	22- 9-14
Cressy	a. c.	12,000	Sunk by German sm. U-29, North Sea	22- 9-14
Aboukir	a. c.	12,000	Sunk by German sm. U-29, North Sea	22- 9-14
Hawke	a. c.	7,350	Sunk by German sm. U-9, North Sea	16-10-14
Good Hope	a. c.	14,100	Sunk by German forces in Pacific	1-11-14
Monmouth	a. c.	9,800	Sunk by German forces in Pacific	1-11-14
Pathfinder	l. c.	2,940	Sunk by German sm., North Sea	5- 9-14
Amphion	l. c.	3,360	Sunk by mine, North Sea	6- 8-14
Pegasus	l. c.	2,135	Sunk by German c. Konigsberg at Zanzibar	20- 9-14
Hermes	l. c.	5,600	Sunk by German sm. U-27, North Sea	30-10-14
Speedy	t. g.	810	Sunk by mine, North Sea	3- 9-14
Niger	t. g.	810	Sunk by German sm. while at anchor	11-11-14

NAME	TYPE	TONNAGE	REMARKS	DATE
Bullfinch	d.	370	Sunk in collision with Dutch merchant ship	18- 8-14
——	d.	——	Ran ashore, Scotch coast	27-12-14
AE-1	sm.	725-810	Accidentally sunk off Australian coast	14- 9-14
E-3	sm.	725-810	Rammed and sunk by German vessel	18-10-14
D-5	sm.	550-600	Sunk by German mine, North Sea	3-11-14
D-2	sm.	550-600	Reported lost. Details not known	1-12-14
E-10	sm.	725-810	Missing. North Sea	——
Oceanic	ax. c.	17,274	Ran aground off north coast of Scotland in storm	8- 9-14
Rohilla	ax. c.	7,400	Ran aground off Whitby, completely wrecked	30-10-14
Viknor	ax. c.	5,386	Lost off Irish coast	14- 1-15

FRENCH WARSHIP LOSSES

Zelee	g. b.	636	Sunk by German cruisers at Tahiti	22- 9-14
No. 347	t. b.	97	Sunk in collision with each other	9-10-14
No. 338	t. b.	97	Sunk in collision with each other	9-10-14
——	t. b.	——	Reported lost	- 1-15
No. 219	t. b.	87	Sunk off Nieuport	- 1-15
Curie	sm.	392	Sunk by Austrians at Pola	23-12-14
Saphir	sm.	386- ?	Sunk at the Dardanelles	15- 1-15

RUSSIAN WARSHIP LOSSES

Pallada	a. c.	7,775	Sunk by German sm., Baltic	11-10-14
Jemtchug	c.	3,130	Sunk by Emden, Penang	28-10-14
Donnetz	g. b.	1,224	Sunk by Turks, Black Sea. Raised by Russians	31-10-14
Kubanetz	g. b.	1,200	By gunfire, Odessa	29-10-14
Putschino	t. b.	——	By gunfire	30-10-14
Prut	ax. c.	5,440	Scuttled to avoid capture	29-10-14
Riasan	tr.	3,522	Captured by Emden	6- 8-14
Oleg	m. l.	1,125	Sunk, Black Sea	24-12-14
Athos	m. l.	1,743	Sunk, Black Sea	24-12-14
Portugal	h. s.	——		

JAPANESE WARSHIP LOSSES

Shirotaye	d.	380	Ran ashore, Tsingtau	4- 9-14
No. 33	t. b.	82	Sunk by mines while mine-sweeping off Tsingtau	11-11-14
Name unknown	t. b.	——	By mine	——
Takachiho	tr. s.	3,700	Torpedoed by German d., S-90, off Tsingtau	17-10-14

In addition to the two above, five special service ships were sunk during the operations off Tsingtau.

LOSSES OF GERMANY AND HER ALLIES

GERMAN WARSHIP LOSSES

Yorck	a. c.	9,350	Sunk by German mine near Wilhelmshaven	3-11-14
Scharnhorst	a. c.	11,420	Sunk by English forces off Falklands	8-12-14
Gneisenau	a. c.	11,420	Sunk by English forces off Falklands	8-12-14
Friedrich Karl	a. c.	8,858	No official report. Press reports that she was lost in the Baltic	-12-14
Bluecher	a. c.	15,550	Sunk by British forces off Doggerbank	24- 1-15
Mainz	p. c.	4,280	Sunk by British forces, North Sea	28- 8-14
Koeln	p. c.	4,280	Sunk by British forces, North Sea	28- 8-14
Ariadne	p. c.	2,618	Sunk by British forces, North Sea	28- 8-14
Hela	p. c.	2,005	Sunk by British sm., E-9, North Sea	13- 9-14
Augsburg	p. c.	4,280	By gunfire	7- 8-14
Emden	p. c.	3,592	Sunk by Australian c. Sydney, Indian Ocean	9-11-14

NAME	TYPE	TONNAGE	REMARKS	DATE
Leipzig	p. c.	3,200	Sunk by English forces off Falklands	8-12-14
Nurnberg	p. c.	3,396	Sunk by English forces off Falklands	8-12-14
Berlin	c. c.	17,324	Interned, Norway	16-11-14
Patagonia	c. c.	——	Seized by Argentina, violation of neutrality	——
Eber	c. c.	1,000	Interned, Bahia	- 9-14
Kormoran II	c. c.	3,508	Interned, Guam	15-12-14
Magdeburg	l. c.	4,478	Ran ashore in fog in Baltic; blown up by own crew after engagement with Russians	27- 8-14
Geier	l. c.	1,630	Interned in Honolulu	8-11-14
Karlsruhe	l. c.	4,822	No official report as to this vessel's destruction. Press reports state she was blown up by internal explosion while cruising in the Atlantic	-11-14
Moewe	g. b.	640	Sunk by her own crew at Dar-es-Salaam when English vessels appeared	14- 8-14
Wissman	g. b.	about 300	Captured by English on Lake Nyasa	20- 8-14
Planet	g. b.	640	Sunk by her own crew at Yap Island on approach of Japanese fleet	7-10-14
Kormoran	g. b.	1,604	Sunk by Germans in Kiau-Chau Bay before surrendering Tsingtau to Allies	6-11-14
Iltis	g. b.	886	Sunk by Germans in Kiau-Chau Bay before surrendering Tsingtau to Allies	6-11-14
Tiger	g. b.	886	Sunk by Germans in Kiau-Chau Bay before surrendering Tsingtau to Allies	6-11-14
Luchs	g. b.	886	Sunk by Germans in Kiau-Chau Bay before surrendering Tsingtau to Allies	6-11-14
Jaguar	g. b.	886	Sunk by Germans in Kiau-Chau Bay before surrendering Tsingtau to Allies	6-11-14
Tsingtau	g. b.	168	Interned, China	17- 8-14
Vaterland	g. b.	168	Interned, China	17- 8-14
V-187	d.	689	Sunk by British forces, North Sea	28- 8-14
S-115	d.	413	Sunk by English destroyers, North Sea	17-10-14
S-117	d.	413	Sunk by English destroyers, North Sea	17-10-14
S-118	d.	413	Sunk by English destroyers, North Sea	17-10-14
S-119	d.	413	Sunk by English destroyers, North Sea	17-10-14
S-90	d.	396	Driven ashore and wrecked by own crew off Tsingtau after having torpedoed the Japanese ship Takachiho	20-10-14
Taku	d.	276	Sunk by Germans in Kiau-Chau Bay before surrendering Tsingtau to Allies	6-11-14
S-124	d.	463	Accidentally rammed and sunk by a merchant vessel in the Baltic	22-11-14
S-126	d.	487	By sm.	6-10-14
S-116	t. b.	477	Sunk by English sm. E-9, North Sea	6-10-14
U-15	sm.	about 450	Sunk by British c. Birmingham	10-10-14
U-18	sm.	——	Destroyed by English d. Garry off Scotch coast	23-11-14
U-3	sm.	——	Rammed	9- 8-14
Prince Adalbert	ax. c.	6,030	Captured by British c.	- 8-14
Sudmark	ax. c.	5,113	Captured by British c.	15- 8-14
Kaiser Wilhelm der Grosse	ax. c.	13,952	Sunk by British c. Highflyer	27- 8-14
Bethania	ax. c.	7,548	Captured by British c.	7- 9-14
Spreewald	ax. c.	3,899	Captured by British c.	12- 9-14
Cap Trafalgar	ax. c.	18,710	Sunk by British ax. c. Carmania	14- 9-14
Max Brock	ax. c.	4,579	Captured by British c.	- 9-14
Itolo	ax. c.	299	Sunk by French g. b. at Kamerun	24- 9-14
Rhios	ax. c.	150	Sunk by French g. b. at Kamerun	24- 9-14
Soden	ax. c.	150	Captured by English c. off Kamerun River	1-10-14
Gneisenau	ax. c.	8,185	Sunk by Belgians prior to evacuation of Antwerp	8-10-14

NAME	TYPE	TONNAGE	REMARKS	DATE
Graecia	ax. c.	2,753	Captured by English c.	10-10-14
Markomannia	ax c.	4,405	Sunk by British c. in Indian Ocean	16-10-14
Navarra	ax. c.	5,794	Sunk by English ax. c. in Atlantic	11-11-14
Greif	ax. c.	——		
Comet	ax. c.	977	Captured by Australian forces	18-10-14
Karnac	ax. tr.	4,437	Interned, Chile	-11-14
Konigin Luise	m. l.	2,163	Sunk by English d., North Sea	5- 8-14
Rufin	m. l.	——	Sunk by Germans in Kiau-Chau Bay before surrendering Tsingtau to Allies	6-11-14
——	m. l.	——	Reported by the French captured outside of Havre disguised as French collier	——
Kingani	a. m. v.	——	Captured by the British on Lake Tanganyika	——
Locksum	n. t.	1,020	Interned, Honolulu	8-11-14

AUSTRIAN WARSHIP LOSSES

Zenta	c.	2,264	Sunk by French fleet off Antwerp	16- 8-14
Kaiserin Elisabeth	c.	3,937	Sunk by her own crew at Tsingtau	7-11-14
No. 19	t. b.	78	Struck by a mine and sank at entrance to Pola	18- 8-14
——	t. b.	——	Sunk by mine in Adriatic	——
——	t. b.	——	Sunk by mine in Adriatic	——

TURKISH WARSHIP LOSSES

Messudieh	b.	10,000	Sunk by British sm. B-11, Dardanelles	14-12-14
Mejidieh	c.	3,330	Sunk by Russian mine	——
Burak Reis	g. b.	502	Scuttled to avoid capture	31-10-14
Issa Reis (type)	g. b.	500	Scuttled	1-11-14
Bezemialen	tr.	——	Sunk, Black Sea	14-11-14
Bachriachmar	tr.	——	Sunk, Black Sea	14-11-14

CHRONOLOGICAL TABLE

DATE		PAGE
1410.	The Teutonic Knights defeated by the Poles at Tannenberg	140
1807.	Heligoland taken by Great Britain	418
1870.	August 4. Germany began the invasion of France	34
1890.	Heligoland ceded to Germany	418
1902.	November 11. Germany signed an agreement with Luxemburg providing that she would not use the railways of the grand-duchy for any military purpose whatsoever	32
1904.	German plan of traversing Belgium to invade France probably approved	9
1906.	General Helmuth von Moltke appointed Chief of the German General Staff	24
1911.	Germany acquired extension of territory in Congo as compensation for French influence in Morocco	20
	General Joffre appointed Chief of the French General Staff	26
1912.	November 22. Franco-British entente affirmed	405
1913.	New German and French army laws enacted	405
1914.	June 24. Kiel Canal extension completed	417
	July 19. King George reviewed British fleet at Spithead	409
	July 28. Austria-Hungary declared war against Serbia	297
	August 1. The German ambassador suggested to the British government respect of Belgian neutrality as the condition of British neutrality	18
	August 2. German troops entered Luxemburg at Wasserbillig	31
	August 2. Germany presented ultimatum to Belgium	49, 51
	August 3. Germany proposed as the condition of British neutrality not to use Belgian coast for warlike purposes	18
	August 3. Belgium declined French aid in her defense	70
	August 3. The Germans entered France on Luxemburg border	80

DATE PAGE

DATE		PAGE
1914.	October 13. The Belgian government seated at Havre .	231
	October 13–November 11. Battle of Ypres .	239, 250–268, 315
	October 15. The Belgian army joined the Allies on the Yser	232
	October 18–November 3. Battle of the Yser .	238, 242–250
	October 21. The Mariannas and the Marshall Islands seized by the Japanese and the British . .	328–329, 475
	October 28. The *Emden* destroyed a Russian cruiser and a French destroyer in the harbor of Penang . .	464–466
	November 1. The Germans defeated the British off Chile	453–457
	November 5. Cyprus formally annexed by Great Britain	333
	November 5. Turkey entered the war	419
	November 9. The *Emden* destroyed at North Keeling Island	466–469
	November 13. German squadron bombarded the east coast of England	436
	November 14. The Russians renewed the siege of Peremysl	288
	December 2. The Austrians defeated by the Serbians in the Battle of Suvobor	305
	December 6. Lodz evacuated by the Russians . . .	294
	December 8. The Germans defeated in battle off the Falkland Islands	459–462
	December 15. King Peter returned to recaptured Belgrade	305
	December 16. Attack made on Scarborough, Whitby, and Hartlepool	392, 436
	December 19. Great Britain proclaimed her Protectorate over Egypt	332
	December 25. The British raided Cuxhaven	439
1915.	January 1. The British battleship *Formidable* sunk .	439–441
	January 15. Great battle in the North Sea	442–445

INDEX

Of some of the subjects presented in this volume varying phases appear in other volumes of this work and the Indexes to all may be profitably consulted.

Made in United States
North Haven, CT
04 November 2026

21766785R10380